FLORA OF MEDAK DISTRICT
ANDHRA PRADESH, INDIA

ts

FLORA OF MEDAK DISTRICT

ANDHRA PRADESH, INDIA

T. PULLAIAH
C. PRABHAKAR
B. RAVIPRASAD RAO

1998
DAYA PUBLISHING HOUSE
DELHI-110 035

THAMMINENI PULLAIAH (B. 1951-)
CHINTALA PRABHAKAR (B. 1958-)
B. RAVIPRASAD RAO (B. 1964-)

ISBN 81-7035-200-2

Published by
Daya Publishing House
1123/74, Deva Ram Park
Tri Nagar, Delhi-110 035
Phone: 710-3999
Fax: (011) 7198902

Laser type-setting by:
Advance Publishing Concept (APCON)
Ph: 5142551

Printed at
Chawla Offset Printers
New Delhi-110 052

CONTENTS

ABBREVIATIONS AND ACRONYS USED

BR	:	B. Ravi Prasad Rao
CP	:	C. Prabhakar
TP	:	T: Pullaiah
KMS	:	K.M. Sebastine
GNR	:	G. Narasimha rao
auct. non	:	auctorum non; not of authors
cm	:	centimetre
ex	:	published by
FBI	:	Hooker, J.D et al., Flora of British India 7 Vols. (`872-1897)
FFH	:	Khan, S. Forest Flora of Hyderabad State
Fig.	:	Figure
Fl & Fr	:	Flowering and fruiting
Gamble	:	Gamble J.S. and C.E. C. Fischer, Flora of the Presidency of Madras. 3 Vols. (1915-1936)
nom. superfl	:	a name superfluous when it was published; in most case it is a name based on the same type as another earlier specific epithet.
RF	:	Reserve forest
ssp	:	subspecies
var	:	variety
vern	:	vernacular name, local name

1

INTRODUCTION

Medak, one of the 10 districts of Telangana region of Andhra Pradesh, lies between 17° 27′ and 18° 18' Northern latitude and 77° 28′ and 79° 10′ of Eastern longitude. The district derived its name from Medak, the then headquarters town of the taluk of the same name. Medak was originally known as Methukudurgam which subsequently changed into Methuku due to the growth of fine and coarse rice in this area.

The total geographical area of the district is 9,699 Sq. km, accounting for 3.5% of the total area of the state. The district has 1265 villages and 10 towns with a population of 22,64,124 (1991 census). The district is bounded on the North by Nizamabad and Karimnagar districts, on the East by Warangal and Nalgonda districts, on the South by Ranga Reddy district and on the West by Bidar district of Karnataka state.

Medak district was not a homogeneous administrative unit in the past and its component parts were, at varying periods of time, under the control of several dynasties, namely the Mouryas (ca.320 A.D), the Satavahanas (5th Century A.D), the Chalukyas of Badami and the Rashtrakutas (ca.757-928 A.D), the Chalukyas of Kalyani (ca 975-1138 A.D.) the Kakatiyas (ca 1138-1323 A.D.), the Bahamanis (1325-1540 A.D), the Barid Shahis and Qutub Shahis (1540-1650 A.D), Moghuls (1650-1724 A.D) and Asaf-Jhais (Nizams of Hyderabad, (1724-1948 A.D).

In 1901, Medak district consisted of 6 taluks, viz., Medak, Ramayampet, Baghat, Kalabgur, Andole and Tekmal besides some Jagirs. In 1905, Tekmal was merged in Andole and Ramayampet partly in Medak and partly in Kamareddy taluk of Nizamabad district. In the same year the district gained the taluk of Ibrahimpatnam from Mahaboobnagar district and Siddipet taluk from Karimnagar district. However, Ibrahimpatnam taluk was added to Baghat.

Thus in 1905, the district comprised 5 taluks, viz., Medak, Siddipet, Baghat, Kalabgur (Sangareddy) and Andole. During the second decade of this century Vikarabad taluk was newly constituted by merging Hatnura and Daulthabad Jagirs. In 1931, Baghat was constituted into a separate district. During the same decade the district gained the taluk of Yellareddy from Nizamabad district and in the next decade (1941-1950) the Sarf-e-Khas district of Atraf-i-Balda and the district of Baghat were abolished and the new district of Hyderabad was created (1948). Consequently on the abolition of Jagirs in 1950, the Paighas of Narsapur and Gajwel were converted into taluks and added to this district with headquarters at Narsapur and Gajwel respectively. Subsequently owing to the reorganization of states in 1956, the district gained the taluk of Narayankhed and Zaheerabad except Nirna Circle and Nyalkal Circle from Bidar district. The district however lost the taluk of Vikarabad which was transferred to Hyderabad.

The district comprised of 8 taluks in 1971 and these taluks were consituted into 11 taluks in 1981. On 25-5-1985, these taluks are further consituted into 45 Revenue Mandals under 3 Revenue Divisions, viz., Sangareddy, Medak and Siddipet by the State Government.

The Government have reconstituted the 11 *Panchayat Samithies* into 45 *Mandal Praja Parishads* on 15-1-1987.

The Revenue Divisions with their respective Mandals are listed in Table 1.

Table 1: The Revenue Divisions with their Respective Mandals

Name of the	Name of Mandals
1. Sangareddy	1. Sangareddy, 2. Patancheru, 3. Ramachandrapuram, 4. Sadasivpet, 5. Munipally, 6. Kondapur, 7. Zaheerabad, 8. Kohir, 9. Jharasangam10. Raikode, 11. Nyalkal, 12. Narayankhed, 13. Kangti, 14. Kalher, 15. Manoor
2. Medak	16. Medak, 17. Kulcharam, 18. Papannapet, 19. Yeldurthi, 20. Andole, 21. Pulkal, 22. Alladurg, 23. Shankarampet (A), 24. Regode, 25. Tekmal, 26. Ramayampet, 27. Sankarampet (R), 28. Chegunta. 29. Narsapur, 30. Jinnaram, 31. Shivampet, 32. Kowdipally, 33. Hathnura
3. Siddipet	34. Siddipet, 35. Chinnakodur, 36. Nangnoor, 37. Kondapak, 38. Dubbak, 39. Mirdoddi, 40. Gajwel, 41. Jagadevpur, 42. Doulthabad, 43. Warangal, 44. Mulug, 45. Toopran

2

TOPOGRAPHY AND GENERAL FEATURES

Topography

Medak district forms part of the tableland of the Deccan Plateau and is crossed by different ranges of hills. The ground is mostly of plains, gentle slopes and undulating hills. Isolated peaks and rocky clusters lie scattered all over the district. The elevation of the ground in the district is between 500 m - 600 m with occasional hills up to 638 m above Mean Sea Level. The hills that are of considerable size in the forest division are in a state of erosion because of reckless felling and indiscriminate grazing.

Geology

The rock formation in the district is of the oldest type (*Archaen gneisses*) and consists principally of Peninsular granite complex i.e. pink and grey granites and their metamorphic variations. Minor inliers of Dharwar rocks occur as narrow bands in the granite and consists of horn blend schists, chlorite schists and banded or massive ferruginous quartzites. A few such exposures are seen due North and North-East of Siddipet. A part of the Sangareddy taluk in the South-West of the district is covered by the Deccan traps (Basalt flows) formation.

Mineral Resources

Building Material

The granites found in the district yield large quantity of building stone and road-metal.

Quartz

There are numerous quartz veins cutting across the granite all over the district. Quartz useful for glass industry may be obtained from selected deposits.

Clay

In the granites feldspars are colonised in some places giving rise to small deposits of white clay in the form of veins and pockets.

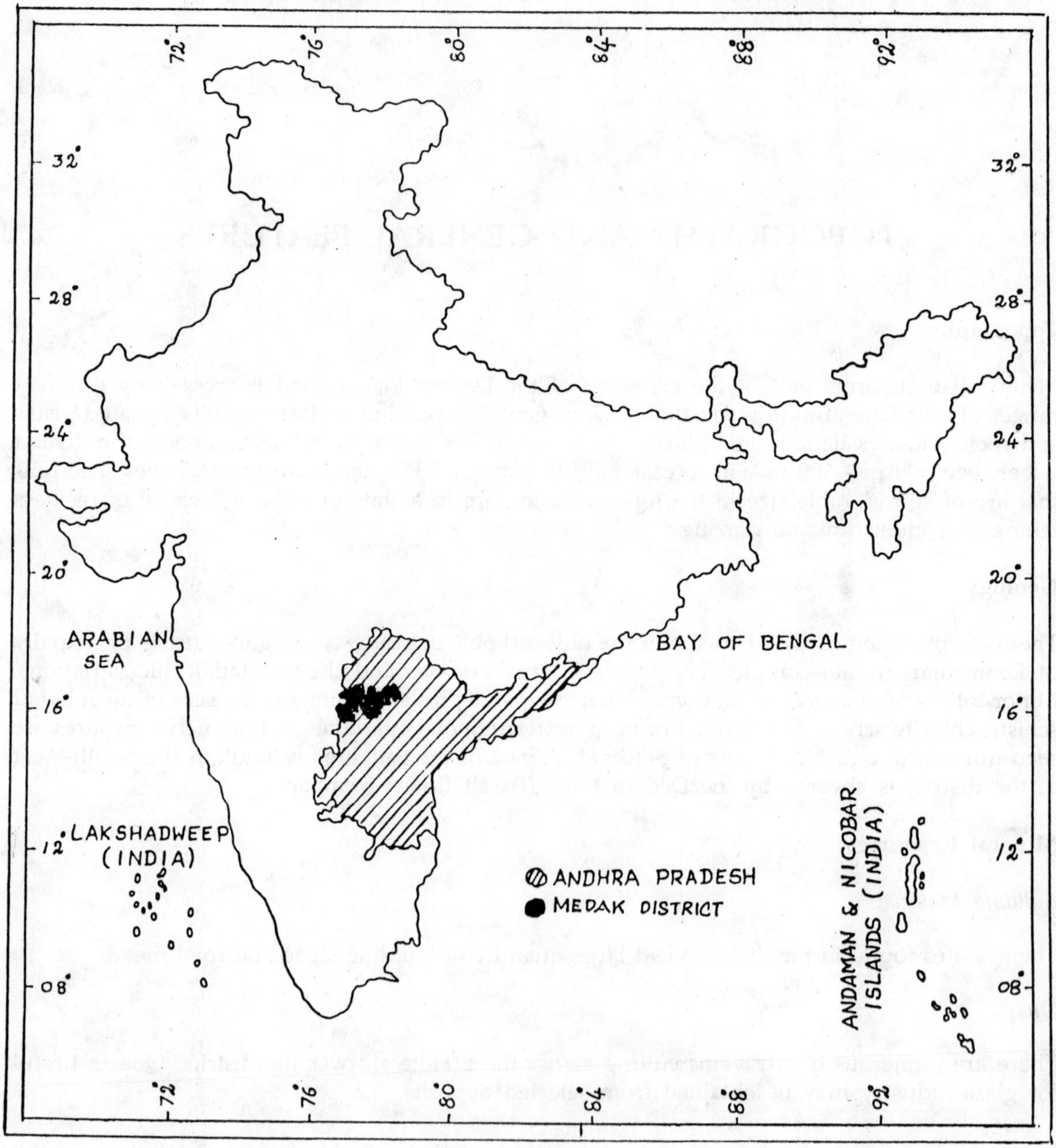

Fig. 2.1: Map of India showing the Position of Andhra Pradesh and Medak District

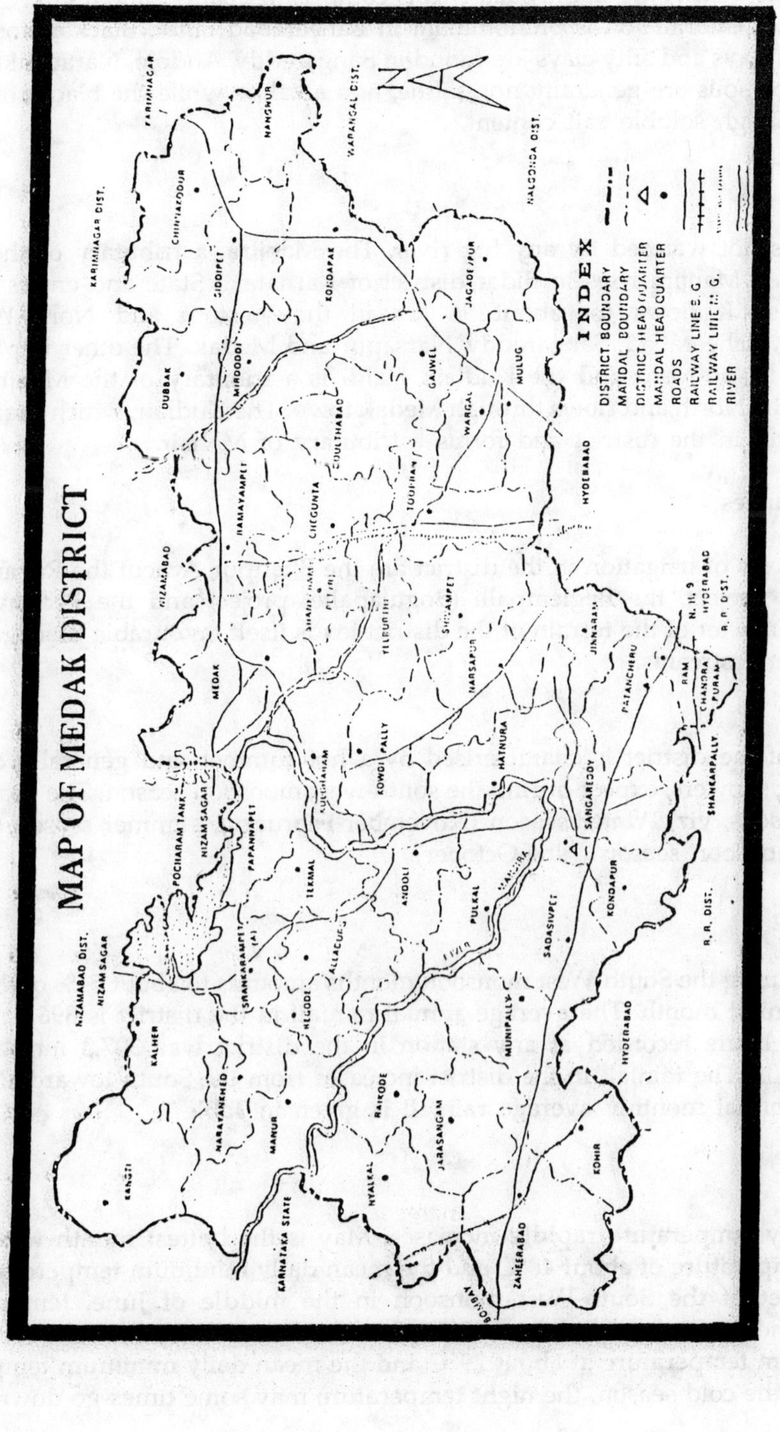

Fig. 2.2: Map of Medak District

Soils

The soils of the district are mainly red earths comprising loamy sands, sandy loams and sandy-clay loams. Red laterite soil is predominant in Zaheerabad taluk. Black cotton soils comprising of clay loams, clays and silty clays are found in Sangareddy, Andole, Narayankhed and Narsapur taluks. The red soils are generally non-saline, non alkaline while the black soils are moderately alkaline with high soluble salt content.

Rivers

The district is not watered by any big river. The Manjira, a tributary of the Godavari, is an important river. Manjira rises in Bidar district of Karnataka State and enters Medak district in the South-East. It flows for about 96 km in the Western and North-Western taluks of Narayankhed, Zaheerabad, Sangareddy, Narsapur and Medak. The other important streams are the Haldi or Pasupuyeru and the Kudlair. Haldi is a tributary of the Manjira and enters the district from the North and flows through Medak town. The Kudlair, which drains Siddipet taluk, is another river in the district and forms a tributary of Mahair.

Irrigation Sources

The chief sources of irrigation in the district are the Bhanpur Ayacut the Rayanpalle project, the Gangakathwa project, the Beglempalli (Bogulapalle) project and the Peddavagu project. The undulating character of the terrain of the district lends itself favourable to irrigation from canals, tanks, wells and streams.

Climate

The climate of the district is characterised by a hot summer and generally dry weather with some pleasing showers, expcet during the south-west monsoon season. The year may be divided into three seasons, viz, Winter season (November-February), summer season (March-May) and South-West monsoon season (June-October).

Rainfall

The rainfall during the South-West monsoon months amounts to about 84% of the annual rainfall. July is the rainiest month. The average annual rainfall in the district is 896.7 mm. The heaviest rainfall in 24 hours recorded at any station in the district was 307.3 mm at Sangareddy in September, 1908. The rainfall in the district increases from the South towards North.

District normal monthly average rainfall is given in Table 2.

Temperature

After February temperature rapidly increases. May is the hottest month with the mean daily maximum temperature of about 40°C and the mean daily minimum temperature of about 26°C. With the onset of the South-West monsoon in the middle of June, temperature decreases appreciably and the weather becomes more pleasant. December is the coldest month with mean daily maximum temperature at about 29°C and the mean daily minimum temperature of about 14°C. During the cold season, the night temperature may some times go down to about 6°C.

Table 2: Monthly Average Rainfall in the District

Seasons	Months	Rainfall (in mm)
South-West mansoon	June	147.1
	July	227.4
	August	191.5
	September	188.2
	October	48.5
	November	18.5
Winter	December	5.2
	January	3.1
	February	13.9
	March	11.6
Summer	April	22.5
	May	19.2
Total		896.7

3

FORESTS AND VEGETATION

The vegetation of the district can be categorised into forest, non forest and aquatic types.

Forest Vegetation

The district forests are of Southern Tropical Dry deciduous type (Champion & Seth,1968) and account for 9.9% of the total geographical area. The forests are grouped into only one division i.e. Medak which includes 6 ranges.

The details of the forest ranges are given in Table 3.

Table 3: Forest Ranges of Medak District

Sl. No.	Name of the Division	Name of the Range	Area (in Sq.Kms)
1.	Medak	Siddipet	178.99
2.	Medak	Ramayampet	165.27
3.	Medak	Medak	250.47
4.	Medak	Narsapur	202.70
5.	Medak	Zaheerabad	91.12
6.	Medak	Narayankhed	71.54

The forests are further classified into Dry mixed deciduous type, Dry deciduous type and Dry savanah type. Locally the forests are subclassified by the forest officials as teak type (Teak over 40%), mixed teak type (Teak 10% to 30%) and mixed type (Teak less than 10%) depending on the abundance of teak in the forests.

Dry mixed deciduous forests

These type of forests are widespread throughout the district and are distributed in the forest blocks of Medak and Narsapur taluks. The composition of these forests are as follows:

TOP STOREY

The stratum consists of species like *Albizia amara, Anogeissus latifolia, Bombax ceiba, Boswellia serrata, Buchanania latifolia, Chloroxylon swietenia, Dalbergia paniculata, Disospyros melanoxylon,*

Givotia moluccana, Hardwickia binata, Lagerstroemia parviflora, Lannea coromandelica, Madhuca indica, Mitragyna parvifolia, Morinda pubescens, Ougeinia oojeinesis, Phyllanthus emblica, Soymida febrifuga, Strychnos nux-vomica, Tectona grandis, Terminalia bellirica, T. coriacea etc. *Terminalia arjuna* is common along the banks of streams.

UNDER STOREY

This storey consists of *Butea monosperma, Cassia fistula, Cassine glauca, Cordia obliqua, Dendrocalamus strictus, Gardenia gummifera, G.latifolia, Holarrhena pubescens, Limonia acidissima, Nyctanthus arbor-tristis, Wrightia tinctoria* etc.

SHRUBS

The common shrubs in these forests are *Alangium salvifolium, Annona squamosa, Cassia auriculata, Catunaregam spinosa, Combretum albidum, Dichrostachys cinerea, Dodonaea angustifolia, Grewia hirsuta, G. tiliifolia, Helicteres isora, Maytenus emarginata, Rhus mysorensis, Vitex negundo* and *Ziziphus* spp.

CLIMBERS

The climbers that are met within the forests are *Abrus precatorius, Ampelocissus latifolia, Aspidopterys cordata, Butea superba, Capparis zeylanica, Cissampelos pareira, Cocculus hirsutus, Cryptolepis buchanani, Derris scandens, Dioscorea pentaphylla, Gymnema sylvestre, Ichnocarpus frutescens, Ipomoea* spp., *Jasminum auriculatum, Olax scandens, Paracalyx scariousus, Rivea hypocrateriformis, Ziziphus oenoplia* etc.

HERBS

The common herbs that are encountered in these forests are *Acalypha indica, Aerva lanata, Ageratum conyzoides, Alysicarpus* spp. *Biophytum sensitivum, Cassia tora, Crotalaria verrucosa, C. retusa, Curculigo orchioides, Desmodium dichotomum, D. gangeticum, Euphorbia cristata, Glinus oppositifolius, Hemigraphis latebrosa, Hibisus lobatus, Indigofera linnaei, I. trita, Polycarpaea corymbosa, Polygala elongata, Pulicaria wightiana, Triumfetta rhomboidea, Uraria picta, Zornia gibbosa* etc.

GRASSES

The common grasses in these forests are *Apluda mutica, Aristida adscensionis, A setacea, Cymbopogon coloratus, C. martinili, Dichanthium annulatum, Eragrostis unioloides, Heteropogon contortus, Oplismenus burmannili, O. compositus* etc.

PARASITES

The parasitic species present in the district are *Cassytha filiformis, Cuscuta reflexa, Dendrophthoe falcata, Scurrula parasitica, Striga asiatica, Viscum nepalense* etc.

PTERIDOPHYTES

Vascular cryptogams are few and usually occur near streams and in rock crevices in forests. *Actinopteris radiata* is common in rock crevices. *Marselia quadrifolia* is seen in ponds and ditches.

Dry Deciduous Scrub Forests

Scrub forests are mostly distributed in forest blocks of Siddipet, Zaheerabad and Narayankhed ranges. Scrub vegetation is characterised by the predominance of *Annona squamosa, Capparis zeylanica, Cassia auriculata, C. occidentails, Diospyros melanoxylon, Gymnosporia spinosa, Dodonaea angustifalia, lxora parviflora, Lantana camara* with trees of *Phoenix loureiri.*

Dry Savannah Forests

These type of forests are distributed in patches in the outer edges of the forest blocks and usually found in Siddipet, Narayankhed and Zaheerabad ranges and parts of Ramayampet and Narsapur. The trees stand far apart singly or in small groups in more or less heavy grass in which certain fire resistant plants persist. Stemless *Phoenix, Phoenix loureiri* is characteristic. Other common species encountered in these forests include *Cassia auriculata, Dodonea angustifalia* and *Lantana camara.*

Non-Forest Vegetation

Waste Lands and Road Sides

The vegetation of waste lands and road sides include *Acalypha ciliata, Acanthospermum hispidum, Achyranthes aspera, Amaranthus tricolor, Boerhavia diffusa, Calotropis gigantea, C. procera, Cassia auriculata, C. occidentalis, C. tora, Cleome gynandra, Corchorus aestuans, C. fascicularis, C. trilocularis, Croton bonplandianum, Datura innoxia, Echinops echinatus, Euphorbia hirta, E. indica, Evolvulus alsinoides, Impatiens balsamina, Indigofera cordifolia, Jatropha gossypifolia, Parthenium hysterophorus, Prosopis chilensis, Solanum surattense, Sida cordata, S. cordifolia, Tephrosia purpurea, T. villosa, Tridax procumbens, Vernonia cinerea, Vicoa indica* etc.

The avenue trees like *Azadirachta indica, Cassia roxburghil, Delonix regia, Ficus benghalensis, Mangifera indica, Pongamia pinnata, Tamarindus indica,* grow in towns and villages and along the sides of roads in the district.

Hedges

The common hedge plants are *Abutilon indicum, Caesalpinia bonduc, Cascabela thevetia, Catunaregam spinosa, Clerodendrum inerme, Grewia hirsuta, Lawsonia inermis, Parkinsonia aculeata, Prosopis chilensis* etc. Chief climbers seen in hedges are *Cissampelos pareira, Cocculus hirsutus, Derris scandens, Pergularia daemia, Tinospora cordifolia, Tylophora indica,* etc.

Weeds

Common weeds of dry and cultivated fields and dry irrigated fields are *Abelmoschus ficulneus, Acanthospermum indicum, Acalypha indica, Ageratum conyzoides, Alysicarpus rugosus, Amaranthus spinosus, A. viridis, Amischophacelus axillaris, Argemone mexicana, Celosia argentea, Cleome gynandra, Chenopodium album, Corchorus aestuans, C. fascicularis, C. trilocularis, Crotalaria laburnifolia, C. retusa, Cynodon dactylon, Cyperus rotundus, Desmodium triflorum, Digera muricata, Eragrostis atrovirens, E. viscosa, Euphorbia geniculata, E. hirta, E. indica, Justicia* spp., *Lagascea mollis, Leucas aspera, Melilotus indica, Merremia emarginata, Panicum repens, Parthenium hysterophorus, Phylanthus amarus, Physalia minima, Portulaca oleracea, Rorippa indica, Sphaeranthus indica, Trianthema portulacastrum, Vigna* spp., etc.

Aquatic Vegetation

The aquatic plants can be classified under the following categories depending on their relations with soil, water and air.

Floating hydrophytes

FREE FLOATING

These plants have no contact with soil. They float on the surface of water and are in contact with air and water only.

e.g : *Pistia stratiotes, Spirodela polyrhiza, Utricularia stellaris.*

ATTACHED WITH FLOATING SHOOTS

These plants are attached to the muddy floor by their roots but their shoots come and float on the surface of water.

e.g.: *Ipomoea aquatica, Ludwigia adscendens.*

ATTACHED WITH FLOATING LEAVES

These plants are attached to the muddy floor. Their stems remain under water in contact with soil and water, while leaves float on the surface of water.

e.g .: *Aponogeton natans, Nymphaea nouchali, Nymphoides hydrophylla.*

Submerged Hydrophytes

SUSPENDED SUBMERGED

These plants remain in water, but have no contact with soil and air.

e.g.: *Hydrilla verticilata, Nechamandra alternifolia.*

ATTACHED SUBMERGED

These plants remain in water but rooted in the soil.

e.g.: *Ottelia alismoides, Potamogeton nodosus, Monochoria vaginalis.*

Emergent Hydrophytes

These plants are attached to soil and are covered with water at their base, but most of their vegetative parts come out of water surface.

e.g.: *Aeschynomene aspera, A. indica, Ammannia baccifera, A. multiflora, Bacopa monnieri, Bergia ammannioides, Cyperus nutans, Echinochioa colona, Marselia quadrifolia, Polygonum barbatum, P. glabrum, Rotala* spp., *Scirpus articulatus, Sopubia delphinifolia* and *Typha angustata.*

Wetland Hydrophytes

These plants are rooted in the soil saturated with water, but also survive in dried conditions in the later part of their life cycle.

e.g.: *Ageratum conyzoides, Bacopa monnieri, Bergia ammannioides, Caesulia axillaris, Centella asiatica, Commelina* spp., *Cyperus* spp., *Dopatrium junceum, Eclipta prostrata, Fimbristylis* spp., *Hemiadelphis polysperma, Hoppea dichotoma, Hygrophila aurciculata, Ipomoea carnea, Lobelia alsinodes, Lindernia spp., Ludwigia perennis, Panicum trypheron, Phyla nodiflora, Sopubia delphinifolia.*

Apart from the species encountered in above vegetation types, number of plants are under cultivation in the district. They are appended at the end of concerned family in the systematic enumeration.

4

PAST AND PRESENT WORK

Past Work

Medak district received scant attention in relation to floristic studies, except with reports on Narsapur reserve forest and its adjoining areas. No attempt has been made for floristic survey on other parts of the district. All the previous reports are exclusively concerned with the flora of erstwhile Nizam's Dominions (old Hyderabad State).

The study of the flora of Hyderabad state dates back to the earlier part of the 19th century when Walker (1849) and Bradley (1849) published their poioneer work, which included agricultural, medicinal and other economically important species of Daulatabad and Warangal districts of the state. Campbell (1898) listed forest plants of Hyderabad State in his *Glimpses of the Nizam's Dominions.* Some information is available on forest plants of the State in the *Imperial Gazetteer of India*. Provincial series, Hyderabad State (1909). Patridge (1911) described 450 species belonging to 69 families of Angiosperms and Gymnosperms in his book "Forest flora of Hyderabad State". Khan (1953) revised this book and added some more information on the vegetation and described 567 wild and cultivated species. Sayeeduddin (1935,1938 1941a, b, 1954) surveyed the Nizam's Dominions, Hyderabad, Deccan and reported a total of 370 species and published a series of papers. All the above reports include very few species with Medak distribution.

The district remained botanically under explored till 1960. Sebastine *et al.* (1960) enumerated 268 species from Narsapur reserve forest of Medak district. Narasimha Rao (1986) reported 616 taxa of Angiosperms including 116 cultivated species, belonging to 395 genera and 111 families from Narsapur taluk of Medak district.

Other chief works on the flora of Telangana region include Santapau (1954). Thothathri (1964), Sebastine and Henry (1966), Subba Rao and Kumari (1967), Ramayya and Rajagopal (1973), Kapoor and Kapoor (1973), Rajagopal (1973), Ramachandrachary (1980), Raju (1984), Ramachandrachary and Ramayya (1986), Raju and Ragan (1988, 1989, 1989), Ravishankar and Rama Rao (1989), Raghava Rao (1989) Pullalah *et al.*, (1992) Pullaiah and Ravi Prasad Rao (1995) and Silar Mohammad (1991).

Present Work

The present work on the flora of Medak, is based on intensive explorations by the authors during the last four years (1991-1995) and also on the critical analysis of collected specimens. In the present investigation a total of 708 species belonging to 414 genera under 119 families have been included.

Materials and Methods

FIELD STUDY

A few exploration trips were conducted during different months of a year covering all ranges of the forests and non-forest areas in the district. During field trips every plant was collected in quadruplicates either with flowering and/or fruiting stage. Each collection of the individual specimen was labelled with field numbers and every attempt was made to study the habit, habitat, colour of the flower, flowering and fruiting season, frequency of distribution and relative abundance. All the above information was recorded in the field itself. Special care was taken for collecting acquatic species, bulbs, corms, tubers etc.

HERBARIUM PREPARATION

For the preparation of Herbarium, methodology described by Santapau (1955) and Jain and Rao (1977) have been followed. The collected specimens were poisoned by dipping the whole plant or twig in saturated solution of mercuric chloride in ethyl alcohol. Immediately they were placed between the blotting papers with the help of forceps and a bunch of 25-50 specimens were tied properly with the help of iron plant pressers. After one day the blotting papers were changed and the specimens were spread properly in such a way that their physical appearance is not changed. These spread specimens were changed once again into dried blotting papers and tied in iron pressers. This process was continued for about 10-15 days, till the specimens were completely dried.

The poisoned, well pressed and dried specimens were pasted with glue (gum arabic) on a standard herbarium sheet measuring 42 x 28 cm. Only one specimen from the collected four was pasted on the herbarium sheet and the rest were kept as stock. Labels with all relevant information such as name of the plant, family, locality, altitude, date of collection and collector's name were affixed on the right hand bottom corner of the mounted sheets.

IDENTIFICATION

Every specimen was carefully studied by dissecting the floral parts of duplicate specimens under binocular stereoscope and compound microscope. Provisional identification was made with the help of Gamble's "Flora of the Presidency of Madras" and also other floras and monographs. Various parts of the dissected flower were mounted on a small adhesive paper board and placed it in a small pocket and in turn it is affixed on the left hand bottom side of the concerned herbarium sheet . Illustrations were drawn after critical study. The identifications were further confirmed after matching the specimens with the authentic specimens housed in the Madras Herbarium (MH). All the specimens are deposited in the herbarium of Botany Department, Sri Krishnadevaraya University, Anantapur.

Plan of Enumeration

In the systematic enumeration all the families are arranged in sequence according to Bentham and Hooker's (1862-83) system of classification with certain exceptions to accommodate recent modifications. All the genera in a family and all the species in a genus are arranged in alphabetical order. All the keys provided are of dichotomous type. The systematic enumeration is given in the following manner:

a. Dichotomous key for the genera under each family.
b. Genus name followed by its author.
c. Key to species under each genus, having 2 or more species.
d. Name of the species with original citation according to latest ICBN (Greuter, 1988).
e. Species description.

The nomenclature of taxa is brought up-to-date according to International Code of Botanical Nomenclature (Greuter, 1988), Uniformity is followed in abbreviating the author names, journals, books etc. Remarks on taxonomic affinities and other relevant information on nomenclature are provided wherever necessary.

Species description is presented in 4 paragraphs.

CITATION

At first the correct and valid name is given followed by its author name and place of publication. The correct name is followed by basionym if necessary and/or synonyms to connect the name with Hooker's "Flora of British India", Gamble's "Flora of the Presidency of Madras" and Khan's "Forest Flora of Hyderabad State". References of relevant monographic work or an important contribution on the concerned taxon is also added in the citation wherever necessary. For the flora of the Presidency of Madras the page number in parentheses represent 1957 edition whereas the page number outside the parentheses represent the original edition (1915-35).

DESCRIPTION

The terminology followed is based on Lawrence (1951) and Radford *et al.* (1974). The sequence is as follows: (a) Habit followed by maximum measurement of its tallness; stem or branch or branchlets nature. (b) Leaves shape (if compound then the nature of leaflets) followed by its range of lengh x breadth, base apex and margin. (c) Flowers position with nature of inflorescence, colour. (d) Fruit type, shape followed by maximum measurement of length × breadth, texture; seed number. All the measurements are given in metric system.

PHENOLOGICAL DATA

Flowering and Fruiting periods and vernacular names, if any are provided.

DISTRIBUTION

Frequency of distribution, locality in which the plant was collected followed by collectors's name with field numbers are given.

The cultivated plants are given at the end of the concerned family.

Floristic Analysis

In the present study a total of 956 field numbers have been collected and identified. These comprise 708 wild and naturalised species belonging to 414 genera and 119 families. Among the 708 species Dicots comprise 525 species, Monocots 179 species and Pteridophytes 4. The results are depicted in Table 4.

Table 4.4: Statistical Data of Flora of the Medak District

		Polypetalae	Gamopetalae	Monochlamydae	Monocots	Pteridophytes	Total
1.	No. of Families	54	25	15	21	4	119
2.	No. of Genera	144	135	38	93	4	414
3.	No. of Species	254	200	71	179	4	708

The ratio of Monocotyledons to Dicotyledons is 1 : 4.47 of families (21 : 94), 1 : 3.40 of genera (93 : 317); 1 : 2.94 of species (179 : 525). The ratio of genera to species in the present study is 1 : 1.7, whereas for the entire Indian region it is 1 : 7. This shows the small proportion of species to number of genera, confirming the general rule that smaller area comprise smaller range of floristic composition.

The total number of families, genera and species of Dicotyledons, Moncotyledons and Pteridophytes with their percentage to the total number of taxa is given in Table 5.

Table 4.5: Statistical Analysis of the Vegetation of the Medak District

	Dicotyledons		Monocotyledons		Pteridophytes		Grand Total	Overall %
	Total	%	Total	%	Total	%		
Families	94	78.99	21	17.64	04	3.36	119	100
Genera	317	76.57	93	22.46	04	0.96	414	100
Species	525	74.15	179	25.28	04	0.56	708	100

In order to get an insight into the relations of the Flora of Medak district with the neighbouring floras a comparative analysis of ten dominant families in Medak district, Flora of the Presidency of Madras and Flora of India is presented, in order the their frequency, in Table 6.

Table 4.6: Comparative Analysis of Dominant Families of the Medak District with other Floras

Rank	Medak District (Present Study)	No. of Species	Flora of the Presidency of Madras (Gamble)	No. of Species	Flora of India	No. of Species
1	Leguminosae	104	Leguminosae	432	Poaceae	1225
2	Poaceae	83	Poaceae	388	Orchidaceae	990
3	Cyperaceae	49	Rubiaceae	226	Leguminosae	775
4	Asteraceae	37	Acanthaceae	202	Asteraceae	754
5	Euphorbiaceae	31	Euphorbiaceae	199	Rubiaceae	495
6	Acanthaceae	22	Orchidaceae	199	Cyperaceae	449
7	Rubiaceae	20	Asteraceae	189	Euphorbiaceae	419
8	Lamiaceae	18	Cyperaceae	172	Lamiaceae	393
9	Convolvulaceae	17	Lamiaceae	128	Acanthaceae	379
10	Amaranthaceae	15	Asclepiadaceae	93	Scrophulariaceae	356

The genera having 5 or more than 5 species are *Cyperus* and *Eragrostis* with 12 species followed by *Crotalaria* and *Fimbristylis* (11); *Indigofera* (10); *Cassia* and *Ipomoea* (09); *Desmodium, Euphorbia, Phyllanthus* and *Schoenoplectus* (07); *Acacia, Alysicarpus, Ficus, Hedyotis, Heliotropium, Justicia* and *Leucas* (06); *Commelina* and *Grewia* (05).

5

KEY TO FAMILIES

Note: When a character applies to upto two genera/species in a family, these are given in brackets at the end of the concerned lead. The number of times a family keyed out is indicated by roman numerals in brackets after family name. In case of exception, the name of the genus/species is given in brackets at the end of the concerned lead.

1. Plants producing seeds (Angiosperms):
    2. Leaves usually net-veined, stipulate or exstipulate; flowers 2-polymerous; cotyledons usually two, lateral, tap root usually persistent:
        3. Perianth 2-seriate: differentiated into petals and sepals, petals free or perianth 1-many-seriate, segments not differentiated into petals and sepals, free or connate or perianth absent:
    4. Parasites or autophytes with tendrils:
        5. Parasites:
            6. Leafy; stamens 4-5 **Loranthaceae**
            6. Leafless; stamens 4-12;
                7. Branchlets flattened, noded; stamens 4 **Viscaceae**
                7. Branchlets wiry, not noded; stamens 12 **Lauraceae**
        5. Autophytes with tendrills:
            8. Inflorescence tip tendrillar (*Antigonon*) **Polygonaceae (I)**
            8. Inflorescence tip not tendrilar:
            9. Stamens 8, fruit inflated (*Cardiospermum*) **Sapindaceae (I)**
            9. Stamens 4-5; fruit not inflated
                10. Androgynophore present; flowers solitary **Passifloraceae**
                10. Androgynophore absent; flowers in cymes **Vitaceae**
    4. Autophytes without tendrils:
    11. Plants insectivorous with glandular hairs **Droseraceae**
    11. Plants not insectivorous, glandular hairs absent:
        12. Leaves absent, reduced to spines **Cactaceae**
        12. Leaves present, not reduced to spines:
            13. Fruit a legume or lomentum; placentation marginal:
                14. Flowers actinomorphic, mostly armed **Mimosaceae(I)**
                14. Flowers zygomorphic, mostly unarmed:
                    15. Corolla descendingly imbricate, the posterior petal outermost **Fabaceae**

15. Corolla ascendingly imbricate, the posterior petal innermost **Caesalpiniaceae**
12. Fruit not a legume or lomentum; placentation other than marginal:
16. Plants acquatic:
17. Plants stoloniferous, floating; leaves orbicular and dentate **Nymphaeaceae**
17. Plants not stoloniferous, submerged; leaves linear, entire **Ceratophyllaceae**
16. Plants terrestrial, rarely in moist places:
18. Trees (except *Annona squamosa* & *Ficus hispida*):
19. Stems sharply prickled; stamens polyadelphous **Bombacaceae**
19. Stems and stamens not as above:
20. Fruit a samara:
21. Leaves compound **Simaroubaceae**
21. Leaves simple **Ulmaceae**
20. Fruit other than samara:
22. Leaves simple:
23. Inflorescence a syconium (*Ficus*) **Moraceae (I)**
23. Inflorescence other than syconium:
24. Leaves opposite:
25. Stamens 5 **Santalaceae**
25. Stamens numerous:
26. Fruit a berry **Myrtaceae**
26. Fruit a capsule (*Lagerstroemia*) **Lythraceae**
24. Leaves alternate:
27. Flowers exceeding 3 cm across; petals bright-yellow or white:
28. Stamens monadelphous (*Thespesia*): **Malvaceae(I)**
28. Stamens free or connate at base:
29. Flowers bright yellow **Cochlospermaceae**
29. Flowers white **Barringtoniaceae**
27. Combination of characters not as above:
30. Flowers bright yellow or red; stamens 10-many:
31. Stamens 10-15 (*Helicteres*) **Sterculiaceae (I)**
31. Stamens more than 40 **Tiliaceae (I)**
30. Combination of characters not as above:
32. Perianth 1-seriate:
33. Stamens 4 (*Streblus*) **Moraceae (II)**
33. Stamens 5-many:
34. Stamens 5 (*Anogeissues & Terminalia*) **Combretaceae(I)**
34. Stamens 6-many **Flacourtiaceae**
32. Perianth 2-seriate (except in some Euphorbiaceae):
35. Flowers in panicles **Anacardiaceae (I)**
35. Flowers solitary or in clusters or in cymes:
36. Calyx tube adnate to ovary, ovary inferior **Alangiaceae**
36. Calyx not adnate to ovary, ovary superior:
37. Flowers unisexual **Euphorbiaceae (I)**
37. Flowers bisexual:
38. Stamens 5
38. Stamens numerous **Annonaceae**

22. Leaves compound:
  39. Stamens 10, compeltely connate into a tube **Meliaceae**
  39. Stamens otherwise:
    40. Leaves odd-pinnate
      41. Ovary on a long gynophore (*Crataeva*) **Capparaceae**
      41. Ovary sessile:
        42. Stamens 8 (*Lannea*) **Anacardiaceae (II)**
        42. Stamens 10-many:
          43. Leaves gland-dotted **Rutaceae**
          43. Leaves not gland-dotted **Burseraceae**
    40. Leaves even-pinnate:
      44. Leaves 2-6 foliolate (*Sapindus*) **Sapindaceae (II)**
      44. Leaves 20-30 foliolate **Flindersiaceae**

18. Shrubs or herbs:
  45. Leaves reduced to scales, amplexicaul at base, enclosing the stem **Tamaricaceae**
  45. Leaves well developed, not reduced to scales:
    46. Leaves compound (except *Cleome monophylla):*
      47. Leaves 2 foliolate **Balanitaceae**
      47. Leaves 3-many-foliolate:
        48. Fruits armed **Zygophyllaceae**
        48. Fruits unarmed:
          49. Flowers 4-merous; stamens 6-many **Cleomaceae**
          49. Flowers 5-merous; stamens 5+5 **Oxalidaceae**
  46. Leaves simple:
    50. Leaves armed with prickles **Papaveraceae**
    50. Leaves unarmed:
      51. Leaves opposite or whorled:
      52. Sepals 2, leaves fleshy **Portulacaceae**
        52. Sepals 4-5 or absent, leaves other than fleshy:
          53. Climbing shrubs:
            54. Samara 2-3-winged **Malipighiaceae**
            54. Drupe 4-winged (*Combretum)* **Combretaceae (II)**
          53. Erect or prostrate:
            55. Flowers unisexual **Euphorbiaceae (II)**
  55. Flowers bisexual:
    56. Leaves sharply serrate **Elatinaceae**
    56. Leaves entire:
      57. Fruit a glandular anthocarp **Nyctaginaceae**
      57. Fruit other than glandular anthocarp:
        58. Calyx covered with bulbous-based bristles, stamens 8-10 **Melastomaceae**
        58. Calyx and stamens not as above:
        59. Flowers 4-merous: **Lythraceae**
        59. Flowers 3-or 5-merous:
          60. Perianth dichlamydeous **Caryophyllaceae**
          60. Perianth monochlamydeous:
            61. Flowers in sessile clusters **Amaranthaceae (J)**

61. Flowers solitary or in umbellate cymes:
62. Tepals free **Aizoaceae**
62. Tepals connate **Molluginaceae**
51. Leaves alternate:
63. Perianth monochlamydeous:
64. Perianth irregular; limb 1-or 2-lipped, exceeding 2.5 cm in length **Aristolochiaceae**
64. Perianth regular; limb absent:
65. Fruit a winged capsuale (*Dodonaea*) **Sapindaceae (III)**
65. Fruit other than a winged capsule:
66. Flowers unisexual:
67. Climbers **Menispermaceae**
67. Non-climbers (except *Tragia*):
68. Inflorescence cyathium or racemes or panicles, perianth not spurred **Euphorbiaceae (III)**
68. Flowers in globose heads, perianth spurred **Urticaceae**
66. Flower bisexual (except *Amaranthus*):
69. Leaves stipulate **Polygonaceae (II)**
70. Stamens connate at base **Amaranthaceae (II)**
70. Stamens free:
71. Stem and leaves succulent **Basellaceae**
71. Stem and leaves not succulent **Chenopodiaceae**
63. Perianth dichlamydeous:
72. Flowers unisexual; stamens 10-15 (*Chrozophora*) **Euphorbiaceae (IV)**
72. Flowers bisexual or polygamous; stamens 3-9, if 10 then plants aquatic or armed:
73. Stamens 6, tetradynamous **Brassicaceae**
73. Stamens otherwise:
74. Stamens 8-many, monadelphous:
75. Stamens 8; anthers oblong **Polygalaceae**
75. Stamens numerous; anthers reniform **Malvaceae (II)**
74. Combination of characters not as above:
76. Leaves pinnatifid or reniform; fruit a cremocarp **Apiaceae**
76. Leaves and fruits otherwise:
77. Lower sepal 2-3 cm in length, spurred **Balsaminaceae**
77. Lower sepal not spurred:
78. Petals unequal; pedicel jointed; stamens 5, anthers connate **Violaceae**
78. Combination of characters not as above:
79. Ovary seated on 1-4 cm long gynophore **Capparaceae (II)**
79. Ovary not as above:
80. Plants erect, unarmed:
81. Styles 3 **Turneraceae**
81. Styles solitary:
82. Stamens free **Tiliaceae (II)**
82. Stamens connate or inserted on petals:
83. Stamens united into a staminal column **Leeaceae**
83. Stamens not united into a staminal column:

84. Stamens 5, connate (*Waltheria & Melochia*) **Sterculiaceae (II)**
84. Stamens 4, 8 or 10 inserted on petals **Onagraceae**
80. Plants straggling or climbing, if erect then armed:
85. Stamens 5, all perfect:
86. Fruit a capsule **Celastraceae**
86. Fruit a drup or samara **Rhamnaceae (II)**
85. Stamens 8-10, with or without staminodes: **Olacaceae**
3. Perianth 2-seriate; differentiated into petals and sepals, petals connate:
87. Insectivores or parasites; leaves absent:
88. Insectivorous, aquatic; stamens 2 **Lentibulariaceae**
88. Parasitic, terrestrial; stamens 5 **Cuscutaceae**
87. Autophytes (except *Striga*); leaves present (except *Caralluma*):
89. Plants with tendrils **Cucurbitaceae**
89. Plants without tendrils:
90. Inflorescence a capitulum; calyx reduced to pappus **Asteraceae**
90. Inflorescence other than capitulum; calyx not reduced to pappus:
91. Fruit a legume or lomentum **Mimosaceae (II)**
91. Fruit other than a legume or lomentum:
92 Aquatic; leaves peltate, floating **Menyanthaceae**
92. Aquatic or terrestril; leaves otherwise:
93. Plants scapigerious, leaves radical **Plantaginaceae**
93. Plants non scapigerous, leaves cauline:
94. Leaves alternate, atleast the lower ones:
95. Trees:
96. Calyx persistent:
97. Stamens free from the corolla **Ebenaceae**
97. Stamens inserted on the corolla (*Cordia & Ehretia*) **Cordiaceae (I)**
96. Calyx deciduous **Sapotaceae**
95. Shrubs or herbs:
98. Staments 8-10; capsules 3-valved (*Jatropha*) **Euphorbiaceae (V)**
98. Combination of characters not as above:
99. Fruit a drupe:
100. Latex milky (*Cascabela*) **Apocynaceae (I)**
100. Latex watery or absent:
101. Stamen filaments not exceeding 0.4 mm in length (*Coldenia & Heliotropium*) **Boraginaceae (I)**
101. Stamen filaments exceeding 2 mm in length (*Carmona*) **Cordiaceae (II)**
99. Fruit other than a drupe:
101. Calyx tube exceeding 1 cm in length, conspicuously glandular **Plumbaginaceae**
102. Calyx tube otherwise:
103. Fruit a berry or prickly capsule **Solanaceae**
103. Fruit a capsule, not prickly:
104. Stamens 2-4 **Scrophulariaceae (I)**
104. Stamens 5:

105. Branchlets winged: leaves serrate **Lobeliaceae**
105. Combination of characters not a above **Convolvulaceae**
94. Leaves opposite or whorled, atleast the lower ones:
106. Ovary inferior; leaves with intra- or inter-petiolar stipules **Rubiaceae**
106. Ovary superior; leaves stipulate or not:
107. Latex milky; fruit a follicle:
108. Anthers connate with the stigma; pollinia and corona present; stigma 5-gonous **Ascelepiadaceae**
108. Anthers connivent around the stigma; pollinia and corona absent; stigma not 5-gonous **Apocynaceae (II)**
107. Latex watery or absent:
109. Climbing shrubs or trees; stamens 2 **Oleaceae**
109. Combination of characters not as above:
110. Trees:
111. Leaves pinnately compound **Bignoniaceae**
111. Leaves simple or digitately compound:
112. Fruit a berry **Loganiaceae**
112. Fruit a drupe **Verbenaceae (I)**
110. Shrubs or herbs:
113. Flowers actinomorphic; stamens 4-5, equal:
114. Stems hispid: stamens connivent into a cone (*Trichodesma*) **Boraginaceae (II)**
114. Stems glabrous or pubescent; stamens otherwise:
115. Fruits beaked, exceeding 1.5 cm in length **Pedaliaceae**
115. Fruits not as above **Gentianaceae**
113. Flowers zygomorphic; stamens 2 or 4, didynamous:
116. Fruit a capsule; placentation axile:
117. Fruit opening elastically from the apex of 2 loculicidal valves; seeds with retinacula **Acanthaceae**
117. Fruit not opening elastically; seeds lacking retinacula **Scrophulariaceae (II)**
116. Fruit a drupe, pyrene or nutlets, placentation basal:
118. Style gynobasic; ovary 4-lobed **Lamiaceae**
118. Style terminal; ovary entire **Verbenaceae (II)**
2. Leaves usually parallel-veined, exstipulate; flowers usually 3-merous; cotyledons 1, terminal or undifferentiated; mature root system adventitious. Stem without a central pith, vascular bundles scattered, withut cambium;
119. Aquatic; plants thalloid, not exceeding 8 mm across **Lemnaceae**
119. Aquatic or terrestrial; plants otherwise
120. Plants stoloniferous, free-floating **Araceae**
120. Plants otherwise:
121. Aquatic, submerged:
122. Perianth 2-seriate, 6-lobed:
123. Flowers in spathes; leaves < 1 cm across (except *Ottelia*). **Hydrocharitaceae**
123. Combination of Characters not as above:
124. Flowers in racemes or spikes; leaves > 5 cm across **Pontederiaceae**
124. Flowers in umbels, involucrate with spathaceous bracts **Alismataceae**

122. Perianth 1-seriate; 2-4-lobed:
125. Leaves radical; tepals 2 **Aponogetonaceae**
125. Leaves cauline; tepals 4 **Potamogetonaceae**
121. Aquatic, emergent or terrestrial:
126. Flowers in spilkelets:
127. Stems hollow (excepted *Dendrocalamus*), noded; phyllotaxy 1/2; flowers subtended by 2 glumaceous bracts **Poaceae**
127. Stems solid, noded or not; phyllotaxy 1/3; flowers subtended by single bract **Cyperaceae**
126. Flowers not in spikelets:
128. Climbers **Dioscoreaceae**
128. Non-climbers (except *Asparagus & Gloriosa*):
129. Inflorescence a densely-pappose cylindrical brown spike **Typhaceae**
129. Inflorescence otherwise
130. Leaves fan-like or 1-pinnate **Arecaceae**
130. Leaves simple, not fan-like:
131. Perianth absent **Pandanaceae**
131. Perianth present:
132. Leaves exceeding 75 cm in length, apex spiny **Agavaceae**
132. Leaves not as above:
133. Leave radical, not exceeding 3 mm across **Eriocaulaceae**
133. Combination of characters not as above:
134. Flowers actinomorphic: perfect stamens 3-6:
135. Bracts exceeding 4 cm in length **Amaryllidaceae**
135. Bracts absent, if present not as above:
136. All perianth segments corolline:
137. Flowers bright yellow **Hypoxidaceae**
137. Flowers otherwise **Liliaceae**
136. Outer perianth segments calycine **Commelinaceae**
134. Flowers zygomorphic; perfect stamens 1-2:
138. Median petal modified into a lip; staminodes absent **Orchidaceae**
138. Median petal otherwise; staminodes petaloid **Cannaceae**
1. Plants not producing seeds (Pteridophytes)
139. Aquatic; leaves 4-foliolate **Marsiliaceae**
139. Terrestrial; leaves otherwise:
140. Stem heterophyllous **Selaginellaceae**
140. Stem homophyllous:
141. Leaves fan-shaped **Actinopteridaceae**
141. Leaves otherwise **Cheilanthaceae**

6

SYSTEMATIC ENUMERATION

ANNONACEAE

1. Flowers greenish-yellow; fruits syncarpic **Annona**
1. Flowers purple; fruits not syncarpic **Miliusa**

ANNONA Linnaeus

Annona squamosa L. Sp. Pl. 537. 1753; FBI 1:78; Gamble 20 (14); Khan in FFH 52

Trees or Shrubs, about 5 m tall; branchlets glabrous. Leaves oblong-lanceolate, 2-4 × 0.5-1.5 cm, glabrescent, base cuneate, apex obtuse-acute, margin entire. Flowers axillary, solitary or 2-4 in clusters, greenish-yellow. Fruits ovate-globose, 7 cm across, glaucous, seeds many.

Fls. & Frs.: March - October. *Vern.*: Sithaphalam.

Common in villages, forests and waste lands. Ramayampet, **BR & CP** 11510.

MILIUSA Leschenault

Miliusa tomentosa (Roxb.) Sinclair in Gard. Bull. Straits. 14: 378. 1955. *Uvaria tomentosa* Roxb. Pl. Corom. t. 35. 1795. *Saccopetalum tomentosum* (Roxb.) Hook. f. & Thoms. Fl. Ind. 152. 1855; FBI 1:88; Gamble 22 (16); Khan in FFH 53.

Trees, about 20 m tall; branchlets tomentose. Leaves elliptic-ovate, 8-16 × 4.5-8 cm, tomentose, base rounded, apex acute-obtuse, margin entire. Flowers axilary, solitary or in fascicles, purple. Fruitlets subglobose, 2 x 1.5 cm; seeds 2.

Fls. & Frs.: March - may *Vern.* : Chiladudi.

Rare in forests. Narsapur RF, **KMS** 8011 and **GNR** 4391.

Polyalthia longifolia (Sonn.) Thw. is planted as an avenue tree. *Annona reticulata L.* is cultivated for edible fruits.

MENISPERMACEAE

1. Leaves peltate; stamens connate **Cissampelos**

1. Leaves not peltate; stamens free:
   2. Leaves hirsute; flowers in short panicles **Cocculus**
   2. Leaves glabrous; flowers in racemes **Tinospora**

CISSAMPELOS Linnaeus

Cissampelos pareira L., Sp. Pl. 1031. 1753. var. **hirsuta** (Buch. - Ham. ex DC.) Forman in Kew Bul. 22 : 356. 1968. *C. hirsuta* Buch-Ham. ex DC., Syst. Nat. 1 : 535. 1817. *C. pareira L.*, Sp. Pl. 1031. 1753. *p.p.*; FBI 1 : 103; Gamble 30 (21); Khan in FFH 55.

Climbing shrubs; branchlets tomentose. Leaves reniform cordate, 5-6 × 5-7 cm, pubescent, base cordate, apex retuse, mucronate, margin entire. Male flowers in subcorymbose cymes, female flowers in racemes, greenish-yellow. Drupes ovate, 0.5 x 0.4 cm; seed 1.

Fls. & Frs. : September - March. *Vern. :* Adavibanka teega.

Common on bushes and trees in forests. Narsapur RF, **TP & CP** 14021 & 14038.

COCCULUS A. P.de Condolle *nom. cons.*

Cocculus hirsutus (L.) Diels in Engl., Pflanzenr. 46 : 236. 1910; Gamble 29 (21). *Menispermum hirsutum* L., Sp. Pl. 341. 1753. *Cocculus villosus* (Lam.) DC., Syst. Nat. 1: 525. 1817; FBI 1 : 101; Khan in FFH 55. *Menispermum villosum* Lam., Encycl. 4 : 97. 1797.

Climbing shrubs; banchlets rusty-tomentose. Leaves ovate, 0.5-3 x 0.3-1.7 cm, chartaceous, base obtuse, apex mucronate, margin entire. Flowers axillary, male in panicles, female in fascicles, green. Drupes globose, compressed, 0.3 cm across; seed 1.

Fls. & Frs. : Round the year. *Vern. :* Dusarateega

Common over bushes and trees and in field hedges. Gangapur RF, **BR & CP 11445;** Pochammaralu, **BR & CP 11483;** Venkatapuram, **TP & CP 14002.**

TINOSPORA *Miers nom. cons.*

Tinospora cordifolia (Wild). Hook. f. & Thoms., Fl. Ind. 184. 1855; FBI 1: 97; Gamble 26 (19); Khan in FFH 55. *Menispermum cordifolium* Wild., Sp. Pl. 4: 826. 1806.

Climbing shrubs, branchlets striated, glabrous. Leaves cordate, 2.5-5 x 2-4.5 cm, glabrous, base cordate, apex acuminate, margin entire. Flowers axillary, in racemes, green. Drupes globose, 0.5 cm across, seed 1.

Fls. & Frs. : Round the year. *Vern.* : Tippatheega.

Common on field hedges. Medak, **BR & CP** 11484; Pathur thanda, **TP & CP 12103.**

NYMPHAEACEAE

Nymphaea Linnaeus *nom. cons.*

Nymphaea nouchali Burm.f., Fl. Ind. 120. 1768. *N. stellata* Wild., SP. Pl. 2 : 1154. 1799; FBI 1 : 114, Gamble 33 (24); Khan in FFH 57.

Stoloniferous aquatic herbs. Leaves orbicualr, 12-20 x 10-20 cm, glabrous, base deeply cordate, margin sharply dentate. Flowers solitary, white-rose. Berries globose, 3 cm across; seeds many.

Fls. & Frs. : September - February. *Vern.* : Kaluva.

Common in Ponds and tanks. Shivampet, **TP & CP** 14001; Dosalagunda lake, **KMS** 6793; Bonthapalle, **GNR** 4440.

PAPAVERACEAE

Argemone Linnaeus

Argemone mexicana L., Sp. Pl. 508. 1753. FBI 1 : 117, Gamble 35 (25); Khan in FFH 58.

Erect prickly herbs, about 80 cm tall; branchlets with yellowish sap. Leaves pinnatifid, 4-12 x 1-4.5 cm, glaucous, base subamplexicaul, margin spinulose-dentate. Flowers terminal, solitary, bright yellow. Capsules oblong, 4.5 x 2 cm, prickly; seeds many.

Fls. & Frs. : February - May. *Vern.* : Yerrikusuma.

Common weed of dry fields and waste lands. Mombojipally, **BR & CP** 11449.

BRASSICACEAE (*nom.alter.* CRUCIFERAE)

1. Slender herbs; sepals not pouched at base **Rorippa**
1. Stout herbs; sepals pouched at base **Brassica**

Brassica Linnaeus

Brassica juncea (L.) Czern., Consp. Pl. Charc. 8 1859; FBI 1 : 157; Gamble 38 (28). *Sinapsis juncea* L., Sp. Pl. 668. 1753.

Erect herbs, about 75 cm tall; branchlets glabrous. Leaves lyrate, 2-7.5 x 1-3.7 cm, glabrous, base attenuate, apex obtuse-subacute, margin irregularly dentate. Flowers terminal, in racemes, yellow. Siliqua linear, 6 x 0.3 cm, subglabrous; seeds many.

Fls. & Frs. : September - February. *Vern.* : Avaalu.

Cultivated, ofen as an escape, Medak, **BR & CP** 11457.

Rorippa Scopoli

Rorippa indica (L.) Hiern, Cat. Afr. Pl. Welw. (Addit. & Corr.) 1: 26. 1896. *Sisymbrium indicum* L., Mant. Pl. 93, 1767. *Nasturtium indicum* (L.) DC., Prodr. 1: 139. 1824; FBI 1: 134. Gamble 37 (27).

Erect herbs, about 30 cm tall, branchlets glabrous. Leaves lyrate, pinnatifid, 6-10 x 3-4 cm; pinnae ovate, glabrous, base attenuate, apex aucte, margin dentate. Flowers terminal, in racemes, yellow. Siliqua 1.5 x 0.2 cm, glabrous; seeds many.

Fls. & Frs. : November - March.

Occasional in cultivated fields and moist localities. Brahmanapalle, **GNR** 4833.

CAPPARACEAE

CAPPARIS Linnaeus

1. Unarmed, trees — **C. grandis**
1. Armed, climbing shrubs:
   2. Flowers in corymbs or in umbels — **C. sepiaria**
   2. Flowers in supra axillary rows — **C. zeylanica**

Capparis grandis L.f., Suppl. Pl. 263. 1781; FBI 1 : 176; Gamble 46 (33); Khan in FFH 62.

Trees, about 5m tall; branchlets velvety-tomentose. Leaves elliptic-obovate, 2.5-5 x 1.5-2.5 cm, thin-coriaceous, base truncate, apex obtuse-acute, margin entire. Flowers axillary or terminal, in racemes or umbels, creamish-white, Berries globose, 2 cm across, seeds 3-6.

Fls. & Frs. : February - May. Vem.: Regoti.

Occasional in dry deciduous forests. Pocharam RF, **BR & CP** 11604.

Capparis sepiaria L. Syst. Nat 2; 1071. 1759; FBI 1: 177; Gamble 46 (32).

Prickly shrubs, about 2 m tall; branchlets terete, grey-pubescent or densely fulvous. Leaves oblong-elliptic or obovate, 1.5 x 0.8-1.5 cm, coriaceous, base cuneate, apex notched, margin entire. Flowers terminal, in corymbose umbels, white. Berries globose, 1 cm across; seeds 2.

Fls. & Frs. : April - July.

Occasional in field hedges. Ranjhole, **CP** 11599.

Capparis zeylanica L., Sp. Pl. ed. 2.720. 1762; Gamble 46(33); Khan in FFH 62. *C.horrida* L.f., Suppl. Pl. 264. 1781; FBI 1 : 178.

Climbing shrubs; branchlets grey-pubescent. Leaves elliptic-ovate, 2.5-4 x 1.2-2 cm, coriaceous, base cuneate, apex retuse, margin entire; spines recurved. Flowers supra axillary, in rows, whitish-rose. Berries ellipsoid, 3.5 x 2.5 cm, glabrous; seeds many.

Fls. & Frs. : February - May *Vern.* : Adavinimma.

Common in forests and hedges. Mombojipally, **BR & CP** 11453.

CLEOMACEAE

CLEOME Linnaeus

Note: A.P. De Condolle (Prodr.1.242.1824) treated *Cleome* L, *Gynandropsis* DC. and *Polanisia* Rafin. as distinct genera. Schultes f. (Syst.1829) merged all the three genera into one genus *Cleome* L. Based

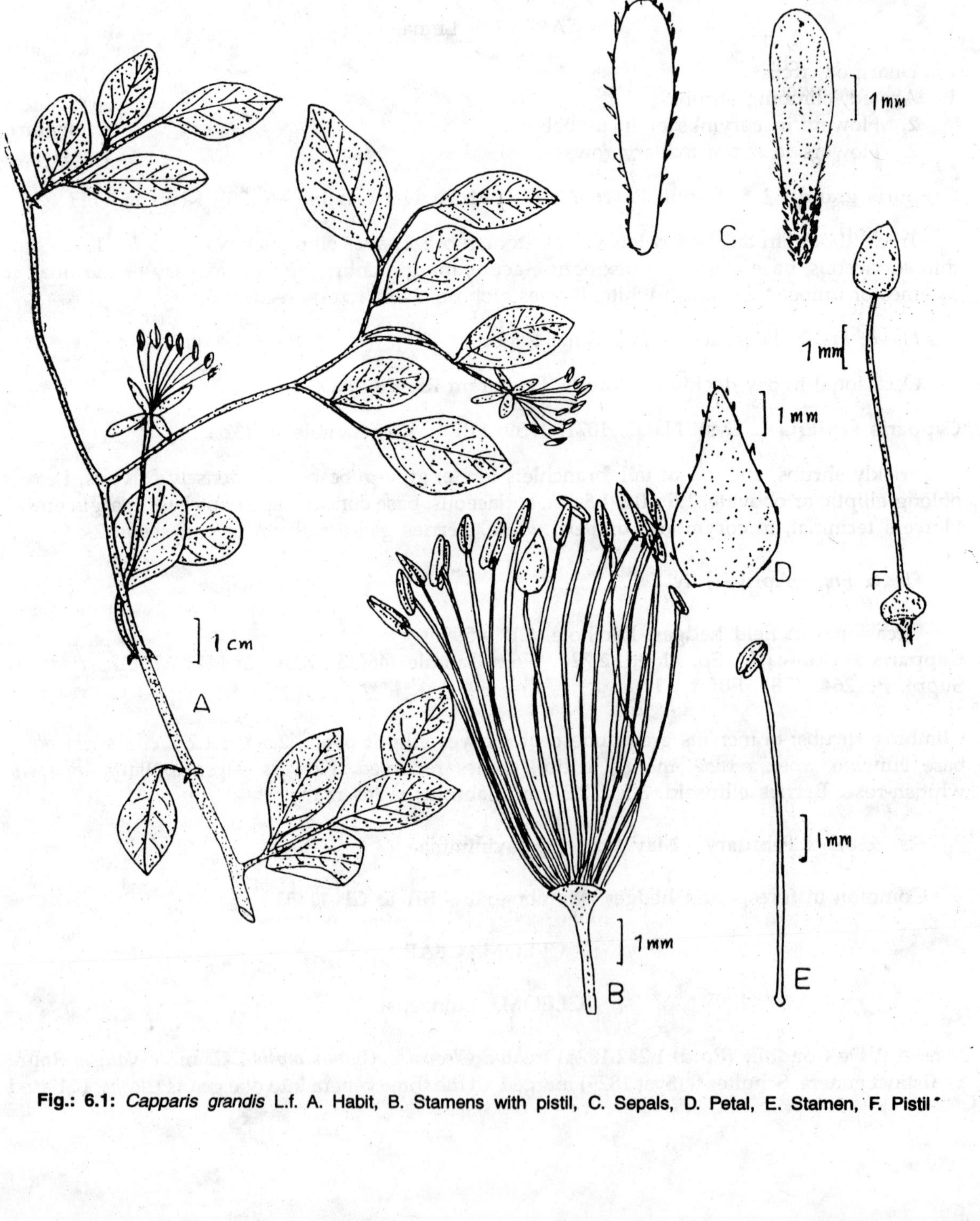

Fig.: 6.1: *Capparis grandis* L.f. A. Habit, B. Stamens with pistil, C. Sepals, D. Petal, E. Stamen, F. Pistil

on morphological characters lltis (Brittonia 10: 1958 & 12: 1960) and Jacobs (Fl. Males. 6: 100.1960) have treated *Gynandropsis* under *Cleome* L. as *C. gynandra*. In view of good evidence from various disciplines the conception of lltis (*loc. cit.*) and Jacobs (*loc. cit.*) in treating *Gynandropsis* under *Cleome* is adopted in the present work.

1. Leaves simple **C. monophylla**
1. Leaves 3-5-foliolate:
    2. Ovary long stalked **C. gynandra**
    2. Ovary sessile:
        3. Stems with glandular prickles; stamens 6 **C. aspera**
        3. Stems viscid; stamens many **C. viscosa**

Cleome aspera J. Koenig ex DC., Prodr. 1: 241. 1824; FBI 1: 169; Gamble 41 (29).

Spreading herbs; branchlets ascending, minutely-prickled. leaves 3-foliolate, leaflets oblanceolate, 0.5-2 x 0.2-.0.3 cm, hairy, base cuneate, apex subacute, margin entire. Flowers axillary, solitary, yellow. Capsules cylindric, 3 x 0.2 cm, glabrous, obliquely striate; seeds many.

Fls. & Frs.: Round the year

Occasional in waste lands. Pochammaralu, CP 11759.

Cleome gynandra L, Sp. Pl. 671. 1753. *Gynandropsis pentaphylla* DC., Prodr. 1: 238. 1824; FBI 1: 171; Gamble 40 (29).

Erect herbs, about 50 cm tall; branchlets glandular-pubescent. Leaves 3-5-foliolate; leaflets oblanceolate, 1.3 x 0.5 - 1.7 cm, glabrous, base cuneate, apex acute, margin entire-shortly dentate. Flowers terminal, in corymbose racemes, white -pink. Capsules linear, 8 x 0.2 cm, terete, glandular; seeds many.

Fls. & Frs.: Round the year, Vern.: Vaminta.

Common in cultivated fields and waste lands. Narsingi, **BR & CP** 11532; Lingareddypally, **CP** 11715.

Cleome monophylla L., Sp. Pl.672. 1753; FBI 1;168; Gamble 41 (29).

Erect herbs, about 50 cm tall; branchlets pubescent. Leaves oblong-lanceolate, 2-4 x 0.5-2 cm, pubescent, base truncate, apex acute, margin entire, ciliate. Flowers axillary, solitary, pink. Capsules cylindric, 6 x 0.2, cm, hairy, seeds many.

Fls. & Frs.: July - January.

Common in cultivated fileds and waste lands. Marrikunta, **CP** 11386; **NSF, BR & CP** 11657.

Cleome viscosa L., Sp. Pl.672.1753; FBI 1: 170; Gamble 41 (29).

Erect herbs, about 50 cm tall; branchlets gandular viscid. Leaves 3-5-foliolate: leaflets obovate-elliptic, hairy, base cuneate, apex acute-obtuse, margin entire. Flowers axillary, solitary, yellow. Capsules cylindric, 6.5 x 0.2 cm, hairy; seeds many.

Fls & Frs.: Round the year. Vern.: Kukka vaminta.

Common weed of cultivated fields and waste lands. Sivanoor, **BR & CP** 11546.

VIOLACEAE

HYBANTHUS N.J. Jacquin *nom. cons.*

Hybanthus enneaspermus (L) F. Muell., Fragm. 10: 81. 1876. *Viola enneasperma* L., Sp. Pl. 937.1753. *V. suffruticosa* L., Sp. Pl.937.1753. *Ionidium suffruticosum* (L.) Ging. ex DC., Prodr. 1: 311. 1824; FBI 1: 185; Gamble 49 (35).

Erect herbs, about 25 cm tall; branchlets puberulous. Leaves lanceolate, 1-2.5 x 0.2-0.5 cm, glabrescent, base attenuate, apex acute, margin serrate. Flowers axillary, solitary, purple-red. Capsules subglobose, 0.7 x 0.6 cm, sparsely hairy; seeds 8-10.

Fls. & Frs.: August - December. Vern.: Ratnapurusha.

Common weed of drylands, fields and forests. Pillutla, **BR & CP** 11264: Thornal, CP 11399.

COCHLOSPERMACEAE

COCHLOSPERMUM Kunth ex A.P. de Condolle *nom. cons.*

Cochlospermum religiosum (L.) Alston in Trimen, Handb. Fl. Ceylon (Suppl.) 6: 14.1931. *Bombax religiosum* L., Sp. Pl.552. 1753. *Cochlospermum gossypium* DC., Prodr. 527. 1824; FBI 1: 190; Gamble 50(36); Khan in FFH 64.

Trees, about 10m tall; branchlets pubescent. Leaves 3-5-lobed, 5-18 x 7-14 cm, glabrous above, grey-tomentose below: lobes ovate-lanceolate, base cordate, apex acute, margin entire. Flowers terminal, in corymbose panicles, golden yellow. Capsules obovate, 6x4 cm, 5-valved, woolly; seeds many.

Fls. & Frs.: March - June. Vern.: Kondagogu.

Occasional in forests. Narsapur RF, **TP & CP** 14049.

FLACOURTIACEAE

FLACOURTIA Commerson ex L' Heritier

Flacourtias indica (Burm.f.) Merr., Interpr. Rump. Herb Amb.377. 1917. *Gmelina indica* Burm. f., Ind. 132. t. 39. f. 5. 1768. *Flacourtia ramontchi* L'Herit., Strip. Nov. 3 : 59. 1785; FBI 1: 193; Gamble 54 (39); Khan in FFH 65. *F. sepiaria* Roxb., Pl. Corom.t. 68. 1795; FBI, Gamble & Khan *loc. cit.*

Armed trees, about 8 m tall; spines axillary. Leaves elliptic - ovate, 1.5-3 x 1-1.7 cm, glabrous above, pubescent below, base obtuse-rounded,apex obtuse, margin serrate. Flowers in short racemes, greenish-yellow. Drupes globose, 1.2 cm across; seeds 8-14.

Fls. & Frs.: February - May. Vern.: Kanregu.

Occasional in forests. Gangapur RF, **BR & CP** 11433; Pocharam RF, BR & CP 11614.

POLYGALACEAE

POLYGALA Linnaeus

1. Flowers in 1-2.5 cm long axillary racemes **P. arvensis**
1. Flowers in 5-10 cm long terminal racemes **P. elongata**

Polygala arvensis Willd., Sp. Pl 3: 876. 1802; Chandrabose and Nair in Proc. Ind. Acad Sci. (Plant Sci.) 90; 123. 1981. *P. chinensis* auct. non L: FBI 1: 204; Gamble 58(42).

Erect herbs, about 50 cm tall, branchlets sparsely pubescent. Leaves elliptic-oblong or obovate, 0.5-3.5 x 0.3-1 cm, glabrous-pubescent, base obtuse, apex obtuse-apiculate, margin entire. Flowers axillary, in racemes, yellow. Capsules ovate, 0.5 x 0.4 cm, margins pubescent; seeds 2.

Fls. & Frs.: Round the year.

Common weed of waste lands. Narsapur RF, **TP & CP** 14023.

Polygala elongata Klein ex Willld., Sp. Pl. 3: 879. 1802; FBI 1: 203: Gamble 59(42); Chandrabose and Nair in Proc.Ind. Acad. Sci. (Plant Sci.) 90: 125. 1981.

Erect herbs, about 30 cm, tall; branchlets sparsely pubescent. Leaves lanceolate-elliptic, 2-3.5 x 0.4-0.7 cm, glabrous, base obtuse, apex rounded, margin entire. Flowers solitary or in racemes. yellow. Capsules oblong, 0.5 x 0.2 cm, densely pubescent; seeds 2.

Fls. & Frs.: June - December.

Common in waste lands and open forests. Pillutla RF, **BR & CP** 11262; Nagulabanda, CP 11329; Venkatapuram, **TP & CP** 14009; Narsapur, **TP & CP** 14022.

CARYOPHYLLACEAE

POLYCARPAEA Lamarck *nom. cons.*

1. Flowers in terminal cymes; internodes densely tomentose **P. aurea**
1. Flowers in axillary and terminal cymes; internodes with scattered hairs **P. corymbosa**

Polycarpaea aurea Wight & Arn., in Nat. Hist. Ser. 1.3: 91. 1839; Gamble 65 (47) *P. corymbosa* Lam. var *aurea* (Wight & Arn.) Wight, Ill 2: 44.t. 110. 1850; FBI 1: 245.

Undershurbs; branchlets whitish. Leaves linear, needle-like. Flowers in compound racemes, whitish-yellow.

Fls. & Frs.: August - December.

Occasional in rocky areas. Gummadidala, **GNR** 4459.

Polycarpaea corymbosa (L.) Lam., Tabl.Encycl. 2: 129. 1792;FBI 1: 245; Gamble 65 (46).*Achyranthes corymbosa* L., Sp. Pl. 205.1753.

Erect herbs, about 20 cm tall; branchlets densely villous. Leaves linear, 0.5-1 x .01 cm, glabrescent, needle-like. Flowers axillary and terminal, in dense corymbose cymes, silverywhite. Capsules oblong, 1 x 0.5 mm, seeds 3-7.

Fls. & Frs.: December - March,. Vern.: Bommasari.

Occasional in forests and dry fields. Narsapur RF, **BR & CP** 11228; Pochammaralu, **BR & CP** 11476.

PORTULACACEAE

PORTULACA Linnaeus

1. Stems with hairs at nodes **P. suffrutiosa**
1. Stems without hairs at nodes **P. oleracea**

Portulaca oleracea L., Sp. Pl. 445. 1753; FBI 1: 246; Gamble 66 (47). var. **oleracea.**

Prostrate herbs; stem and leaves fleshy, glabrous. Leaves obovate-spathulate, 0.4-1.5 x 0.1-0.7 cm, base cuneate, apex obtuse-rounded, margin entire, Flowers terminal, in clusters, yellow. Capsules ovate, 0.6 x 0.3 cm, glabrous: seeds many.

Fls. & Frs.: Round the year Vern.: Pappukura.

Common in cultivated fields and moist localities. NSF, **BR & CP** 11666.

Protulaca oleracea L. Sp. Pl. 445. 1753; FBI 1: 246;Gamble 66(47). var. **linearifolia** Sivarajan et Manilai in New Botanist 4: 30. 1977 & J Bombay Nat.Hist. Soc. 78: 258.1981.

Leaves linear, 0.3-1.2 x 0.1-0.3 cm, seeds 55-75 per capsule, very small. All other characters similar to the above variety.

Fls. & Frs.: Round the year.

Common in cultivated fields. Ramayampet, **BR & CP** 11512; NSF, **BR & CP** 11667.

Portulaca suffruticosa Wight ex Wight & Arn. Prodr. 356.1834: FBI 1: 247; Gamble 66 (48).

Erect herbs, about 10 cm tall; branchlets from base, terete. Leaves fleshy, 0.5-2 x 0.1 cm, puberulous, base cuneate, apex subacute, margin entire. Flowers terminal, solitary or 2-3 in clusters, violet-pink, Capsules subglobose-oblong, 0.5 cm across; seeds many.

Fls. & Frs.: November - March.

Common in moist places. Narsapur tank, **TP & CP** 12010.

TAMARICACEAE

TAMARIX Linnaeus

Tamarix ericoides Rottler in Ges. Naturf. Freunde Berlin Neue Schriften 4: 214. 1803; FBI 1: 249; Gamble 68(49); Khan in FFH 67. *Trichaurus ericoides* (Rottler) Arn. ex Wight & Arn., Prodr. 40.1834.

Shrubs, about 1.5 m tall; branchlets jointed. Leaves ovate-lanceolate, scale like, 0.1-.0.3 cm, grey-pubescent, base amplexicaul, apex abruptly acuminate. Flowers axillary and terminal, in compact racemes,rose, Capsules conical, 2 x 0.6 cm, glabrous; seeds many.

Fls. & Frs.: December - February.

Common in riverbeds and sandy tankbeds. Pochammaralu, **BR & CP** 11496.

ELATINACEAE

BERGIA Linnaeus

Bergia ammannioides Roxb. ex Roth, Nov. Pl. Spec.219. 1821; FBI 1: 251; Gamble 69(50).

Erect herbs, about 20 cm tall; branchlets wiry, glandular-pubescent, Leaves Oblanceolate-elliptic, 0.3-0.8 x 0.1-0.3 cm, pubescent, base cuneate, apex acute, margin sharply serrate. Flowers at nodes. in condensed clusters, reddish-brown. Capsules globose, 0.2 cm across, glabrous; seeds many.

Fls. & Frs.: November-March.

Common in moist places. Medak, **BR & CP** 11424; Gangapur **TP & CP** 12097.

MALVACEAE

1. Fruit a schizocarp :
    2. Epicalyx present:
        3. Leaves not lobed **Malvastrum**
        3. Leaves lobed :
            4. Mericarps winged **Pavonia**
            4. Mericarps not winged **Urena**
    2. Epcialyx absent:
        5. Leaves cordate; carpels 2-or more-seeded **Abutilon**
        5. Leaves ovate-lanceolate; carpels 1- seeded **Sida**
1. Fruit a capsule:
    6. Calyx woody **Thespesia**

6. Calyx thin :
  7. Epicalyx persistent; calyx not spathaceous **Hibiscus**
  7. Epicalyx deciduous; calyx spathaceous **Abelmoschus**

ABELMOSCHUS Medikus

Note: Hochreutiner (1924) erected *Abelmoschus* Med. as a distinct genus formerly treated under *Hibiscus* L.) based on the calyx character, which is shortly 5-toothed, spathaceous and adnate to the corolla and falls with the latter after flowering. These characters of calyx are specific to *Abelmoschus* Med. and no where found in any Malvaceae memebr (Borssum Walkes, 1966).

Abelmoschus ficulneus (L) Wight & Arn., Prodr. 53.1834; Borssum Walkes in Blumea 14: 101. 1966. *Hibiscus ficulneus* L., Sp. Pl.695. 1753: FBI 1: 340; Gamble 97(70).

Spreading herbs, about 75 cm tall; branchlets minutely prickled. Leaves 3-lobed, orbicular; lobes obovate-spathulate, 1.5 - 3.5 x 1-3 cm, hispid, base cordate, apex obtuse-rotund, margin crenate-serrate. Flowers axillary, solitary or in racemes, white. Capsules oblong-ovate, 1.5 x 0.5 cm, tomentose; seeds many.

Fls. & Frs.: September - February.

Occasional in waste lands and cultivated fields. Pathur, **TP & CP** 12126.

ABUTILON P. Miller

1. Branchlets hispid; seeds hairy **A. hirtum**
1. Branchlets tomentose; seeds glabrous **A. indicum**

Abutilon hirtum (Lam.) Sweet. Hort. Brit. 53.1826; Gamble 91 (65); Borssum Walkes in Blumea 14: 168. 1966. *Sida hirta* Lam., Encycl 1; 7 1783. *S. graveolens* Roxb. ex Hornem, Hort. Hafn Suppl. 77.1819. *Abutilon graveolens* (Roxb. ex Hornem) Wight & Arn., Prodr. 56. 1834. var. *hirtum* (Lam.) Masters in FBI 1: 327, 1874.

Herbs, about 1 m tall; branchlets hispid. Leaves ovate-suborbicular, 4.5-8 x 5.5-9 cm, tomentose, base cordate, apex acuminate, margin serrate-dentate, Flowers axillary, solitary, orange-yellow. Schizocarps globose, 1 cm across, pubescent; mericarps 10-20.

Fls. & Frs.: June - March. *Vern*: Pala benda.

Ocasional in field hedges. Narsapur, **BR & CP** 14099.

Abutilon indicum (L) Sweet, Hort Brit. 54.1826; FBI 1:326; Gamble 91 (102). ssp. **indicum.**; Borssum Walkes in Blumea 14: 171. 1966. *Sida indica* L., Cent. Pl 2: 26. 1756.

Woody herbs, about 2 m tall; branchlets tomentose. Leaves cordate-orbicular, 2-7 x 2-4.5 cm, pubescent, base cordate, apex acuminate, margin dentate. Flowers axillary, solitary, yellow. Schizocarps globose, 1 x 2 cm, tomentose; mericarps 15-20.

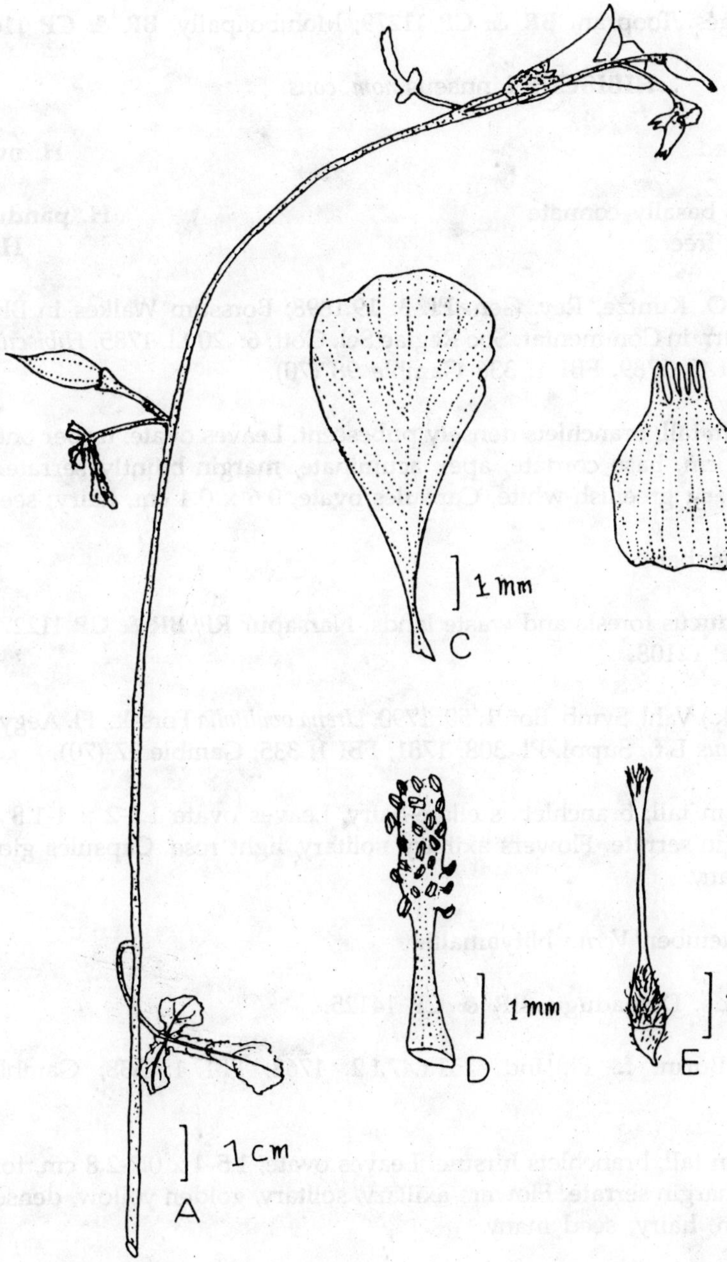

Fig. 6.2: ***Abelmoschus ficulneus*** **(L.) Wt. & Arn. A. Twig, B. Calyx with split open, C. Petal, D. Staminal column, E. Pistil**

Fls. & Frs.: June - December. Vern.: Thuthurabenda.

Common in field hedges. Toopran, **BR & CP** 11279; Mombojipally, **BR & CP** 11635.

HIBISCUS Linnaeus *nom. cons.*

1. Leaves simple not lobed **H. ovalifolius**
1. Leaves lobed :
   2. Epicalyx segments basally connate **H. panduriformis**
   2. Epicalyx segments free **H. lobatus**

Hibiscus lobatus (Murr.) O. Kuntze, Rev. Gen. Pl. 3: 19.1898; Borssum Walkes in Blumea 14: 77. 1966. *Solandra lobata* Murr. in Commentat. Soc Regiae Sci. Gott. 6: 20 t.l. 1785. *Hibiscus solandra* L'Herit, Strip. Nov. 1.103 t.49 1789; FBI 1: 336 Gamble 98 (70).

Erect herbs, about 50 cm tall; branchlets densely pubescent. Leaves ovate, upper ones deeply 3-lobed, 3- 4.5 x 0.5-1.3 cm, base cordate, apex acuminate, margin bluntly serrate. Flowers axillary and terminal, solitary, greenish-white. Capsules ovate, 0.6 x 0.4 cm, hairy; seeds many.

Fls. & Frs.: July - December.

Occasional in dry deciducus forests and waste lands. Narsapur RF, **BR & CP** 11227; Medak, **CP** 11731; Pathur, **TP & CP** 12108.

Hibiscus ovalifolius (Forssk.) Vahl, Symb. Bot. 1: 50. 1790. *Urena ovalifolia* Forssk., Fl. Aegypt.-Arab. 124. 1775. *Hibiscus micranthus* L.f., Suppl. Pl. 308. 1781; FBI 1: 335; Gamble 97 (70).

Undershrubs, about 1 m tall; branchlets stellate-hairy, Leaves ovate 1.5-2 x 1-1.8 cm, base rounded, apex acute, margin serrate. Flowers axillary, solitary, light rose. Capsules globose, 0.6 cm across, hairy; seeds many.

Fls. & Frs.: March-September. *Vern.*: Nityamalli.

Common in waste lands. Domadugu, **BR & CP** 14125.

Hibiscus panduriformis Burm. f., Fl. Ind. 151.t.47.f.2. 1768; FBI 1: 338; Gamble 98(70) (*panduraeformis*).

Undershrubs, about 1 m tall; branchlets hirsute. Leaves ovate, 1.5-4 x 0.7-2.8 cm, tomentose, base cordate, apex acute, margin serrate. Flowers axillary, solitary, golden yellow, densely hairy. Capsules ovate, 1.5 x 1 cm, hairy; seed many.

Fls. & Frs.: October - January.

Rare in fallow lands. Patancheru, **BR & CP** 11310.

MALVASTRUM A. Gray *nom cons.*

Malvastrum coromandelianum (L.) Garcke in Bonplandia 5 : 295. 1857; Gamble 88 (64); Borssum Walkes in Blumea 14:152. 1966. *Malva coromandeliana* L., Sp. Pl. 687. 1753. *M. tricuspidata* R.Br. in W.T. Aiton, Hortus Kew. ed. 2. 4:210. 1812 (*nom. superfl.*). *Malvastrum tricuspidatum* (R.Br.) A. Gray, Pl. Wight. 1: 16. 1852; FBI 1: 321.

Erect herbs, about 75 cm tall; branchlets strigose. Leaves ovate-lanceolate, 2-4 x 1-2.2 cm, densely hairy, base truncate- rounded, apex acute, margin dentate, Flowers axillary, solitary or 2-3 in clusters, yellow. Schizocarps globose, 0.3 cm across, hairy; mericarps 10-14, bristled.

Fls. & Frs. : Round the year.

Common weed of cultivated fields. Narsapur, **BR & CP** 11236.

PAVONIA Cavanilles *nom. cons.*

1. Leaf-segments obovate; mericarps narrowly winged **P. zeylanica**
1. Leaf-segments ovate; mericarps not winged **P. odorata**

Pavonia odorata Willd. Sp. Pl. 3:837. 1800; FBI 1:331; Gamble 93 (67).

Erect herbs, about 90 cm tall; branchlets viscid. Leaves ovate- orbicular, 3-lobed; lobes ovate, 1-5 x 0.6 - 4.5 cm, base cordate, apex acute, margin dentate. Flowers axillary, solitary, pinkish-white. Schizocarps globose, 0.5 cm across, pubescent; seeds many.

Fls. & Frs.: Round the year.

Occasional along road sides in forests. Way to Pegarikutta, **KMS** 6614; Narsapur, **GNR** 3911, 4015 & 4181

Pavonia zeylanica (L) Cav., Diss 3:134, 1787; FBI 1:330; Gamble 93(67), *Hibiscus zeylanicus* L., Sp. Pl. 697. 1753.

Erect herbs, about 1 m tall; branchlets viscid. Leaves orbicular, 3-lobed, 0.8-2 x 0.6-2 cm, pubescent, base cordate, apex acute-obtuse, margin dentate. Flowers axillary, solitary, pink. Schizocarps globose, 0.2 cm across, glabrous, enclosed in persistent involucre; seeds many.

Fls. & Frs.: August - February. Vern.: Karubenda.

Common weed of waste lands and road sides. Burugupally, **BR & CP** 11617; Thornal, **CP** 11703.

SIDA Linnaeus

1. Leaves glabrous, lanceolate **S. acuta**

1. Leaves pubescent or tomentose, other than lanceolate:
   2. Leaves ovate, cordate:
      3. Stem trailing **S. cordata**
      3. Stem erect **S. cordifolia**
   2. Leaves obovate, cuneate **S. rhombifolia**

Sida acuta Burm. f., Fl. Ind. 147. 1768; FBI 1:323;Gamble 90(64); Borssum Walkes in Blumea 14: 186. 1966.

Erect herbs, about 75 cm tall; branchlets glabrous. Leaves lanceolate, 1-3 x 0.3-1 cm, glabrous, base obtuse, apex acute, margin serrate. Flowers axillary, solitary, yellow, Schizocarps globose, 0.3 cm across, glabrous; mericarps 5-6.

Fls. & Frs.: August - December. Vern. : Visha boddi

Common weed of cultivated fields and waste lands. Narsapur, **BR & CP** 11229; Rampur, **BR & CP** 11653.

Sida cordata (Burm.f.) Borssum Walkes in Blumea 14;182. 1966. *Melochia cordata* Burm.f., Fl. Ind. 143. 1768. *Sida veronicifolia* Lam., Encycl. 1: 5. 1783; Gamble 89 (64). *S. humilis* Cav., Diss. 5: 277. 1788; FBI 1: 322.

Trailing herbs; branchlets hispid. Leaves ovate, 1-1.5 x 1-1.5 cm, hairy, base cordate, apex acute, margin dentate. Flowers axillary, solitary or in small racemes, yellow. Schizocarps depressed-globose, 0.3 cm across, hairy; mericarps 5.

Fls. & Frs.: June - November Vern.: Gayapuaku.

Common weed of waste places. Akkannapet, **BR & CP** 11561.

Sida cordifolia L. Sp. Pl. 684. 1753; FBI 1: 324; Gamble 89 (64); Borssum Walkes in Blumea 14: 199. 1966.

Undershrubs, about 1 m tall; branchlets hairy. Leaves ovate, 1- 3.5 x 0.5-2.4 cm, tomentose, base cordate, apex obtuse, margin crenate-denatate. Flowers axillary, solitary or few in clusters, yellow. Schizocarps depressed-globose, 0.4 cm across, tomentose; mericarps 7-12.

Fls. & Frs. : July - February. *Vern.* : Chirubenda.

Rare in forests and waste lands. Narsapur town, **BR & CP** 11238.

Sida rhombifolia L., Sp. Pl. 684. 1753; emend. Masters in FBI 1:323.1874. incl.var. *obovata, retusa, rhomboidea;* Gamble 90 (65); Borssum Walkes in Blumea 14: 193. 1966.

Erect herbs, about 50 cm tall; branchlets hairy. Leaves obovate, 1.5-3x0.5-2 cm, pubesecent, base cuneate, apex obtuse- truncate, margin dentate. Flowers axillary, solitary, yellow. Schizocarps depressed- globose, 0.3 cm across, hairy; mericarps 8-10.

Fls. & Frs. : February - October.

Common in moist places and waste lands. Narsapur-Hyderabad Road, BR & CP 11212; Lingareddypally, CP 11713.

THESPESIA Solander ex Correa *nom. cons.*

Thespesia lampas (Cav.) Dalz. ex. Dalz & Gibs. Bombay Fl. 19. 1861; FBI 1: 345. *Hibiscus lampas* Cav. Diss 3: 154. t.56. f.2 1787; Gamble 98 (71).

Shrubs, about 2.5 m tall; branchlets stellate- tomentose. Leaves; lower ones palmately 3-lobed; upper ones ovate, 5-10 x 3-7.5 cm, tomentose above, woolly below, base cordate, apex acuminate, margin entire. Flowers axillary, solitary or in racemes, yellow. Capsules ovate, woody; seeds many.

Fls. & Frs. : August - December. *Vern.* : Kondapatti.

Occasional in forests. Narsapur RF, **TP & CP** 14046.

URENA Linnaeus

Urena lobata L., Sp. Pl. 692. 1753; FBI 1: 329; Gamble 92(66) spp. **sinuata** (L.) Borssum Walkes in Blumea 14:142. 1966. *Urena sinuata* L., Sp. Pl. 692. 1753.

Woolly herbs, about 1 m tall, branchlets hairy. Leaves orbicular, 4-5-lobed, 2-3 x2-3.5 cm, hispid above, woolly below, base subcordate, apex obtuse-acute, margin serrate-crenate. Flowers axillary, solitary or 2-3 in clusters, pink. Schizocarps globose, 0.5 cm across; mericarps 5.

Fls. & Frs. : August - December. Vern. : Nallabenda.

Occasional in field hedges and road sides. Mombojipally, **BR & CP** 11638.

Several species of *Gossypium* L. are cultivated for cotton. *Abelmoschus esculentus* (L) Moench and *Hibiscus cannabinus* L. are cultivated for fruits and leaves respectively.

BOMBACACEAE

BOMBAX Linnaeus *nom. cons.*

Bombax ceiba L., Sp. Pl. 511. 1753. *Bombax malabaricum* DC., Prodr, 1:479. 1824; FBI 1:349; Gamble 99 (71); Khan in FFH 73.

Trees, about 20 m tall; branchlets prickled. Leaves 5- 7- foliolate; leaflets elliptic-oblong, 5.5-7.2 x 4-7 cm, glabrous, base cuneate, apex acuminate, margin entire. Flowers on leafless branchlets, solitary, scarlet red. Capsules oblong, 10 x 5 cm, 5-valvate; seeds many.

Fls. & Frs. : March - May. *Vern.*: Adavi burgu.

Occasional in forests. Narsapur RF, **BR & CP** 14102.

Cieba pentandra (L.) Gaertn. is grown for its silk cotton.

STERCULIACEAE

1. Trees **Helicteres**
1. Herbs:
    2. Flowers pink; calyx obscurely lobed **Melochia**
    2. Flowers yellow; calyx deeply lobed **Waltheria**

HELICTERES Linnaeus

Helicteres isora L., Sp. Pl. 963.1753; FBI 1:365; Gamble 107 (77); Khan in FFH 77.

Trees, about 6 m tall; branchlets stellate tomentose. Leaves obovate-orbicular, 4-12 x 3-8 cm, stellate-tomentose, base subcordate-subacute, apex obtuse, margin irregularly serrate. Flowers axillary, in cymes, red. Follicles 5, spirally twisted, 6.5 x 0.6 cm, stellate-tomentose; seeds many.

Fls. & Frs. : September - March.

Occasional in forests. Narsapur RF, **TP & CP** 14034.

MELOCHIA Linnaeus

Melochia corchorifolia L., Sp. Pl. 675. 1753; FBI 1: 374; Gamble 110 (79).

Erect herbs, about 50 cm tall; branchlets glabrous. Leaves ovate-lanceolate, obscurely 3-lobed, 1-1.5 x 0.3-0.5 cm, glabrous, base obtuse, apex acute, margin serrte. Flowers terminal, in capitate clusters, pink. Capsules depressed-globose, 5-valved, 0.3 cm across. sparsely hairy; seed 1 per cell.

Fls. & Frs. : September - December.

Occasional weed of forests and waste places. Narsapur RF, **TP & CP** 11994.

WALTHERIA Linnaeus

Waltheria indica L., Sp. Pl. 673; FBI 1: 374; Gamble 111 (79). *W. americana* L., Sp. Pl.673. 1753.

Errect herbs, about 60 cm tall; branchlets stellate- tomentose. Leaves ovate-elliptic, 3-4.5 x 2.4 cm, softly pubescent, base subcordate-rounded apex obtuse, margin crenate-serrate. Flowers axillary, in capitate clusters, yellow. Capsules obconical, 2- valved, 0.2 cm, enclosed in calyx; seed 1.

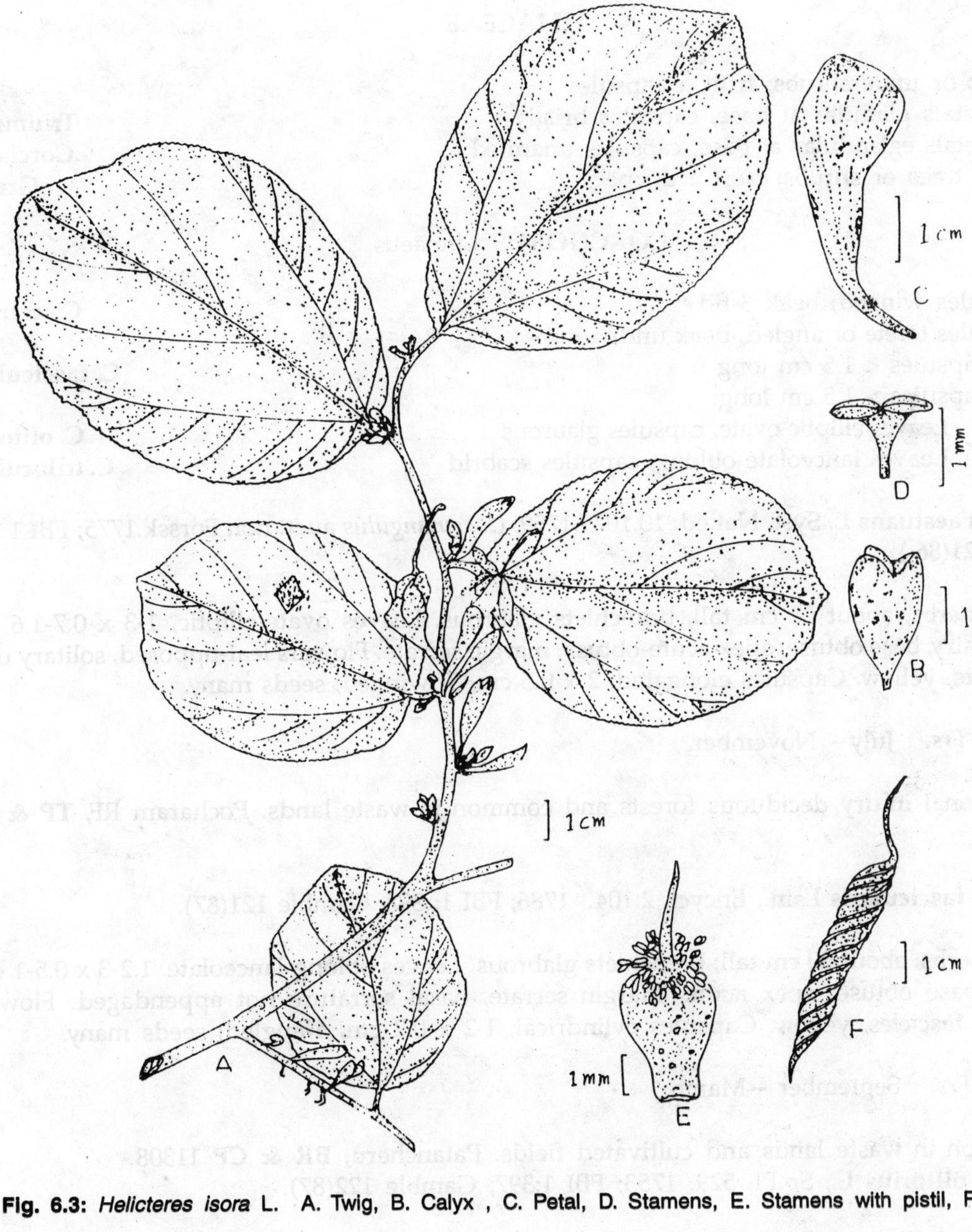

Fig. 6.3: *Helicteres isora* L. A. Twig, B. Calyx , C. Petal, D. Stamens, E. Stamens with pistil, F. Pistil

Fls. & Frs. : September - December. *Vern.* : Nallabenda.

Common in forests and waste places. Ponnnala, **CP** 11369; Narsingi, **BR & CP** 11539.

TILIACEAE

1. Herbs or undershrubs; fruit a capsule:
   2. Petals gladular at base; capsules bristly **Triumfetta**
   2. Petals eglandular at base; capsules unarmed **Corchorus**
1. Small trees or shrubs; fruit a drupe **Grewia**

CORCHORUS Linnaeus

1. Capsules winged; beak 3-fid **C.aestuans**
1. Capsules terete or angled; beak undivided:
   2. Capsules < 1.5 cm long **C. fascicularis**
   2. Capsules > 1.5 cm long:
      3. Leaves elliptic-ovate, capsules glabrous **C. olitorius**
      3. Leaves lanceolate-oblong, capsules scabrid **C. trilocularis**

Corchorus aestuans L. Syst. Nat.ed. 10.1079.1759. *C. acutangulus* auct. non Forssk.1775; FBI 1:398; Gamble 121(86.)

Erect herbs, about 50 cm tall, branchlets glabrous. Leaves ovate-elliptic, 1-3 x 0.7-1.6 cm, sparsely hairy, base obtuse, apex acute-obtuse, margin serrate. Flowers leaf-opposed, solitary or 2-3 in clusters, yellow. Capsules elongated, 2 x 0.5 cm, 3- winged; seeds many.

Fls. & Frs. : July - November.

Occasional in dry deciduous forests and common in waste lands. Pocharam RF, **TP & CP** 12057.

Corchorus fascicularis Lam., Encycl. 2:104. 1786; FBI 1: 398, Gamble 121(87).

Erect herbs, about 50 cm tall; branchlets glabrous. Leaves oblong-lanceolate. 1.2-3 x 0.5-1 cm, glabrous, base obtuse, apex acute, margin serrate, basal serrature not appendaged. Flowers axillary, in fascicles, yellow. Capsules cylindrical, 1.2 x 0.2 cm, 4-angled; seeds many.

Fls. & Frs. : September - March.

Common in waste lands and cultivated fields. Patancheru, **BR & CP** 11308.

Corchorus olitorius L., Sp.Pl. 529. 1753; FBI 1:397; Gamble 122(87).

Erect herbs, about 1 m tall; branchlets glabrous. Leaves elliptic-ovate,2-4 x 0.5-1.7 cm, glabrescent, base obtuse, apex acuminate, margin sharply serrate. Flowers leaf-opposed, 1-2 in short cymes, yellow. Capsules cylindric, turgid, 5 x 0.3 cm, 10- ribbed, glabrous; seeds many.

Fls. & Frs. : July - November.

Occasional in moist places and waste lands. Patancheru, **BR & CP** 1309; Ramayampet, **BR & CP** 11501.

Corchorus trilocularis L., Mant. Pl. 77:1767; FBI 1:397; Gamble 122 (87).

Erect herbs, about 75 cm tall; branchlets glabrescent. Leaves lanceolate-oblong. 1.5-3.5 x 1-1.4 cm, glabrous, base cuneate, apex acute, margin serrate- crenate. Flowers axillary, solitary or 2-3 in clusters, yellow. Capsules linear, 6 x 0.2 cm, 3-4-angled; seeds many.

Fls. & Frs. : August - January.

Occasional in waste lands and cultivated fields. Gangapur, **TP & CP** 12081.

GREWIA Linnaeus

1. Leaves 5-or more-ribbed:
   2. Leaves ovate-oblong stipules foliaceaus **G. tiliifolia**
   2. Leaves obovate-orbicular,stipules lanceolate **G. orbiculata**
1. Leaves 3-5-ribbed:
   3. Shrubs or small trees:
      4. Buds oval; drupes glabrous **G. polygama**
      4. Buds cylindric; drupes tomentose **G. flavescens**
   3. Straggling shrubs **G.hirsuta**

Grewia flavescens A.L. Juss. in Ann. Mus. Natl. Hist. Nat. 4: 91, 1804; Gamble 119(85).. *G. carpinifolia* auct. non A.L. Juss: FBI 1: 387.

Shrubs, about 3 m tall; branchlets stellate-tomentose. Leaves elliptic-oblong, 4-10 x 2.5-4.5 cm, stellately hairy, base truncate, apex abruptly acute, margin serrate. Flowers axillary, in cymose clusters, yellow. Drupes globose, obscurely 2-lobed, 1 cm across, tomentose.

Fls. & Frs. : July - Novembr.

Occasional in forests and field hedges. Venkatapuram, **TP & CP** 14003.

Grewia hirsuta Vahl, Symb. Bot. 1: 34. 1790; FBI 1: 391; Gamble 119(86).

Straggling shrubs, about 3 m tall; branchlets tomentose. Leaves ovate-lanceolate, 2.4-8 x 1-3 cm, stellate-pubescent above, woolly below, base cordate, apex acute, margin 2-dentate. Flowers axillary, in umbellate cymes, yellow. Drupes globose, obscurely 4- lobed, 1 cm across, hirsute.

Fls. & Frs. : March - December. *Vern.* : Tadiki.

Common in forest areas. Pillutla RF, **BR & CP** 11254; Pocharam RF, **CP** 11750; Narsapur RF, **TP & CP** 11969.

Fig. 6.4: *Grewia flavescences* Juss. A. Twig, B. C. & D. Sepals, E. & F. Petals, G. Staminal column, H. & J. Stamens, K. Pistil

Grewia orbiculata Rottler in Ges. Naturf. Freunde Berlin Neue Schriften 4: 205. 1803; FBI 1: 386. *Grewia rotundifolia* A.L. Juss. in Ann. Mus. Natl. Hist. Nat. 4: 92 t,. 50.f.3.1804; Gamble 118 (84).

Straggling shrubs, about 8 m tall; branchlets tomentose. Leaves obovate-orbicular, 3-9 x 2-6 cm, pubescent above, glaucous tomentose below, base obliquely cordate, apex obtuse-acuminate, margin crenate-serrate. Flowers axillary, in umbellate cymes, yellow. Drupes subglobose, obscurely 2-lobed, 0.6 cm across, tomentose.

Fls. & Frs: April - October.

Occasional in dry deciduous forests. Narsapur, **GNR** 4300.

Grewia polygama Roxb., Fl. Ind. 2: 588. 1832; FBI 1: 391; Gamble 119 (83).

Shrubs, about 2 m tall; branchlets pubescent. Leaves oblong, 1.5-4 x 1-2.5 cm, stellate-pubescent above, tomentose below, base rounded, apex acute, margin irregularly serrate. Flowers axillary, in cymes, white. Drupes obscurely 4-lobed, 1.5 cm across. glabrous.

Fls. & Frs. : July - November.

Occasional in waste lands and fields. Zaheerabad, **CP** 11589.

Grewia tiliifolia Vahl, Symb. Bot. 1:35. 1790; FBI 1:386; Gamble 118 (84) (*tiliaefolia*); Khan in FFH 79.

Trees, about 8 m tall, branchlets tomentose. Leaves ovate- oblong, 5-10 x 4.8 cm, thin-coriaceous, glabrous, base oblique- cordate, apex acute, margin serrate- dentate. Flowers axillary, in cymes, yellow. Drupes globose, 2- lobed, 1 cm across, glabrous.

Fls. & Frs. : April - December. *Vern.* : Thada.

Occasional in forests. Narsapur RF, **TP & CP** 14032.

TRIUMFETTA Linnaeus

1. Flowers in axillary clusters or racemes:
   2. Caplsules globose **T. rhomboidea**
   2. Capsules ovate **T. pentandra**
1. Flowers in terminal racemes **T. rotundifolia**

Triumfetta pentandra A. Rich in Gullemin & Perr., Fl. Seneg.Tent. 1: 93 t. 19. 1831; Gamble 120(86). *T. neglecta* Wight & Arn., Prodr. 75. 1834; FBI 1: 396.

Undershrubs, about 1 m tall; branchlets brown, stellate tomentose. Leaves lower ones palmately-lobed, upper ones rhomboid-elliptic, 1-5.2 × 0.5-6.5 cm, strigose above, tomentose below, base cuneate, apex acuminate, margin crenate-serrate. Flowers axillary, in racemes, yellow. Capsules ovate, 0.4 x 0.2 cm., bristles hispid: seeds beaked.

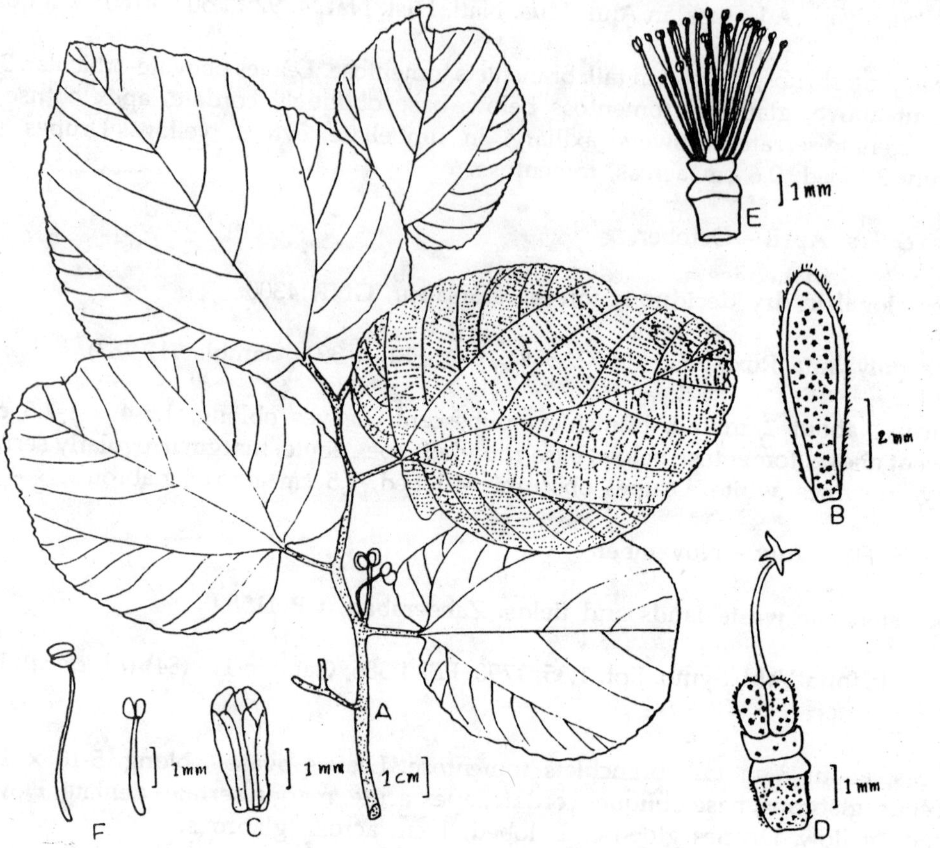

Fig. 6.5: *Grewia tillifolia* Vahl. A. Twig, B. Sepal, C. Petal, D. Pistil, E. Staminal column, F. Stamens

Fls & Frs. : August - December.

Common in waste lands. Narsapur-Hyderabad Road, **BR & CP** 11214.

Triumfetta rhomboidea Jacq., Enum. Syst. Pl.22. 1760; FBI 1:395; Gamble 120(86). *Bartramia indica* L., Sp. Pl. 389. 1753. (non *Triumfetta indica* Lam. 1792). *T. bartramia* L., Syst. Nat. ed. 10.2 1044. 1759, *nom. illegit*.

Undershrubs, about 2 m tall; branchlets pubescent. Leaves, lower ones palmately lobed, upper ones lanceolate-rhomboid, 2-4 × 0.5-3.5 cm, stellate- pubescent, base truncate, apex acuminate, margin irregulrly serrate. Flowers axillary, in clusters, yellow- orange. Capsules globose, 0.3 cm across, scabrid, bristles glabrous; seeds smooth.

Fls. & Frs. : August - December. *Vern.* : Bankathuthara.

Common in forests. Sardhana, **BR & CP** 11627.

Triumfetta rotundifolia Lam., Encycl. 3: 421. 1792; FBI 1: 395; Gamble 120 (86).

Erect herbs, about 1 m tall; branchlets stellate-tomentose. Leaves orbicular-rhomboid, 1.5-3 × 1-2.5 cm, stellate-pubescent above, tomentose below, base cuneate-obtuse, apex acute, margin iregularly crenate-serrate. Flowers terminal, in interuppted racemes, orange. Capsules globose, 0.5 cm across, pubescent, seeds beaked.

Fls. & Frs. : August - January.

Occasional in hedges and waste lands. Manjira Barrage, **BR & CP** 14097.

MALPIGHIACEAE

ASPIDOPTERYS A.H.L. Jussieu

Aspidopterys cordata Adr. Juss. in Ann. Sci. Nat. Bot. Ser.2.13. 1840; FBI 1: 421; Gamble 130 (92).

Climbing shrubs; branchlets woolly. Leaves broadly ovate- orbicular, 10-15 × 10-14 cm, tomentose, base deeply cordate, apex abruptly acuminate, margin entire. Flowers axillary, in panicles, yellowish-green. Samara orbicular-elliptic, 4 x 3 cm, wings 1.

Fls. & Frs. : September - February.

Occasional in forests. Narsapur RF, **TP & CP** 14028.

ZYGOPHYLLACEAE

TRIBULUS Linnaeus

Tribulus terrestris L., Sp. Pl. 387, 1753; FBI 1:423; Gamble 130(92).

Prostrate herbs; branchlets hairy. Leaves even pinnate; leaflets 4-5 pairs, oblong, 1-1.5 × 0.5-0.7 cm, base oblique, apex mucronate-acute, margin entire. Flowers axillary, solitary, golden yellow. Schizocarps globose, 1 cm across, 5-angled; Cocci 5; seeds 5.

Fls. & Frs. : Round the year. *Vern.* : Palleru.

Occasional in wastelands and cultivated fields. Manjira Barrage, **BR & CP** 14093.

OXALIDACEAE

1. Diffuse creeping herbs; leaves 3-foliolate **Oxalis**
1. Erect herbs; leaves many-foliolate **Biophytum**

BIOPHYTUM A.P. de Condolle

Biophytum sensitivum (L.) DC., Prodr. 1:690. 1824; FBI 1:436; Gamble 133 (95). *Oxalis sensitiva* L., Sp. Pl. 434. 1753.

Erect herbs, about 20 cm tall; branchlets glabrous. Leaves even-pinnate, to 9 cm, all crowned at the top of the stem; leaflets 8-12 pairs, terminal pair largtest, ovate-obovate, 0.4-0.8 × 0.3- 0.5 cm, base obtuse, apex apiculate, margin enitre. Flowers in umbels, yellow. Capsules oblong, 0.6 × 0.2 cm; seeds many.

Fls. & Frs. : July - December.

Frequent in forests. Gangapur RF, **TP & CP** 12073.

OXALIS Linnaeus

Oxalis corniculata L., Sp. Pl. 435. 1753; FBI 1:436; Gamble 132 (94).

Creeping herbs; branchlets softly pilose. Leaves digitately 3- foliolate; leaflets obcordate, 0.4-1 × 0.7-1.2 cm, pilose, base cuneate, apex emarginate, margin entire. Flowers axillary, in simple umbels, yellow. Capsules oblong 2.3 × 0.2 cm, seeds many.

Fls. & Frs. : December - March. *Vern.* : Ambotikura.

Occasional in moist and shady places. Medak, **BR & CP** 11469.

BALSAMINACEAE

IMPATIENS Linnaeus

Impatiens balsamina L., Sp. Pl.938 1753; FBI 1:453; Gamble 142 (101).

Succulent herbs, about 1 m tall; branchlets glabrescent. Leaves lanceolate, 8-13 × 1.5-2.5 cm, sparsely pubescent, base attenuate, apex acuminate, margin sharply serrate. Flowers axillary, solitary or in clusters, rose. Capsules oblong, 1 × 0.4 cm; seeds many.

Fls. & Frs. : October - January. *Vern.* : Muddagorinta.

Common in moist waste lands and villages. Medak, **CP** 11729.

RUTACEAE

1. Petiole not winged **Aegle**
1. Petiole winged:
   2. Stamens 10-12; disc annular **Limonia**
   2. Stamens 8-10; disc columnar **Naringi**

AEGLE Correa *nom. cons.*

Aegle marmelos (L.) Correa in Trans. Linn. Soc. London 5:223.1800; FBI 1:516; Gamble 161 (115); Khan in FFH 89. *Crataeva marmeios* L., Sp. Pl. 444.1753.

Armed trees, about 6 m tall; branchlets pubescent. Leaves 3-foliolate, to 5 cm; leaflets ovate, 2.5 - 5.5 × 1.4-2.5 cm, glabrous, base obtuse, apex acuminate, margin subcrenulate. Flowers axillary, in panicles, pale green. Barries globose, 4 cm across; seeds many.

Fls. & Frs. : March - Setember. *Vern.*: Maredu.

Common in dry deciduous forests, cultivated near temples. Pocharam sanctuary, **BR & CP** 11608; Pathur, **TP & CP** 12129.

LIMONIA Linnaeus

Note : Swingle (Journ.Wash. Acad.Sci 4:328. 1914) is of the opinion that the generic name *Limonia* L. (1764) is illegitimate "being a mere variant of *Limonium* Miller (1764)" of Plumbaginaceae. He therefore took the next available name *Feronia* Cor. for the genus which has been subsequently followed by many. Airy Shaw (Kew Bull. 1939 : 293. 1939) states that *Limonia* is not an orthographic variant of *Limonium* of Plumbaginaceae and the legitimate name *Limonia* L., must be retained.

Limonia acidissima L., Sp. Pl. ed. 2.554.1762; B.C. Stone & D.H. Nicolson in Taxon 27:551-552. 1978. *Feronia limonia* (L) Swingle in Journ. Wash. Acad. Sci. 4: 328. 1914. *Schinus limonia* L., Sp. Pl. 1: 389. 1753. *Limonia elephantum* (Correa) Panigrahi in Taxon 26: 577. 1977. *Feronia elephantum* Correa in Trans. Linn. Soc. London 5:225. 1800; FBI 1: 516; Gamble 160 (114).

Trees, about 15 m tall; branchlets spiny, tomentose. Leaves 3-7-foliolate; leaflets oblong-obovate, 2-4 x 1-2 cm, glabrous, base cuneate, apex emarginate, margin entire. Flowers axillary and terminal, in panicles, pinkish-white. Berries globose, 6 cm across; seeds many.

Fls. & Frs. : March - August. *Vern.* : Tora-elka.

Common in dry deciduous forests. Narsapur RF, **TP & CP** 14062.

Note : Panigrahi (*loc. cit.*) has argued for the rejection of the name *Limonia acidissima* L. as ***nomen ambiguum***. However Stone and Nicolson (*loc. cit.*) have argued for retaining the original name. The concept of Stone and Nicolson is followed in the present work.

NARINGI Adanson

Naringi crenulata (Roxb.) Nicolson in Saldanha & Nicolson, Fl. Hassan 387. 1976. *Limonia crenulata* Roxb. Pl. Cor. t. 86. 1798; Gamble 157 (112). *Limonia acidissima* auct. non L.; Khan in FFH 86.

Trees, about 8 m tall; branchlets glabrous, spines axillary. Leaves in clusters; leaflets 3 pairs, obolong-ovate, 2-6.5 x 1-3 cm, glabrous, base obtuse, apex subacute, margin crenulate; petiole and rachis jointed, winged. Flowers axillary, in subumbellate racemes, white. Berries globose, 0.8 cm across; seeds 3-4.

Fls. & Frs. : April - June.

Occasional in forests. Narsapur RF, **BR & CP** 14106.

Citrus grandis Osbeck., C. medica L. are cultivated for their fruits. *Murraya koenigii* Spreng. is grown for its leaves.

FLINDERSIACEAE

CHLOROXYLON A.P. de Condole *nom. cons.*

Chloroxylon swietenia DC., Prodr., 1: 625. 1824; FBI 1: 569; Gamble 152 (109); Khan in FFH 99.

Trees, about 10 m tall; branchlets pubescent. Leaves even-pinnate, to 15 cm; leaflets 10-15 pairs, oblong-obovate, 2-3 x 1-1.5 cm, glabrous, base oblique, apex obtuse, margin entire. Flowers axillary or terminal, in panicles, cream. Capsules oblong, 2 x 0.6 cm; seeds many.

Fls. & *Frs.* : November -May. *Vern.* : Billudu.

Common in dry deciduous forests. Gangapur RF, BR & **CP** 11439; Narsapur RF, CP 11580.

SIMAROUBACEAE

AILANTHUS Desfontaines *nom. cons.*

Ailanthus excelsa Roxb., Pl. Corom. 24. t. 23. 1795; FBI 1:518; Gamble 163 (116); Khan in FFH 91.

Trees, about 20 m tall; branchlets densely fluffy. Leaves odd-pinnate; leaflets 8-12 pairs, elliptic-lanceolate, 8-20 x 4.5-8 cm, tomentose, base truncate, apex acuminate, margin bluntly dentate. Flowers axillary and terminal, in panicles, yellow. Samaras lanceolate, 6 x 1.2 cm, seed 1.

Fls. & *Frs.* : January - May. *Vern.* : Peddamanu.

Common in villages. Domadugu, **BR & CP** 14123.

BALANITACEAE

BALANITES Delile

Balanites aegyptiaca (L.) Del. Descr. Egypt. Hist. Nat. 2.221. t.28.f 1. 1813. *Ximenia aegyptiaca*, L., Sp. 1194. 1753. *Balanites roxburghii* Planch. in Ann. Sci. Nat. Bot. Ser. 4.2. 258. 1854; FBI 1:522; Gamble 164 (117); Khan in FFH 91.

Armed trees, about 6 m tall; branchlets glabrous; spines sharp. Leaves 2-foliolate; leaflets elliptic-ovate, 2-3 x 1-1.4 cm, glabrous, base cuneate, apex obtuse, margin entire. Flowers axillary, in cymes, green. Drupes oblong, 5-grooved, 5 x 3 cm, Woody.

Fls. & Frs. : March - May *Vern.* : Gari.

Common in open forests. Narsapur, **GNR** 4404.

BURSERACEAE

BOSWELLIA Roxburgh

Boswellia serrata Roxb. ex Coleb. in Asiat. Res. 9; 379. t. 5. 1807; FBI 1: 528; Gamble 168 (120); Khan in FFH 93.

Trees, about 12 m tall; branchlets pubescent. Leaves odd-pinnate; leaflets 15-27, oblong-lanceolate, 2-7 × 0-5.3 cm, glabrous, base oblique, apex rounded, margin coarsely serrate. Flowers axillary, in pubescent panicles, white. Drupes trigonous, 1.5 × 1 cm, pyrenes 3.

Fls. & Frs : February - June. *Vern.* : Anduk.

Common in dry deciduous forests. Narsapur RF, **TP & CP** 14061.

MELIACEAE

1. Leaves 2-3-pinnate **Melia**
1. Leaves 1-pinnate :
    2. Leaves odd-pinnate, sharply serrate **Azadirachta**
    2. Leaves even-pinnate, entire **Soymida**

AZADIRACHTA A.H.L. Jussieu

Azadirachta indica Adr. Juss. in Mem. Hist. Nat. 19: 221.t.2. f.5. 1830; Gamble 177 (127). *Melia azadirachta* L., Sp. Pl. 385. 1753; FBI 1: 544; Khan in FFH 96.

Trees, about 15 m tall; branchlets glabrous. Leaves odd-pinnate, to 25 cm; leaflets 5-7 pairs, asymmetrically lanceolate, 2-6 x 1-1.5 cm, base oblique-cuneate, apex acuminate, margin serrate. Flowers axillary, in compact panicles, white. Drupes oblong 1.5 x 0.5 cm; glabrous.

Fls. & Frs. : February - September. *Vern.* : Vepa

Common in villages and along road sides. Narsingi, **BR & CP** 11527.

MELIA Linnaeus

Melia azedarach L., Sp. Pl. 384. 1753; FBI 1:L544; Gamble 176 (126); Khan in FFH 97.

Trees, about 8 m tall; branchlets glabrous. Leaves bipinnate, to 22 cm; leaflets lanceolate, 2-4 x 1-1.5 cm, glabrous, base oblique, apex acuminate, margin serrate. Flowers axillary, in panicles, lilac. Drupes globose oblong, 0.8 x 0.6 cm, glabrous.

Fls. & Frs. : March - October. *Vern.* : Turakavepa.

Common in villages, along road sides. Mombojipally, **BR & CP** 11630.

SOYMIDA A.H.L. Jussieu

Soymida febrifuga (Roxb.) Adr. Juss. in Mem. Mus. Hist. Nat. 19.251. 1830; FBI 1:567; Gamble 185 (133); Khan in FFH 97.

Trees. about 10 m tall; branchlets glabrous. Leaves even-pinnate; leaflets 3-6 pairs, ovate-oblong, 6-7.5 x 2-3 cm, coriaceous, glabrous, base oblique-truncate, apex obtuse-acute, margin entire. Flowers axillary or terminal, in panicles, white. Capsules obovate, 7 x 5 cm, 5-celled, woody; seeds many.

Fls. & Frs. : February - June *Vern.* : Somi.

Common in dry deciduous forests. Narsapur RF, **CP** 11574.

OLACACEAE

1. Unarmed; calyx enclosing drupes to half the length **Olax**
1. Armed; calyx not enclosing drupes **Ximenia**

OLAX Linnaeus

Olax scandens Roxb., Pl. Corom. t. 102. 1799; FBI 1:575; Gamble 190 (136); Khan in FFH 101.

Climbing shrubs; branchlets sparsely pubescent. Leaves oblong-ovate, 4-7 x 2-3.5 cm, glabrous, base cuneate, apex obtuse-acute, margin entire. Flowers axillary, in racemes, creamish-white. Drupes globose-oblong, 1 x 0.5 cm.

Fls. & Frs. : Round the year. *Vern.* Murkimalle.

Common in forests. Gangapur RF, **BR & CP** 11446; Burugupally, **CP** 11766.

XIMENIA Linnaeus

Ximenia americana L. Sp. Pl. 1193. 1753; FBI 1:574; Gamble 189 (135); Khan in FFH 100.

Armed stragglers, about 4 m tall; branchlets pubescent. Leaves ovate, 1.5-3.5 x 1.5-3 cm, membranous, base cuneate, apex obtuse, margin entire. Flowers axillary, in racemes, creamish-white. Drupes globose-oblong, 1 x 0.6 cm.

Fls. & Frs. January - July. *Vern.* : Nakira.

Common in dry deciduous forests. Gangapur RF, **BR & CP** 11440; Narsapur RF, **CP** 11581 and **TP & CP** 11965.

CELASTRACEAE

1. Trees **Cassine**
1. Erect shrubs or woody climbers:
   2. Armed shrubs; flowers in cymes **Maytenus**
   2. Unarmed woody climbers; flowers in panicles **Celastrus**

CASSINE Linnaeus

Cassine glauca (Rottb.) O. Kuntze, Rev. Gen. Pl. 1:144. 1891. *Mangifera glauca* Rottb. in Nye Samml. Vid. Selsk. Skrift. 2:534. 1738. *Elaeodendron glaucum* (Rottb.) Pers., Syn., Pl. 1:24. 1805; FBI 1:623; Gamble 211 (152). *E. roxburghii* Wight & Arn., Prodr. 157. 1834; Khan in FFH 103.

Trees, about 10 m tall; branchlets glabrous. Leaves elliptic-oblong, 7.5-12 x 2.5-5 cm, glabrous, base obtuse, apex acute, margin serrate-crenate. Flowers axillary and terminal, in corymbose cymes, green. Drupes oblong, 1 x 0.6 cm.
Fls. & Frs. : June - December. *Vern.* : Butankus.

Common in dry deciduous forests. Narsapur RF, **BR & CP** 14105.

CELASTRUS Linnaeus

Celastrus paniculatus Willd., Sp. Pl. 1:1125. 1797; FBI 1:617; Gamble 208 (150); Khan in FFH 102.

Woody stragglers, branchlets glabrous. Leaves ovate-orbicular, 4-7 x 4-6.5 cm. glabrescent, base obtuse-cuneate, apex shortly acuminate, margin dentate. Flowers terminal, in panicles, pale green. Capsules globose, 0.6 cm across; seeds 6.

Fls. & Frs. : March - October.

Occasional in forests. Narsapur RF, **TP & CP** 14036.

Note : This taxon resembles Euphorbiaceae members in fruiting stage due to its 3-celled capsule.

MAYTENUS Molina *emend.* Bosc.

Maytenus emarginata (Wild.) Ding Hou in Fl. Males. Ser. 1.6: 241. 1962. *Celastrus emarginatus* Willd., Sp. Pl. 1:1128. 1798. *C. montanus* Roth in Roemer & Schultes in L., Syst. Veg. 5: 427. 1819. *Gymnosporia montana* (Roth) Benth., Fl. Austral. 1: 400, 1863; Gamble 209 (150). *G. emarginata* (Willd.) Thw., Enum. Pl. Zeyl. 409. 1864; FBI 1 : 621; Gamble 210 (151). *Celastrus senegalensis* Lam., Encycl. 1:661. 1785; Khan in FFH 103.

Armed shrubs, about 3 m tall; branchlets glabrous. Leaves orbicular, 3-6 x 2.5-4 cm, glabrous, base obtuse-attenuate, apex rounded - emarginate, margin serrate-crenate. Flowers axillary and terminal, in dense dichotomous cymes, dull-white. Capsules globose, 0.5 cm across; seeds 1-3.

Fls. & Frs. : July - February. *Vern.* : Danti.

Common in waste lands and Scrub jungles. Medak, **BR & CP** 11415.

Note: *Maytenus senegalensis* (Lam.) Exell. which was once considered as the valid name for the Indian taxon is now found out to be a different African plant and hence the next available name *M. emarginata* is now used for the species.

RHAMNACEAE

1. Unarmed lianas; fruit samara **Ventilago**
1. Armed shrubs or trees; fruit drupe **Ziziphus**

VENTILAGO J. Gaertner

Ventilago denticulata Willd. in Ges. Naturf. Freunde Berlin Neue Schriften 3:417. 1801; Oza in Journ. Bombay Nat. Hist Soc. 63: 455. 1966. *V. calyculata* Tul. in Ann. Sci. Nat. Bot. Ser. 4.8:124. 1857; FBI 1:631; Gamble 218 (156). *V. maderaspatana* Roxb., Fl. Ind. 1:629. 1820; non. Gaertn. 1788; Khan in FFH 104.

Unarmed lianas; branchlets striated, tomentose. Leaves elliptic-oblong, 2-7.5 × 1-4 cm, glabrescent, base rounded, apex sub-acute, margin create-dentate. Flowers in clusters on panicles, green. Samara with globose nut, 0.8 × 0.5 cm; seed 1.

Fls. & Frs. : August - December. *Vern.* : Sulgur teega.

Common in forests. Pocharam RF, **BR & CP** 11607.

ZIZIPHUS P. Miller

1. Stems with hairs at nodes **P. suffruticosa**
1. Leaves ovate-lanceolate; flowers in umbellate cymes **Z. oenoplia**
1. Leaves elliptic-oblong; flowers in axillary clusters:
   2. Leaf base cordate; peduncle dinstinct; drupes tomentose **Z. xylopyrus**
   2. Leaf base rounded; peduncle reduced; drupes glabrous **Z. mauritiana**

Ziziphus mauritiana Lam., Encycl. 3: 319. 1789; var. **mauritiana**. *Z. jujuba* (L.) Gaertn., Fruct. 1; 203. 1788. non Miller 1768; FBI 1:632; Gamble 219 (157); Khan in FFH 105. *Rhamnus jujuba* L., Sp. Pl. 194. 1753.

Armed trees, about 6 m tall; branchlets rusty tomentose. Leaves elliptic-oblong. 2-3 × 1-1.5 cm, pubescent above, tomentose below, base rounded, apex retuse, margin obscurely serrate. Flowers axillary, in dense clusters, greenish-yellow. Drupes globose, 1 cm across, glabrous.

Fls. & Frs. : Round the year. *Vern.* : Gangaregu.

Common in forests and in waste lands. Pochammaralu, **BR & CP** 11482.

Ziziphus oenoplia (L) Miller, Gard. Dict. ed. 8. n.8.1768; FBI 1:634; Gamble 220 (158); Khan in FFH 106. *Rhamnus oenoplia* L, Sp. Pl. 194. 1753.

Armed climbing shrubs; branchlets tomentose. Leaves ovate- lanceolate, 2.5-4 × 1-2 cm, glabrescent above, rusty-tomentose below, base oblique, apex acuminate, margin obscurely serrate. Flowers axillary, in short umbellate cymes, green. Drupes obovate, 1 × 0.4 cm, woody.

Fls. & Frs. : July - January. *Vern.* : Pariki.

Common in field hedges and in forests. Marrikunta, **CP** 11384; Ramayampet, **BR** & **CP** 11508.

Ziziphus xylopyrus (Retz.) Willd., Sp. Pl. 1:1104. 1798; FBI 1:634. 1875; Gamble 220 (158); Khan in FFH 106. *Rhamnus xylopyrus* Retz., Obs Bot. 2:11. 1781.

Armed shrubs, about 5 m tall; branchlets tomentose; Leaves sub-orbicular-elliptic or oblong, 4-6.5 × 3.5-5 cm, glabrous above, tomentose below, base obliquely cordate, apex retuse-rounded, margin obscurely serrate. Flowers axillary, in paniculate cymes, green. Drupes globose, 2 cm, across, tomentose.

Fls. & Frs. : March - September. Vern. : Gotika.

Occasional in forests. Pocharam RF, **BR & CP** 11606; Narsapur RF, **TP & CP** 11981.

VITACEAE

1. Tendrils on peduncles **Ampleocissus**
1. Tendrils leaf-opposed:
   2. Leaves simple; berries 1-seeded **Cissus**
   2. Leaves 3-7-foliolate; berries 2-4-seeded **Cayratia**

AMPELOCISSUS

Ampelocissus latifolia (Roxb.) Planch. in Journ. Vigne Amer. 374. 1884; Gamble 230 (165). *Vitis latifolia* Roxb., Fl. Ind. ed. 2:661, 1832; FBI 1:652; Khan in FFH 107.

Climbing shrubs; branchlets fistulose, glabrous. Leaves orbicular, 5-lobed or 5-angled, 5-16 x 7-18 cm, glabrous, base cordate, apex acuminate, margin irregularly-toothed. Flowers in thyrsoid cymes, reddish-brown. Berries globose, 0.6 cm across; seeds - 2.

Fls. & Frs. : July - November. *Vern.* : Pulatheega.

Common in dry deciduous forests. Burugupally, **CP** 11767; Narsapur RF, **TP & CP** 14032.

CAYRATIA A.L. Jussieu *nom. cons.*

1. Leaves 3-foliolate **C. trifolia**

1. Leaves 5-7-foliolate:
    2. Stipules auriculate; berries ovate **C. auriculata**
    2. Stipules not auriculate; berries subglobose **C. pedata**

Cayratia auriculata (Roxb.) Gamble, Fl.Pres. Madras 237 (170). 1918. *Cissus auriculata* Roxb., Fl. Ind. 1: 412, 1820. *Vitis auriculata* Wall. ex Wight & Arn., Prodr. 129. 1834, *p.p.* FBI 1: 658.

Climbing shrubs; branchlets pubescent. Leaves 5- foliolate; leaflets obovate-oblong, 6-10 × 4-6.5 cm, glabrous above, pubescent below, base cuneate, apex abruptly acuminate, margin distantly serrate. Flowers axillary, in divaricate cymes, greenish-yellow. Berries ovate, 0.8 × 0.3 cm; seeds 4.

Fls. & Frs. : June - November.

Occasional in forests. Kondapur, **GNR** 4110.

Cayratia pedata (Lam.) Juss. ex Gagnep. in Lecomte, Notul. Syst. (Paris) 1:346. 1911. Gamble 236 (169). *Cissus pedata* Lam. Encycl. 1: 31. 1783. *Vitis pedata* (Lam.) Wall. ex Wight & Arn., Prodr. 128. 1834; FBI 1: 661.

Climbing shrubs, about 8 m tall; branchlets terete, puberulous. Leaves 5-7-foliolate; leaflets ovate-oblong or elliptic, 7-12 x 5-8 cm, base rounded - cuneate, apex acuminate, margin serrate. Flowers axillary, in cymes, greenish-white. Berries subglobose, 0.7 × 0.5. cm; seeds 2-4.

Fls & Frs.: January - November.

Occasional on forest hedges. Narsapur RF, **TP & CP** 11991.

Cayratia trifolia (L.) Domin. in Biblloth. Bot. 89.371. 1917. *Vitis trifolia* L., Sp. Pl. 203. 1753. *Cayratia carnosa* (Lam.) Gagnep. in Lecomte. Notul. Syst.(Paris) 1:347. 1911; Gamble 237 (169). *Vitis carnosa* Wall. ex Wight & Arn., Prodr. 127. 1834; FBI 1: 654. *Cissus carnosa* Lam., Encycl. 1: 31. 1783.

Climbing shrubs; branchlets pubescent, glandular. Leaves 3- foliolate; leaflets ovate-orbicular, 3-5.5 × 1.5-4 cm, base obtuse-unequally cordate, apex acute, margin crenate-dentate. Flowers in dichotomously branched umbellate cymes, pale green, Berries obovate, 0.5 × 0.4 cm; seeds 2.

Fls. & Frs.: July - October.

Common on field hedges. Nagulabanda, **CP** 11338.

CISSUS Linnaeus

1. Erect shrubs **C. pallida**
1. Climbing shrubs:
    2. Branchlets pubescent; flowers greenish-white **C. vitiginea**

2. Branchlets glabrous; flowers red **C. adnata**

Cissus adnata Roxb., Fl. Ind. 1: 423, 1820; Gamble 237 (170). *Vitis adnata* (Roxb.) Wall. Cat. 205. n. 5998. 1831-32; FBI 1: 649.

Climbing shrubs; branchlets glabrous. Leaves ovate, 4-10 × 3-7.5 cm, glabrescent, base cordate, apex acuminate, margin serrate. Flowers leaf-opposed, in umbellate cymes, red. Berries globose, 0.8 x 0.5 cm, seed 1.

Fls. & Frs. : October - December.

Occasional on hedges. Sathaihpalle, **GNR** 4106.

Cissus pallida (Wight & Arn.) Planch. in A. & C.DC., Monogr. Phan. 5: 477. 1887; Gamble 234 (167). *Vitis pallida* Wight & Arn., Prodr. 125. 1834; FBI 1:647; Khan in FFH 108.

Erect shrubs, about 1 m tall; branchlets fistulose, glabrous. Leaves cordate, 5-20 × 6-18 cm, glabrous, base deeply cordate, apex abruptly acuminate, margin serrate. Flowers in panicled cymes, red. Berries pyriform 0.6 × 0.5 cm, mucronate; seed 1.

Fls. & Frs. : July - September. *Vern.* : Kurdandi.

Occasional in forests. Nagulabanda, **CP** 11337.

Cissus vitiginea L., Sp. Pl. 117.1753; Gamble 234 (167). *Vitis linnaei* Wall. ex Wight & Arn. Prodr. 126. 1834; FBI 1: 649.

Climbing shrubs; branchlets densely pubescent. Leaves broadly cordate, 3-8 × 2.5-7.5 cm, thick coriaceous, pubescent, base cordate, apex acuminate, margin dentate. Flowers leaf-opposed, in umbellate cymes. yellowish-white. Berries pyriform-ovate, 1 × 0.5 cm; seed 1.

Fls. & Frs. : Round the year. *Vern.* : Adavidraksha.

Common on field hedges and open forests. Medak, **CP** 11726.

Vitis vinifera L. is occasionally cultivated for its fruits.

LEEACEAE

LEEA Linnaeus *nom. cons.*

Leea macrophylla Roxb. ex Hornem. Hort. Hafn. 1:231. 1813; FBI 1:654. *p.p.*

Herbs, about 1m tall; branchlets glabrous. Leaves ovate, base unequally rounded, apex acute, margin serrate. Flowers in sessile corymbs, greenish-white. Berries 4-6-celled, 0.9 x 0.5; seeds black.

Fls. & Frs. : April - November.

Rare in dry forests as under growth. Narsapur, **GNR** 4432.

SAPINDACEAE

1. Climbing herbs **Cardiospermum**
1. Trees or shrubs:
    2. Shrubs; fruit a capsule **Dodonaea**
    2. Trees; fruit a drupe:
        3. Petals 4-5: drupes rugose **Sapindus**
        3. Petals 0; drupes spiny **Schleichera**

CARDIOSPERMUM Linnaeus

Cardiospermum halicacabum L., Sp.Pl. 1: 366. 1753; FBI 1: 670; Gamble 244 (175).

Climbing herbs; branchlets striated, pubescent. Leaves bipinnate, to 12 cm; leaflets ovate-lanceolate, 2-5.5 × 1-2 cm, glabrous above, pubescent below, base attenuate, apex acute-acuminate, margin irregularly deeply incised. Flowers in dichasial cymes, white. Capsules trigonous with 3 bladder like wings, 1.5 x 1 cm, seeds 3.

Fls. & Frs. : June - March. *Vern.* : Buddakakara.

Common weed of cultivated fields. Marrikunta, CP 11392.

DODONAEA P. Miller

Dodonaea angustifolia L.f., Suppl. Pl. 218. 1782. *D. viscosa* auct. non. Jacq.: FBI 1:697; Gamble 253 (181); Khan in FFH 111.

Shrubs, about 2.5 m tall; branchlets glabrous. leaves lanceolate, 2.5-6 × 0.5-1.5 cm, glabrous, base attenuate, apex acute, margin entire. Flowers axillary or terminal, in panicles, green. Capsules globose- oblong, 1.5 cm across, membranous, 2- winged; seeds 1-3.

Fls. & Frs. : Round the year. *Vern.* : Pulivallu.

Common in Forests. Pochammaralu, **BR & CP** 11478; Thornal, **CP** 11702.

SAPINDUS Linnaeus

Sapindus emarginata Vahl, Symb. Bot. 3:54. 1974; Gamble 250(178). *S. trifoliatus* auct. non L., FBI 1: 682.

Trees, about 12 m tall; branchlets tomentose. Leaves 3-6- foliolate, to 12 cm; leaflets irregularly elliptic-obovate, 4- 7.2 × 2.4-3.6 cm, coriaceous, base cuneate, apex emarginate, margin entire. Flowers terminal , in racemes, greenish-yellow. Drupes ovate-globose, 1 cm across, 3-lobed, rugose; seeds 3.

Fls. & Frs. : September - March. *Vern.* : Kunkudu.

Common in villages. Pochammaralu, **BR** & **CP** 11499.

SCHLEICHERA Willdenow *nom. cons.*

Schleichera oleosa (Lour.) Oken, Allg. Natur. 3(2): 1341. 1841. *Pistacia oleosa* Lour., Fl. Cochinch. 2:615. 1790. *Schleichera trijuga* Willd., Sp. Pl. 4(2); 1096. 1805; FBI 1:681; Gamble 248 (177); Khan in FFH 109.

Trees, about 20 m tall; branchlets striate. Leaves even-pinnate, to 30 cm; leaflets 2-4 pairs, oblong-obovate, 7-16 × 3-5 cm, coriaceous, base cuneate-acute, apex obtuse-acuminate, margin entire. Flowers in much branched racemes, Pale yellow. Drupes ovate, 2.5 × 1.5 cm, spiny.

Fls. & Frs. : February - May. *Vern.* : Rakot.

Occasional in dry deciduous forests. Narsapur RF, **CP** 11584.

ANACARDIACEAE

1. Leaves compound:
   2. Leaflets 3; branchlets spine-tipped **Rhus**
   2. Leaflets >3; branchlets unarmed **Lannea**
1. Leaves simple:
   3. Drupes seated on fleshy hypocarp:
      4 Leaves glabrous below; drupes reniform **Anacardium**
      4. Leaves villous below; durpes ovate **Semecarpus**
   3. Drupes not on fleshy hypocarp:
      5. Flowers bisexual; drupes slightly fleshy **Buchanania**
      5. Flowers polygamous; drupes fleshy **Mangifera**

ANACARDIUM Linnaeus

Anacardium occidentale L., Sp. Pl. 383. 1753; Wight & Arn., Prodr. 168. 1834; FBI 2:20; Gamble 260 (185).

Trees, about 8 m tall; branchlets glabrous. Leaves obovate, 9-11 × 4.5 - 7 cm, thick-coriaceous, glabrous, base cuneate-rotund, apex obtuse, margin entire. Flowers terminal, in panicles, polygamous, greenish-with pink stripes. Drupes reniform, 2 × 1.5 cm; hypocarp pyriform.

Fls. & Frs. : January - May. *Vern.* : Jeedimamidi.

Occasionally cultivated. Choutkoor, **TP** & **CP** 11940.

BUCHANANIA K.P.J. Sprengel

1. Leaves elliptic-oblong; panicles rusty-tomentose **B. lanzan**

1. Leaves obovate-linear; panicles glabrous **B. axillaris**

Buchanania axillaris (Desr.) Ramam. in Saldanha & Nicolson Fl. Hassan 374. 1976. *Mangifera axillaries* Desr. in Lam. Encycl. 3:697. 1792. *Buchanania angustifolia* Roxb. Pl. Corom. t. 262. 1820; FBI 2: 23; Gamble 258 (184); Khan in FFH 115.

Trees, about 8 m tall, branchlets glabrous. Leaves elliptic-oblong. 10-16 × 4-6 cm, glabrous, base obtuse, apex subacute- emarginate, margin entire. Flowers axillary, in sparsely branched panicles, greenish-white. Drupes oblique, 1 × 0.5 cm.

Fls. & Frs. : January - May.

Occasional in forests. Narsapur, **GNR** 4359.

Buchanania lanzan Spreng. in Journ. Bot. (Schrader) 2.234. 1801; Gamble 258 (184). *B. latifolia* Roxb., Fl. Ind. 2:385. 1832; FBI 2:23; Khan in FFH 114.

Trees, about 15 m tall; branchlets sericeous. Leaves elliptic-oblong, 12-16.5 × 6.5-10 cm, coriaceous, base rounded, apex rounded-retuse, margin entire. Flowers in rusty-tomentose panicles, greenish-white. Drupes obliquely lentiform, 0.7 × 0.6 cm, hairy, 2-valved.

Fls. & Frs. : December - April. *Vern.* : Morli, Mori.

Common in dry deciduous forests. Gangapur RF, **BR & CP** 11431.

LANNEA A. Richard *nom. cons.*

Lannea coromandelica (Houtt.) Merr. in Journ. Arnold Arbor. 19: 353. 1938. *Dialium coromandelicum* Houtt., Nat. Hist. Ser. 2:39 t. 5. f.2.1774. *Odina wodier* Roxb., Fl. Ind. 2:293. 1832; FBI 2:29; Gamble 263 (187); Khan in FFH 115.

Trees, about 15 m tall; branchlets rusty stellate-tomentose. Leaves odd-pinnate, to 20 cm; leaflets 4 pairs, oblong- lanceolate, 2-4 x 1-2.4 cm, base oblique, apex attenuate, margin entire. Flowers terminal, in pendulous racemes, pale yellow. Drupes oblong, 1 x 0.4 cm.

Fls. & Frs. : February - April. *Vern.* : Gumpena.

Common in dry deciduous forests and villages. Ghangapur RF, **BR & CP** 11438; Narsingi, **BR & CP** 11540.

MANGIFERA Linnaeus

Mangifera indica L., Sp. Pl. 200. 1753; FBI 2:13; Gamble 259 (185); Khan in FFH 112.

Trees, about 20 m tall; branchlets glabrous. Leaves lanceolate-elliptic. 10-18 × 3.5-5 cm, glabrous, base cuneate-subacute, apex acuminate, margin entire. Flowers terminal, in compact panicles, greenish-yellow. Drupes ovate-oblong, 10 × 6 cm.

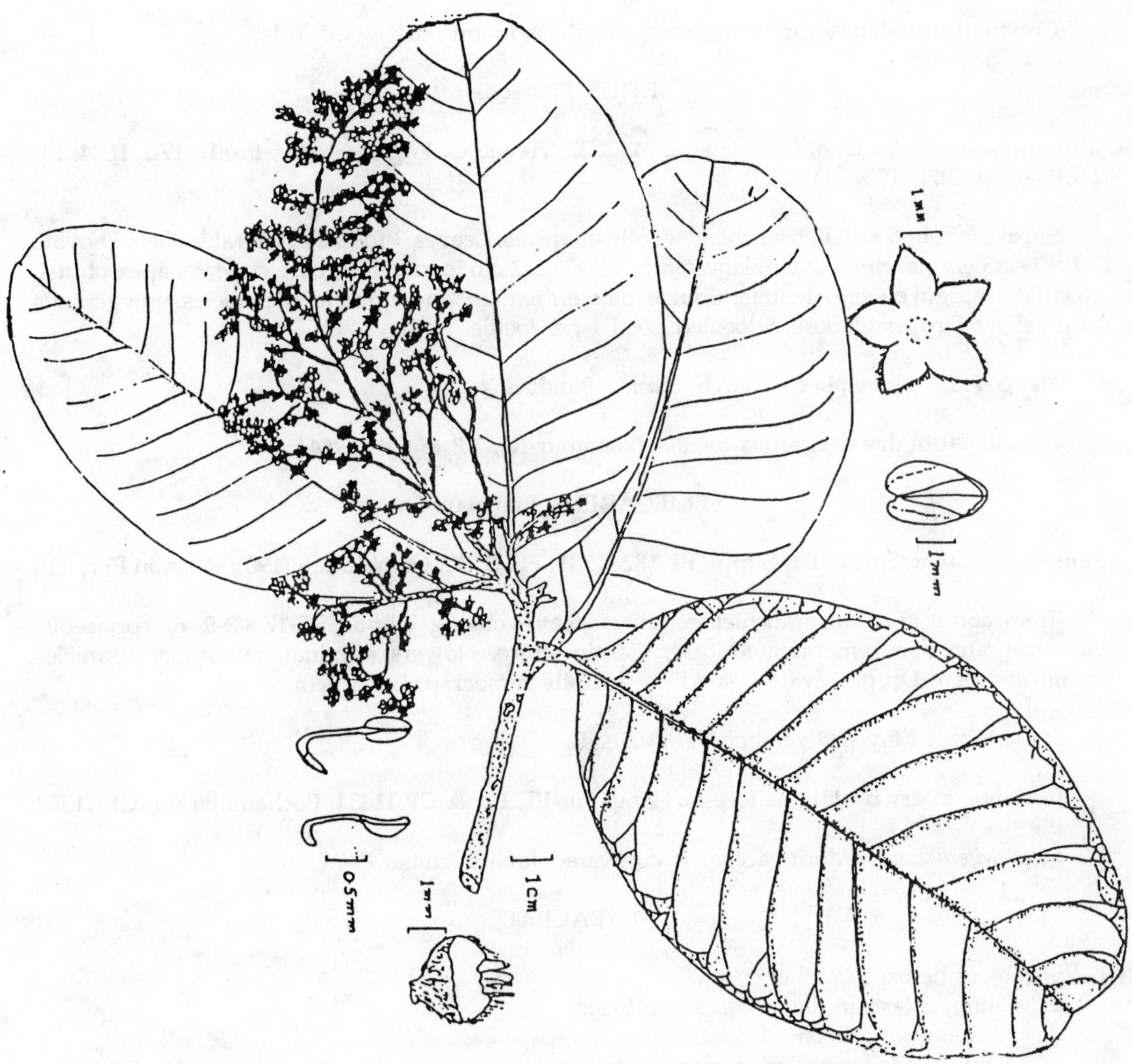

Fig. 6.6: *Buchanania lanzan* Spreng. A. Twig, B. Calyx, C. Petals, D. Pistil, E. Stamens

Fls. & Frs. : February - July . *Vern.* : Mamidi.

Common in villages and along road sides. Laxmapur **BR & CP** 11447.

RHUS Linnaeus

Rhus mysorensis G. Don, Gen. Hist. 2:74. 1832; Heyne ex Wight & Arn., Prodr. 172. 1834; FBI 2:9; Gamble 264 (188).

Shrubs, about 5 m tall; branchlets woolly at apices. Leaves 3-foliolate; lateral leaflets obovate, 1-2 × 1- 1.5 cm, terminal one oblanceolate, 2-3 × 1.5-2 cm, coriaceous, base cuneate, apex obtuse-apiculate, margin crenate-dentate, sinuate. Flowers axillary and terminal, in panicles, polygamous, pale yellow. Drupes globose, 5-loculed; seed 1 per locule.

Fls. & Frs. : November - March. *Vern.* : Sundari.

Occasional in dry deciduous forests. Narsapur RF, **TP & CP** 11964.

SEMECARPUS Linnaeus f.

Semecarpus anacardium L.f. Suppl. Pl. 182. 1781; FBI 2: 30; Gamble 266 (190); Khan in FFH 116.

Trees, about 15 m tall; branchlets sericious. Leaves oblong- elliptic, 5.5-17 × 3-9 cm, coriaceous, base truncate, apex emarginate-retuse, margin entire. Flowers terminal, in compact panicles, creamish-green. Drupes ovate, seated on a fleshy hypocrap, 2.5 × 2cm.

Fls. & Frs. : May - December, *Vern.* : Jeedi.

Common in dry deciduous forests. Gangapur RF, **BR & CP** 11434; Pochammaralu, **CP** 11762.

Moringa oleifera Lam. (Moringaceae) is cultivated for its fruits.

FABACEAE

1. Shrubs or herbs:
    2. Climbers (except *Rhynchosia suaveolens*):
        3. Leaves 3-foliolate:
            4. Pods concealed in membranous calyx **Paracalyx**
            4. Pods not as above:
                5. Pod ends curved in opposite directions **Mucuna**
                5. Pod ends not as above:
                    6. Pods glabrous **Dolichos**
                    6. Pods other than glabrous:
                        7. Flowers on a long naked peduncle **Canavalia**
                        7. Flowers not as above:
                            8. Pods 2-seeded **Rhynchosia**

8. Pods 3-many seeded:
 9. Calyx glandular — **Atylosia**
 9. Calyx eglandular:
  10. Stamens monadelphous — **Teramnus**
  10. Stamens diadelphous — **Pseudarthria**
3. Leaves 5-many-foliolate:
 11 Leaflets even-pinnate — **Abrus**
 11. Leaflets odd-pinnate:
  12. Flowers solitary — **Clitoria**
  12. Flowers racemed — **Derris**
2. Other than climbers:
 13. Leaves 2-foliolate — **Zornia**
 13. Leaves simple or 3-many-foliolate:
  14. Pods moniliform:
   15. Pods included in calyx — **Uraria**
   15. Pods not as above:
    16. Pod joints turgid — **Alysicarpus**
    16. Pod joints flattened:
     17. Leaves 1-3-foliolate — **Desmodium**
     17. Leaves many-foliolate — **Aeschynomene**
  14. Pods not moniliform:
   18. Leaflets lobed — **Vigna**
   18. Leaves or leaflets not lobed:
    19. Flowers yellow:
     20. Anthers uniform:
      21. Pods 1-2-seeded — **Stylosanthes**
      21. Pods many seeded — **Rothia**
     20. Anthers dimorphic:
      22. Pods turgid — **Crotalaria**
      22. Pods flattened — **Goniogyna**
    19. Flowers other than yellow:
     23. Anthers apiculate — **Indigofera**
     23. Anthers not apiculate:
      24. Leaf or leaflets margin dentate:
       25. Leaves 1-foliate — **Psoralea**
       25. Leaves 3-foliate — **Melilotus**
      24. Leaf or leaflets margin entire — **Tephrosia**
1. Trees (except *Butea superba*)
 26. Leaves 3-foliolate
  27. Armed — **Erythrina**
  27. Unarmed
   28. Pods 1-seeded — **Butea**
   28. Pods 2-many-seeded — **Ougeinia**
 26. Leaves 5-many-foliolate:
  29. Leaves odd-pinnate
   30. Leaflets opposite — **Pongamia**

30. Leaflets alternate **Dalbergia**
29. Leaves even-pinnate **Sesbania**

ABRUS Adanson

Abrus precatorious L., Syst. Nat. ed. 12. 2 : 472. 1767; FBI 2 : 176; Gamble 349 (247); Khan in FFH 126. *Glycine abrus* L., Sp. Pl. 753. 1753.

Climbing shrubs; branchlets densely hairy. Leaves even-pinnate; leaflets 10-13 pairs, oblong, 0.5 × 0.2-0.4 cm, hairy, base rounded, apex mucronate, margin entire. Flowers axillary and terminal, in compact racemes, pink. Pods oblong, 4 × 1.5 cm, pubescent; seeds 5-10. scarlet with black end, glabrous.

Fls. & *Frs.* : July - January. *Vern.* : Guruvinda, Erragurja.

Common in forest areas. Pocharam, **BR & CP** 11609; Medak, **CP** 11746.

AESCHYNOMENE Linnaeus

1. Leaflets 25-30 pairs; calyx hispid **A. aspera**
1. Leaflets 15-20 pairs; calyx glabrous **A. indica**

Aeschynomene aspera L., Sp. Pl. 713. 1753; FBI 2 : 151; Gamble 331 (234); Khan in FFH 124.

Under shrubs, about 2 m tall; branchlets glabrous. Leaves odd-pinnate; leaflets oblong, 0.1-0.7 × 0.1-0.2 cm, glabrous, base oblique, apex rotund-obtuse, margin entire. Flowers axillary, in hispid racemes, yellow. Pods oblong, 3.5 × 0.4 cm, glabrous, echinate over seeds, indented along both margins; seeds 5-8.

Fls. & *Frs.* : Round the year. *Vern.* : Niru-jilugu.

Common in moist waste lands. Patancheru, **BR & CP** 11307; Ponnala, **CP** 11371.

Aeschynomene indica L., Sp. Pl. 713. 1753. FBI 2 : 151; Gamble 331 (234).

Erect herbs, about 1 m tall, branchlets terete. Leaves odd-pinnate; leaflets linear, oblong, base obtuse. apex apiculate, margin entire. Flowers axillary, in racemes, whitish-yellow. Pods slightly falcate, 3.2 × 0.5 cm, glabrous; seeds 7-10.

Fls. & *Frs.* : July - December. *Vern.* : Tigajiluga.

Common in moist localities and forests. Choutkoor, **TP & CP** 11938; Gangapur RF, **TP & CP** 12076.

ALYSICARPUS Desvaux *nom. cons.*

1. Calyx shorter or slightly longer than first joint of pod:
   2. Pods not moniliform, compressed **A. procumbens**

2. Pods moniliform, turgid :
    3. Joints of the pod tetragonous, strongly veined **A. vaginalis**
    3. Joints of the pod subglobose, veinless **A. monilifer**
1. Calyx much longer than the first joint of pod :
    4. Joints of the pods not ribbed **A. bupleurifolius**
    4. Joints of the pods transversely ribbed:
        5. Leaves simple, dimorphic **A. scariosus**
        5. Leaves 1-foliolate, not dimorphic **A. rugosus**

Alysicarpus bupleurifolius (L.) DC., Prodr. 2 : 352. 1825; FBI 2 : 158; Gamble 338 (239). *Hedysarum bupleurifolium* L., Sp. Pl. 745. 1753.

Ascending herbs, about 60 cm tall; branchlets glabrous. Leaves lanceolate-oblong, 1-4 × 0.2-0.6 cm, glabrous, base rounded, apex acute, margin entire, Flowers axillary and terminal, in racemes, pink. Pods cylindric, 1 × 0.2 cm, glabrous; seeds 2-4.

Fls. & Frs. : July - December.

Occasional in waste lands. Narsapur tank, **GNR** 4428.

Alysicarpus monilifer (L.) DC., Prodr. 2:353. 1825; FBI 2:157; Gamble 338 (238). *Hedysarum moniliferum* L., Mant. Pl. 102. 1767.

Prostrate herbs, about 25 cm tall; branchlets pilose. Leaves ovate-oblong, 0.5 - 2 × 0.5-1.5 cm, pubescent, base obtuse-subcordate, apex obtuse, margin entire. Flowers axillary, in racemes, pink. Pods moniliform, 2 × 0.2 cm, hairy; seeds 3-7.

Fls. & Frs. : August - November.

Occasional in forests and waste lands. Pocharam RF, **TP & CP** 12060.

Alysicarpus procumbens (Roxb.) Schind. in Repert. Spec. Nov. Reg. Veg. 21.11. 1925. *Hedysarum procumbens* Roxb., Fl. Ind. 3 : 345. 1832; FBI 3 : 345. *Alysicarpus hamosus* Edgew. in Journ. Asiat. Soc. Beng. 21 : 171. 1853; Gamble 338 (238).

Prostrate herbs; branchlets hirsute. Leaves orbicular, 1.5 - 3.2 × 1-2.8 cm, hairy, base rounded, apex mucronate-rounded, margin entire. Flowers axillary or terminal, in racemes, purple. Pods oblong, 1.5 × 0.2 cm, hirsute; seeds 4-6.

Fls. & Frs. July - November.

Occasional in all forests. Pilutla RF, **BR & CP** 11257; Narsapur RF, **TP & CP** 14020.

Alysicarpus rugosus (Willd.) DC., Prodr. 2 : 553. 1825; FBI 2 : 159; Gamble 339 (239). *Hedysarum rugosum* Willd., Sp. Pl. 3 : 1172. 1802.

Diffuse herbs; branchlets glabrous. Leaves oblong, 0.5 - 1.5 × 0.2-1 cm, base obtuse-rounded, apex rounded, margin entire. Flowers axillary and terminal, in compact racemes, pink. Pods moniliform, 2 × 0.3 cm, glabrous; seeds 4-6.

Fls. & Frs. : July - November.

Common in all forests and in waste lands. Sangareddy, **BR & CP** 11295; Patancheru, **BR & CP** 11312.

Alysicarpus scariosus (Rottl. ex Spreng.) Grah. ex Thw. Enum. Pl. Zeyl. 88. 1858. *Hedysarum scariosum* Rottl. ex Spreng. Syst. Veg. 3:319. 1826. *Alysicarpus rugosus* (Willd.) DC. var. *styracifolius* Baker in FBI 2:159. 1876; Gamble 339 (239).

Diffuse herbs; branchlets hairy. Leaves dimorphic, lower elliptic-oblong, 1-2.5 × 0.5-1.2 cm, upper linear-lanceolate, 2-4 × 1-2 cm, base truncate, apex mucronate, margin entire. Flowers axillary and terminal, in compact racemes, pink. Pods moniliform, 1 × 0.2 cm; seeds 4-6.

Fls. & Frs. : June - December. *Vern.* : Chinnakandikaraku.

Occasional weed of cultivated fields and waste lands. Choutkoor, **TP & CP** 11914.

Alysicarpus vaginalis (L.) DC., Prodr. 2 : 353. 1825; FBI 2 : 158; Gamble 338 (238). *Hedysarum vaginale* L., Sp. Pl. 746. 1753.

Prostrate herbs; branchlets pubescent. Leaves obovate-oblong, 0.6 - 2 × 0.4 - 1 cm, sparsely hairy, base subcordate, apex obtuse, margin entire. Flowers axillary, in compact racemes, pink. Pods moniliform, 1.5 × 0.2 cm, downy-pubescent, seeds 5-7.

Fls. & Frs. : July - November.

Occasional in waste lands and cultivated fields. Sivampet, **TP & CP** 14013.

ATYLOSIA Wight & Arnott

Atylosia scarabaeoides (L.) Benth., in Miq., Pl. Jungh. 243. 1852; FBI 2 : 215; Gamble 367 (261); Khan in FFH 133. *Dolichos scarabaeoides* L., Sp. Pl. 726. 1753.

Climbing shrubs; branchlets densely pubescent. Leaves 3-foliolate; leaflets elliptic-ovate, 1-4.5 × 0.5-2.5 cm, pubescent, base cuneate, apex obtuse, margin entire. Flowers axillary, in short racemes, yellow. Pods oblong, 1.5 × 0.5 cm, densely pubescent; seeds 4-5.

Fls. & Frs. : August - December.

Occasional in forests. Narsapur forest nursery, **BR & CP** 11215; Sardhana, **BR & CP** 11628.

BUTEA Roxburgh *nom. cons.*

1. Trees **B. monosperma**
1. Woody climbers **B. superba**

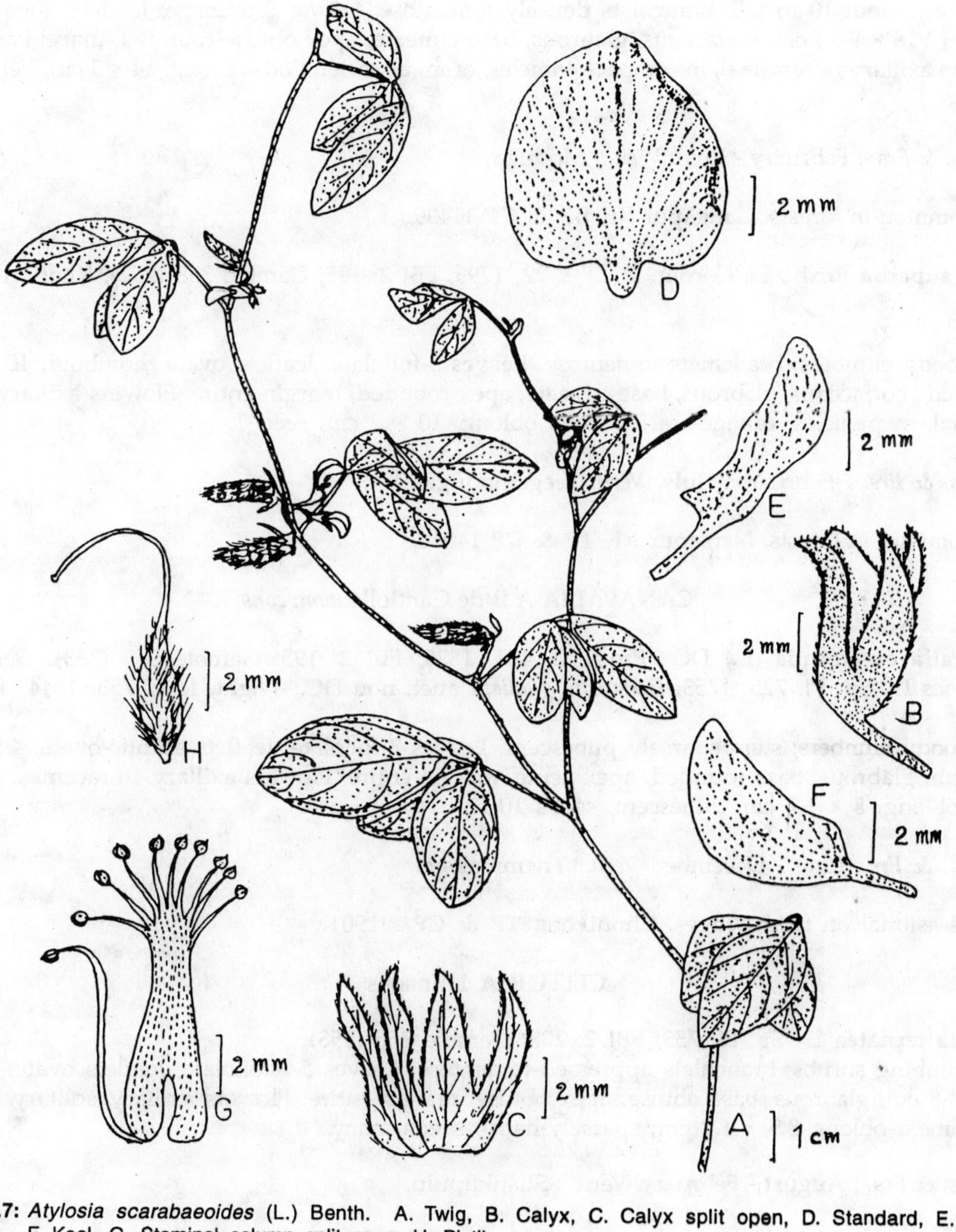

Fig. 6.7: *Atylosia scarabaeoides* (L.) Benth. A. Twig, B. Calyx, C. Calyx split open, D. Standard, E. Wing, F. Keel, G. Staminal column split open, H. Pistil

Butea monosperma (Lam.) Taub. in Engl. Nat. Pflanzenfam. 3 (3): 366. 1894. *Erythrina monosperma Lam., Encycl.*, 1: 391. 1785. *Butea frondosa* J. Koenig ex Roxb. in Asiat. Res. 3: 469. 1792; FBI 2: 194; Gamble 357 (252); Khan in FFH 129.

Trees, about 10 m tall; branchlets densely tomentose. Leaves 3-foliolate; leaflets unequally ovate, 10-18 × 9-15 cm, coriaceous, glabrous, base cuneate, apex obtuse-rounded, margin entire. Flowers axillary or terminal, in compact panicles, orange-scarlet. Pods oblong, 14 × 5 cm, Velvety; seed 1.

Fls. & *Frs.* : February - June. *Vern.* : Moduga.

Common in forests. Gangapur RF, **BR & CP** 11429.

Butea superba Roxb., Pl. Corom. 1: 23. t. 22. 1795; FBI 2: 195; Gamble 358 (252); Khan in FFH 129.

Woody climbers; branchlets tomentose. Leaves 3-foliolate; leaflets ovate-rhomboid, 10-15 × 9-14.5 cm, coriaceous, glabrous, base cuneate, apex rounded, margin entire. Flowers axillary and terminal, in panicles, orange-scarlet. Pods oblong, 10 × 4 cm; seed 1.

Fls. & *Frs.* : February - July. *Vern.* Teegamoduga.

Common in forests. Narsapur RF, **TP & CP** 14044.

CANAVALIA A.P. de Candolle *nom. cons.*

Canavalia ensiformis (L.) DC., Prodr. 2: 404. 1825; FBI 2: 195; Gamble 359 (253). *Dolichos ensiformis* L., Sp. Pl. 725. 1753. *Canavalia gladiata* auct. non DC.: Wight, Ic. t. 753. 1844.

Woody climbers; stem sparsely pubescent. Leaves 3-foliolate; leaflets elliptic-ovate, 4-9.5 × 2 - 6 cm, glabrous, base rounded, apex acute, margin entire. Flowers axillary, in racemes, pink. Pods oblong, 8 × 1.7 cm, pubescent; seeds 10-12.

Fls. & *Frs.*: June - December. *Vern.*: Thammakaya.

Ossasional on field hedges. Choutkoor. **TP & CP** 11901.

CLITORIA Linnaeus

Clitoria ternatea L., Sp. Pl. 753; FBI 2: 208; Gamble 365 (258).

Climbing shrubs; branchlets appressed-tomentose. Leaves 5-7-foliolate; leaflets ovate, 1.5-3 × 0.7-1.8 cm, glabrous, base obtuse, apex obtuse, margin entire. Flowers axillary, solitary, blue. Pods linear-oblong 8.5 × 1.2 cm, sparsely-hairy; seeds many.

Fls. & *Frs.* : August - February. Vern. : Shankupulu.

Common on field hedges as an escape. Narsingi, **BR & CP** 11525.

CROTALARIA Linnaeus

1. Leaves trifoliolate:
   2. Diffuse herbs **C. orixensis**
   2. Erect undershrubs or shrubs :
      3. Pods stipe > 1 cm **C. laburnifolia**
      3. Pods stipe < 0.5 cm **C. pallida**
1. Leaves simple.
   4. Stipules foliaceous **C. verrucosa**
   4. Stipules not foliaceous :
      5. Pods glabrous :
         6. Stipules linear-subulate **C. retusa**
         6. Stipules absent :
            7. Racemes 2-4-flowered **C. prostrata**
            7. Racemes many-flowered **C. albida**
      5. Pods hairy:
         8. Leaves ovate-orbicular.
            9. Standard petal blue within **C. bifaria**
            9. Standard petal yellow within :
               10. Erect herbs **C. hirsuta**
               10. Prostrate herbs **C. angulata**
         8. Leaves oblong **C. pusilla**

Crotalaria albida Heyne ex Roth, Nov. Pl. Spec. 333. 1821; FBI 2: 71; Gamble 295 (208).

Herbs, about 50 cm tall; branchlets pubescent. Leaves lanceolate-obvate, 1-4.5 × 0.6-1.5 cm, pubescent, base cuneate, apex retuse, margin entire. Flowers terminal, in racemes, yellow. Pods oblong, 0.8 × 0.4 cm, glabrous; seeds 4-10.

Fls. & Frs. : June - November.

Occasional in waste lands. Manjira Barrage, **BR & CP** 14092.

Crotalaria angulata Mill. Gard. Dict. ed. 8.n.9. 1768. *C. biflora* (L.) L., Mant. Alt. 570. 1771; FBI 2 : 66; Gamble 292 (206). *Astragalus biflorus* L., Mant. Alt. 273. 1771.

Diffuse herbs; branchlets hairy. Leaves ovate, 1-1.5 × 0.5-1 cm, membranous, base rounded, apex acute, margin entire. Flowers lateral, solitary or in racemes, yellow. Pods suborbicular, 1.8 × 1.2 cm, hairy; seeds 20.

Fls. & Frs. : November - April.

Occasional weed of waste lands. Lingareddypally, **CP** 11714.

Crotalaria bifaria L.f. Suppl. Pl. 322. 1781; FBI 1 : 69; Gamble 293 (207).

Prostrate herbs; branchlets pubescent. Leaves ovate orbicular or lanceolate, 1-4 × 0.5-2 cm, glabrescent above, puberulous below, base obtuse-cuneate, apex subacute, margin entire. Flowers axillary, in racemes, yellow. Pods oblong. 2.5 × 1 cm, puberulous; seeds 10.

Fls. & Frs. : October - December.

Occasional in cultivated fields. Gummadidala, **GNR** 4452.

Crotalaria hirsuta Willd., Sp. Pl. 3 : 978. 1803; FBI 2 : 68; Gamble 293 (207).

Erect herbs, about 50 cm tall; branchlets hairy. Leaves ovate, 3-5 × 2-2.5 cm, membranous, hairy, base cuneate, apex acute, margin entire. Flowers axillary, in racemes, yellow. Pods oblong, 2 × 0.6 cm, hirsute; seeds 8-10.

Fls. & Frs. : July - December.

Occasional in forests. Narsapur town, **KMS** 6674.

Crotalaria laburnifolia L., Sp. Pl. 715. 1753; FBI 2 : 84; Gamble 301 (213).

Shrubs, about 1 m tall; branchlets glabrous. Leaves 3-foliolate; leaflets ovate-elliptic, 3-6 × 1-2.5 cm, glabrous, base cuneate, apex apiculate, margin entire. Flowers lateral and terminal, in lax racemes, deep yellow. Pods cylindric, 4 × 1 cm, glabrous; seeds 10-20.

Fls. & Frs. : September - January.

Occasional in cultivated fields. Narsapur, **BR & CP** 11237; Medak, **CP** 11738.

Crotalaria orixensis Rottler ex Willd. in Ges. Naturf. Freunde Berlin Neue Schriften 4 : 217. 1803; FBI 2 : 83; Gamble 300 (212).

Diffuse herbs, about 60 cm tall; branchlets pubescent. Leaves 3-foliolate; leaflets oblong-obovate, 1-2.5 × 0.5 - 1.2 cm, glabrous above, pubescent below, base cuneate, apex apiculate, margin entire. Flowers axillary and terminal, in lax racemes, yellow. Pods cylindric, stipitate, 2 × 1 cm, glabrous; seeds 8-10.

Fls. & Frs. : August - March.

Occasional in waste lands. Manjira Barrage, **BR & CP** 14096.

Crotalaria pallida Aiton, Hort. Kew. 3 : 20. 1789. *C. striata* DC., Prodr. 2 : 131. 1825; FBI 2 : 84; Gamble 301 (212).

Undershrubs, about 1 m tall; branchlets puberulous. Leaves 3-foliolate; leaflets obovate-elliptic, 3-6.5 × 2-4.5 cm, membranous, base cuneate, apex acute-mucronate, margin entire. Flowers axillary and terminal, in racemes, yellow. Pods oblong, 3 × 0.5 cm, glabrous; seeds 50.

Fls. & Frs. : Round the year.

Occasional in waste lands. Manoharabad, **GNR** 4807.

Crotalaria prostrata Rottler in Willd., Enum. Hort. Berol. 744. 1809; FBI 2 : 67; Gamble 291 (206).

Prostrate herbs; branchlets hispid. Leaves oblong-elliptic, 1.5-2.5 × 0.5-1 cm, appressed-pubescent, base obtuse, apex subacute, margin entire. Flowers leaf-opposed, in racemes, yellow. Pods oblong 1.5 × 0.5 cm, glabrous; seeds 15-20.

Fls. & Frs. : September - January.

Occasional in open forests. Narsapur RF, **TP & CP** 11977.

Crotalaria pusilla Heyne ex Roth, Nov. Pl. Spec. 335. 1821; FBI 2.70; Gamble 296 (209).

Diffuse herbs, about 10 cm tall; branchlets hairly. Leaves oblong, linear, 0.5-1.8 × 0.1- 0.3 cm, hairy, base obtuse, apex acute, margin entire. Flowers lateral or terminal, in racemes, yellow. Pods oblong, 0.6 × 0.3 cm, strigose; seeds **4-8**.

Fls. & Frs. : August - January.

Rare in waste lands Narsapur, **KMS** 6626.

Note : This is the smallest species of the genus in India.

Crotalaria retusa L., Sp. Pl. 715; 1753; FBI 2 : 75; Gamble 293 (207).

Undershrubs, about 1.5 m tall; branchlets pubescent. Leaves oblanceolate, 3-7 × 0.5-2 cm, glabrous above, pubescent below, base attenuate, apex obtuse-retuse, margin entire. Flowers terminal, in racemes, golden yellow. Pods oblong, 4 × 1.2 cm, glabrous; seeds 5.

Fls. & Frs. : August - December.

Common in forests and in waste lands. Choutkoor, **TP & CP** 11921.

Crotalaria verrucosa L., Sp. Pl. 715. 1753; FBI 2 : 77; Gamble 297 (210).

Undershrubs, about 1 m tall; branchlets glabrous; Leaves ovate-rhomboid, 1.5-5 × 1-3.5 cm, pubescent, base attenuate, apex acute, margin entire. Flowers axillary and terminal, in racemes, white-yellow, Pods oblong, 4 × 1.1 cm, sparsely pubescent; seeds 2.

Fls. & Frs. : Round the year.

Common weed of waste lands, Narsapur forest nursery, **BR & CP** 11218; Siddipet, **CP** 11378; Medak, **CP** 11799.

DALBERGIA Linnaeus *nom. cons.*

1. Leaflets 9-17; stamens 10, in two bundles of five each:

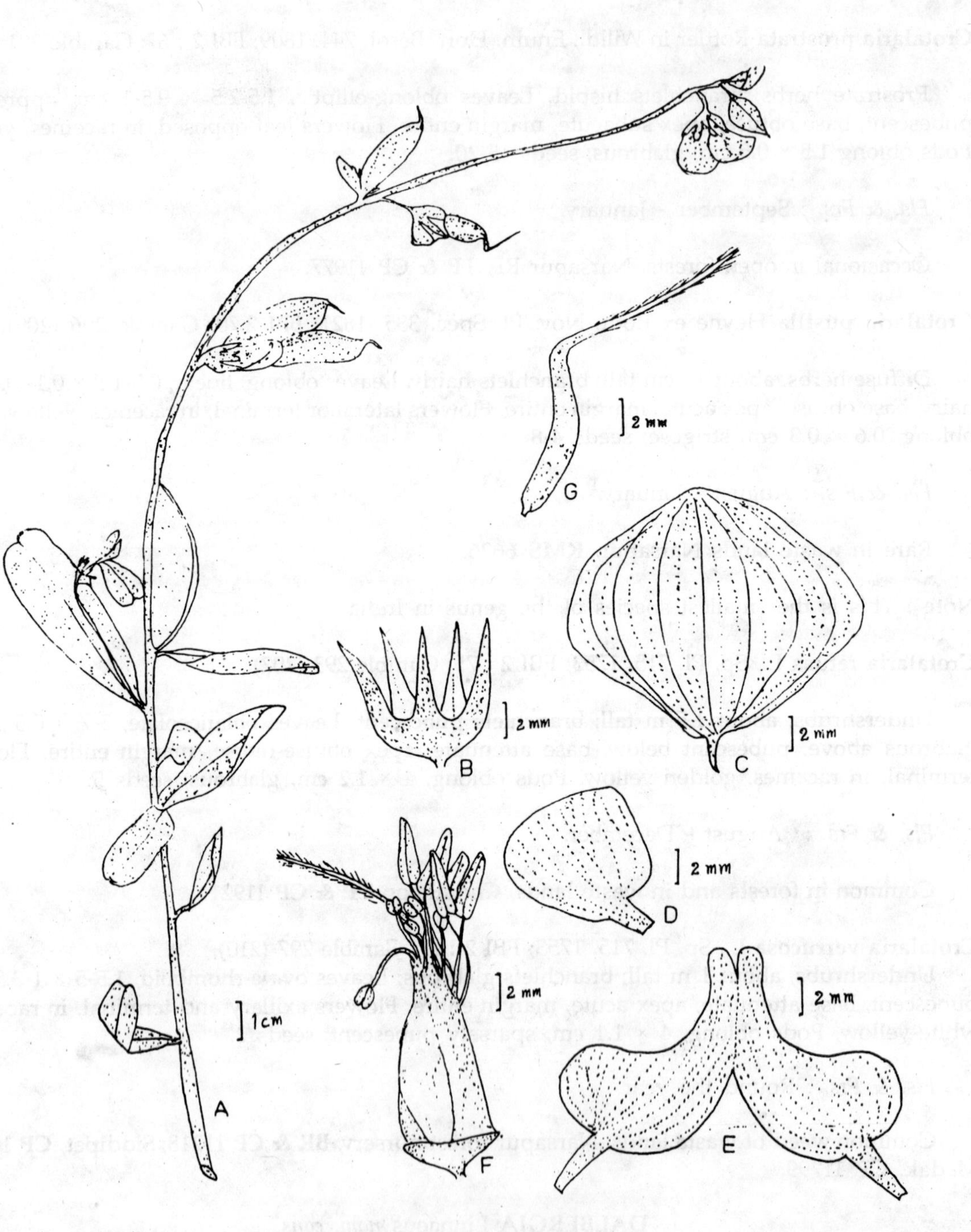

Fig. 6.8: *Crotalaria retusa* L. A. Twig, B. Calyx, C. Standard, D. Wing, F. Keel, F. Staminal column G. Pistil

2. Flowers pink, calyx puberulous, leaves and pods drying grey **D. lanceloria**
2. Flowers white, calyx rusty tomentose, leaves and pods drying black **D. paniculata**

1. Leaflets 5-7; stamens 9, monadelphous:
3. Leaflets with emarginate apex **D. latifolia**
3. Leaflets with acuminate apex **D. sissoo**

Dalbergia lanceolaria L. f. Suppl. Pl. 316. 1781; FBI 2 : 235; Gamble 383 (270); Khan in FFH 135.

Trees, about 15 m tall leaves odd pinnate; leaf lets obovate, 1.8-4.8 × 1-2 cm., chartaceous, glabrous, base cuneate, apex obtuse, margin entire. Flowers pink, in axillarry and terminal panicles. Pods lanceolate, 9 × 2 cm, compressed. Seeds 1-3.

Fls. & Frs.: April-July. *Vern.* : Pachari.

Occasional in forests. Zaheerabad, **CP** 11588.

Dalbergia latifolia Roxb., Pl. Corom. t. 113. 1799; FBI 2 : 231; Gamble 383 (270); Khan in FFH 134.

Trees, about 10 m tall; branchlets glabrous. Leaves odd pinnate; leaflets oblong-orbicular, 3.5-10 × 3-8 cm, coriaceous, glaborus, base rounded, apex slightly notched-emarginate, margin entire. Flowers axillary, in much branched panicles, yellow. Pods oblong, 3 × 1 cm, compressed; seeds 1-3.

Fls. & Frs. : July - April. *Vern.* : Jitegi.

Common in dry deciduous forests. Narsapur RF, **TP & CP** 14068.

Dalbergia paniculata Roxb., Pl. Corom. t. 114. 1799; FBI 2 : 236; Gamble 383 (270); Khan in FFH 136.

Trees, about 20m tall; branchlets glabrous. Leaves odd-pinnate; leaflets oblong, 1.3-2.5 × 1-1.5 cm, base obtuse, apex rounded emarginate, margin entire. Flowers axillary or terminal, in compact panicles, white. Pods lanceolate, 7.5 × 2 cm, glabrous; seeds 1-2.

Fls. & Frs. : April - December. *Vern.* : Sopera.

Common in dry deciduous forests. Narsapur RF, **TP & CP** 11968.

Dalbergia sissoo Roxb., Fl. Ind. 3:223. 1832; FBI 2:231; Gamble 384 (270); Khan in FFH 134.

Trees, about 10 m tall; banchlets glabrous. Leaves odd-pinnate; leaflets obovate, 2.5-4 × 2-4 cm, base obtuse, apex acumiate, margin entire. Flowers axillary or terminal, in panicles, creamish-white. Pods oblong, 7.5 × 1 cm, glabrous; seeds 1-2.

Fls. & Frs. : March - September.

Common along road sides. NSF, **BR & CP** 11648.

DERRIS Loureiro *nom. cons.*

Derris scandens (Roxb.) Benth. in Journ. Linn. Soc. Bot. 4 (Suppl.): 103. 1860; FBI 2:240; Gamble

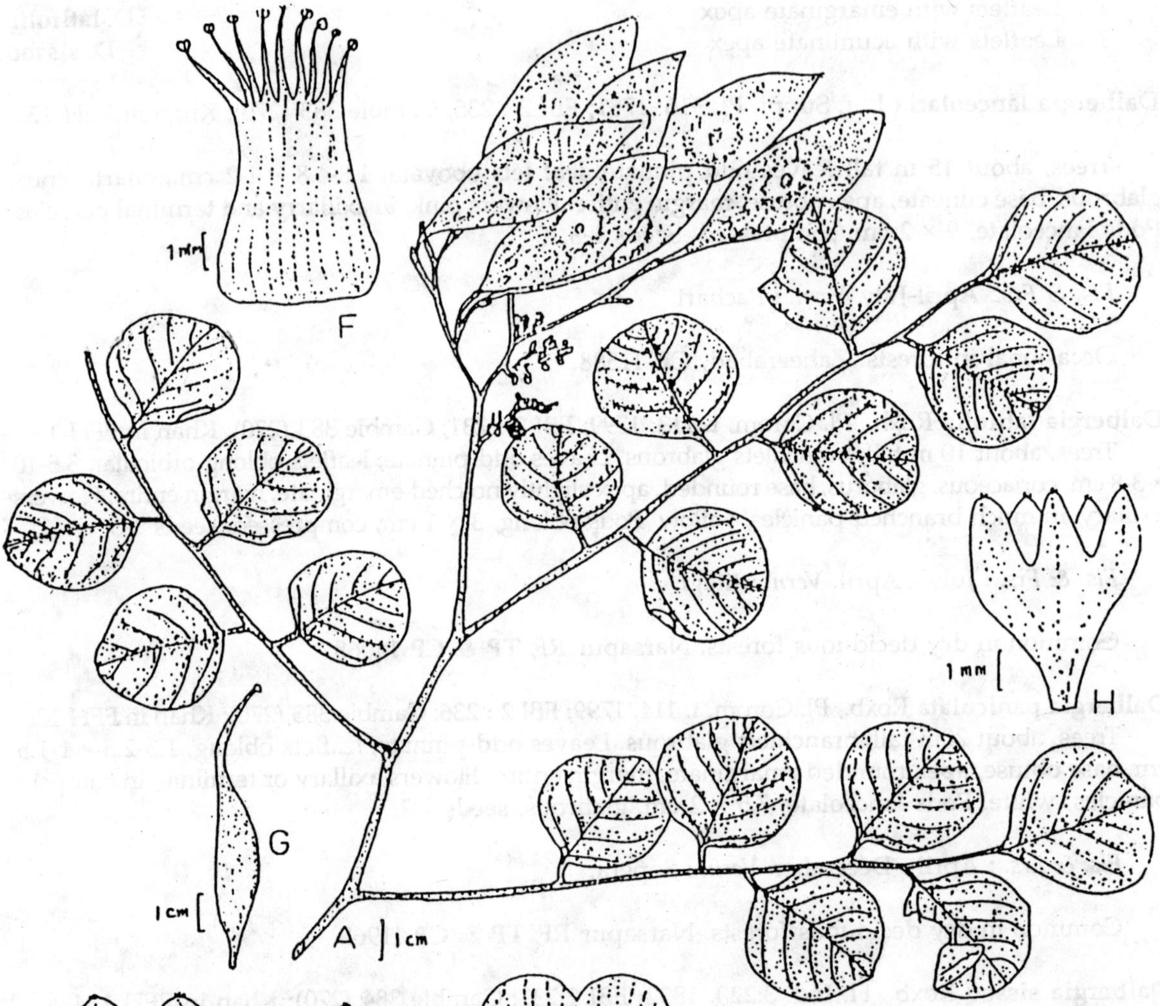

Fig. 6.9: *Dalbergia latifolia* Roxb. A. Twig, B. Calyx split open, C. Standard, D. Wing, F. Keel, F. Staminal column split open, G. Pistil

387 (273); Khan in FFH 138. *Dalbergia scandens* Roxb., Pl. Corom. t. 192. 1805.

Climbing shrubs; branchlets rusty-tomentose. Leaves odd-pinnate; leaflets 5-7, elliptic-oblong, 2.5-5 × 15.2 cm, glabrous, base rounded, apex retuse, margin entire. Flowers axillary and terminal, in racemes, white-pink. Pods oblong, 5 × 0.5 cm, glabrous; seeds 1-5.

Fls. & Frs. : June - November. Vern. : Nallateega.

Common along river banks and streams. Gnanapur, **CP** 11789.

DESMODIUM Desvaux *nom. cons.*

1. Leaves simple **D. gangeticum**
1. Leaves 3-foliolate :
    2. Foliar bracts present **D. pulchellum**
    2. Foliar bracts absent :
        3. Prostrate herbs or shrubs :
            4. Leaves estipulate; flowers in lax racemes **D. biarticulatum**
            4. Leaves stipulate; flowers solitary or in pairs **D. triflorum**
        3. Erect herbs or undershrubs:
            5. Stipules foliaceous; pods indented on both sutures **D. dichotomum**
            5. Stipules linear; pod straight on one suture:
                6. Lateral leaflets less than half as long as the terminal **D. alysicarpoides**
                6. Lateral leaflets half as long as the terminal **D. laxiflorum**

Desmodium alysicarpoides Van Meeuwen in Reinwardtia 6:246. 1962. *Alysicarpus parviflorus* Dalz. J. Bot. Kew Gard. Misc. 3:211. 1851; Gamble 339 (240). *Desmodium parviflorum* (Dalz.) Baker in FBI 2:172. 1876.

Erect herbs, about 30 cm, tall; branchlets glabrous. Leaves 1-3-foliolate, ovate-elliptic, 1-3.5 × 0.5-1.5cm, glabrous, base subcordate, apex obtuse, margin entire. Flowers axillary and terminal, in racemes, pink-violet. Pods oblong, 1.5 × 0.3 cm; seeds 5-6.

Fls. & Frs. : September - December.

Occasional along stream banks. Gummadidala. **GNR 4254.**

Desmodium biarticulatum (L.) F.V. Muell. Fragm. Phyt. Austr. 2. 121. 1861; FBI 2:163; Gamble 344 (243). *Hedysarum biarticulatum* L. Sp. Pl. 747. 1753.

Slender shrubs, about 50 cm tall; branchlets glabrous or downy. Leaves 3-foliolate, leaflets oblong-oblanceolate, 0.5-1.5 × 0.3-0.4 cm, glabrous, base cuneate, apex obtuse, margin entire. Flowers terminal, in lax racemes, bright red. Pods with 2 joints, oblong 1 × 0.2 cm, pubescent; seeds 2.

Fls. & Frs. : April - September.

Occasional in scrub jungles. Reddypalle, **TP & CP** 14076.

Desmodium dichotomum (Willd.) DC., Prodr. 2:336. 1825. *Hedysarum dichotomum* Wild., Sp. Pl. 3: 1180. 1802. *Desmodium diffusum* (Willd.) DC. in Ann. Sci. Nat. (Paris) 4:100. 1825, *nom. illegit.* : FBI 2: 169; Gamble 346 (244).

Diffuse straggling herbs, about 1m tall; branchlets pubescent. Leaves 3-foliolate; leaflets ovate-elliptic, 2-5 × 0.8 - 3 cm, pubescent, base subcordate, apex acute, margin entire. Flowers axillary or terminal, in lax racemes, pink. Pods moniliform, 2 × 0.3 cm, hairy; seeds 4-6.

Fls. & Frs. : June - November.

Occasional in forests and waste lands. Pathur, **TP & CP** 12109.

Desmodium gangeticum (L.) DC., Prodr. 2:327. 1825; FBI 2:168; Gamble 345 (244). *Hedysarum gangeticum* L., Sp. Pl. 746. 1753.

Undershrubs, about 1.5 m tall; branchlets pubescent. Leaves elliptic-obovate, 4-9 × 2-4 cm, pubescent, base truncate, apex acute, margin entire. Flowers terminal, in racemes, violet. Pods moniliform, 2 × 0.2 cm; seeds 3-5.

Fls. & Frs. : August - December.

Occasional in forests. Nagulabanda, **CP** 11344; Muttaypally, **CP** 11780.

Desmodium laxiflorum DC. Ann. Sci. Nat. (Paris) 4:100. 1825; FBI 2:164; Gamble 344 (243). *Hedysarum recurvatum* Roxb. Fl. Ind. 3:358. 1832. *Desmodium recurvatum* (Roxb.) Graham ex Wight & Arn., Prodr. 226. 1834.

Undershrubs, about 1 m tall; branchlets pubescent. Leaves 3-foliolate; leaflets ovate-elliptic, 2-5 × 1.5-3 cm, puberulous above, pubescent below, base subacute, apex acute-obtuse, margin entire. Flowers axillary and terminal, in racemes, blue. Pods linear, stipitate, 2 × 0.3 cm; seeds 5-7.

Fls. & Frs. : November - Ferbruary.

Occasional in forests. Narsapur, **GNR** 3947.

Desmodium pulchellum (L.) Benth., Fl. Hongk. 83. 1861; FBI 2:162; Gamble 344 (243); Khan in FFH 125. *Hedysarum pulchellum* L., Sp. Pl. 747. 1753.

Shrubs, about 1 m tall; branchlets tomentose. Leaves 3-foliolate; leaflets ovate, lateral leaflets : 3.5-7 × 1.5-4 cm, median leaflets : 5-12.5 × 2.3-6.7 cm, pubescent, base subcrodate, apex acute, margin entire. Flowers axillary, hidden by foliaceous bracts, yellow. Pods moniliform, 1.2 × 0.2 cm, glabrous; seeds 1-3.

Fls. & Frs. : July - November. *Vern. :* Teega-velaga.

Rare in dry deciduous forests. Gangapur RF, **TP & CP** 12095.

Note : The plant is very distinct with the bracts covering the flowers.

Desmodium triflorum (L.) DC., Prodr. 2:334. 1825; FBI 2:173; Gamble 347 (245). *Hedysarum triflorum* L. Sp. Pl. 749. 1753.

Trailing herbs, branchlets elongate-pubescent. Leaves 3-foliolate; leaflets obovate, 0.4-1.5 × 0.4-1 cm, base round, apex emarginate-obtuse, margin entire. Flowers axillary or terminal, solitary or in pairs, purple-pink. Pods moniliform, 2 × 0.3 cm, pubescent; seeds 2-4.

Fls. & Frs. : July - November.

Occasional in all forests and waste lands. Pillutla RF, **BR & CP** 11265.

DOLICHOS Linnaeus

Dolichos trilobus L., Sp. Pl. 726. 1753. *D. falcatus* Klein ex Willd., Sp. Pl. 3:1047; 1802; FBI 2:211; Gamble 366 (259).

Twining herbs; branchlets hairy. Leaves 3-foliolate, terminal leaflets ovate-rhomboid, 1.4-2.2 × 0.8-1.5 cm, lateral leaflets ovate, 0.7-2 × 0.5-1.3 cm, sparsely pubescent, base truncate, apex acute-apiculate, margin entire. Flowers axillary, in lax racemes, yellow. Pods linear, 7 × 1 cm, glabrous; seeds 4-6.

Fls. & Frs. : August - December.

Common in field hedges. Mambojipally, **CP** 11632; Gangapur, **TP & CP** 12090.

ERYTHRINA Linnaeus

1. Armed; keel petals connate **E. stricta**
1. Unarmed; keel petals free **E. variegata**

Erythrina stricta Roxb., Fl. Ind. 3:251. 1832; FBI 2:189; Gamble 354 (250); Khan in FFH 129.

Trees, about 8 m tall; branchlets prickled. Leaves 3-foliolate; leaflets rhomboid, 12-15, × 10-16.5 cm, glabrous, base truncate, apex acuminate, margin entire. Flowers terminal, in racemes, scarlet. Pods linear, 12 × 2 cm, glabrous; seeds 5-6.

Fls. & Frs. : December - April. *Vern.* : Mulla-modugu.

Occasional in waste lands. Manjira Barrage, **BR & CP** 14095.

Erythrina variegata L., Herb. Amboin. 10.1754 var. **orientalis** Merr. Interpr. Herb. Amboin. 276.1617

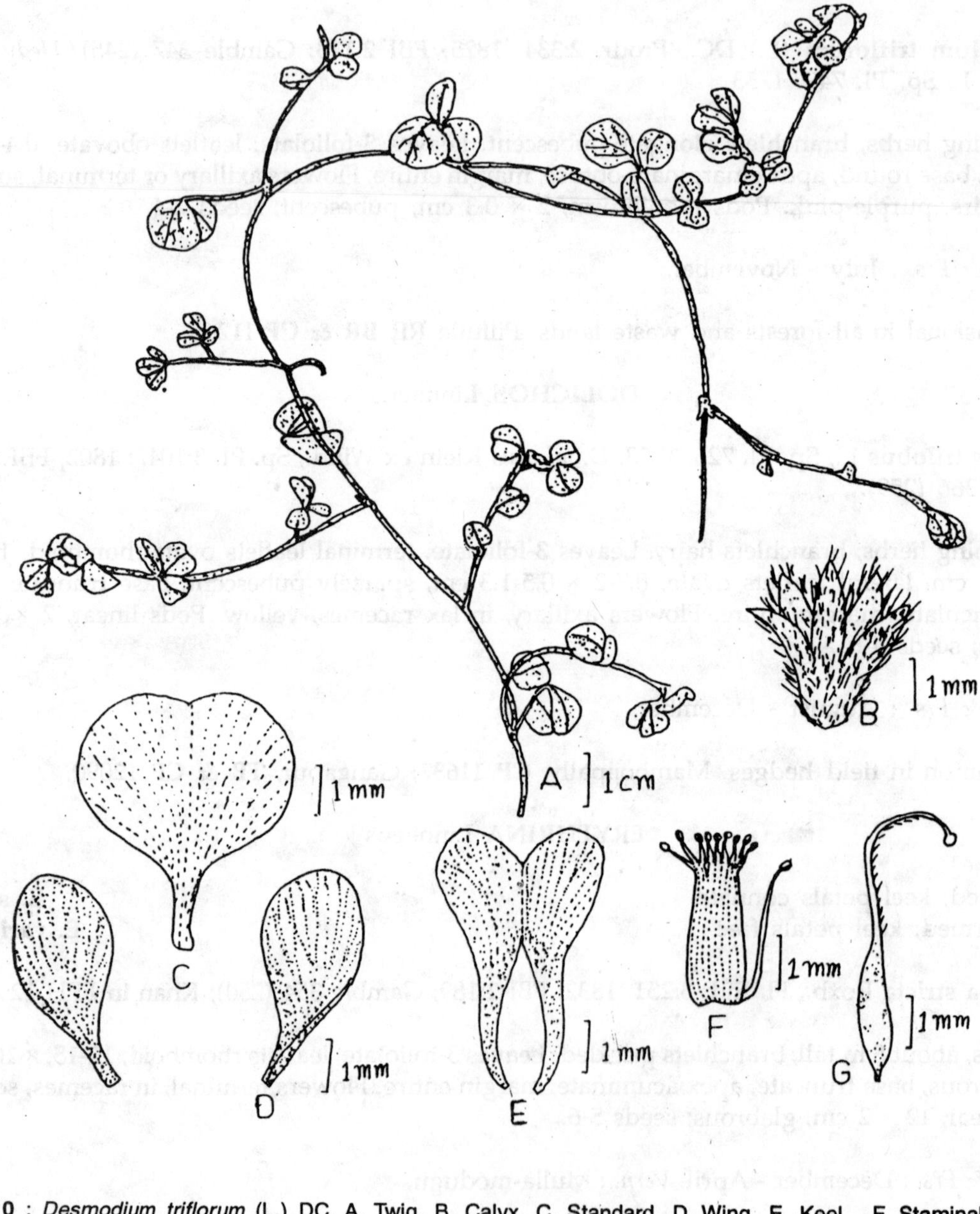

Fig. 6.10 : *Desmodium triflorum* (L.) DC. A. Twig, B. Calyx, C. Standard, D. Wing, E. Keel, F. Staminal column split open, G. Pistil

E. indica Lam., Encycl. 2:391. 1786' Roxb. Fl. Ind. 3. 249. 1832; FBI 2.188; Gamble 353 (249); Khan in FFH 128.

Trees, about 15 m tall; branchlets pubescent. Leaves 3-foliolate; leaflets rhomboid-ovate, 10-15 × 8-10 cm, coriaceous, base obtuse-truncate, apex acuminate, margin entire. Flowers axillary and terminal, in racemes, bright red. Pods falcate; seeds 6-8.

Fls. & Frs. : February - June.

Occasional in hedges. Doulatabad, **GNR** 4783.

GONIOGYNA A.P. de Condolle

Goniogyna hirta (Willd). Ali in Taxon 16:463. 1967. *Hallia hirta* Willd., Sp. Pl. 3(2):1169. 1803. *Heylandia latebrosa* DC., Mem. Leg 6:201. 1825; FBI 2:65; Gamble 280(199).

Prostrate herbs; branchlets hispid. Leaves ovate, 0.5-1 × 0.3-0.6 cm, silky, base subcordate, apex acute, margin entire. Flowers axillary, solitary, pale-yellow. Pods oblong, 0.5 × 0.3 cm, silky; seeds 1-2.

Fls. & Frs. : September - March.

Common weed of waste lands and fields. NSF, **BR & CP** 11660.

INDIGOFERA Linnaeus

1. Leaves simple:
    2. Pods prickly — **I. nummularifolia**
    2. Pods not prickly :
        3. Pods 1-seeded — **I. linifolia**
        3. Pods 2-many seeded — **I. cordifolia**
1. Leaves 3-many foliolate :
    4. Leaves 3-foliolate :
        5. Pods 1-seeded — **I. glandulosa**
        5. Pods many-seeded :
            6. Pods not exceeding 1 cm in length — **I. trifoliata**
            6. Pods exceeding 1 cm in length — **I. trita**
    4. Leaves 5-many-foliolate :
        7. Diffuse herbs — **I. linnaei**
        7. Erect subshrubs or shrubs :
            8. Flowers in sub-capitate clusters — **I. tinctoria**
            8. Flowers on elongated recemes :
                9. Leaflets 5-13 — **I. hirsuta**
                9. Leaflets 17-21 — **I. cassioides**

Indigofera cassioides Rottler ex DC., Prodr. 2:225. 1825. *I. pulchella* Roxb., Fl. Ind. 3:382. 1832; FBI 2:101; Gamble 313 (221); Khan in FFH 122.

Shrubs, about 2 m tall; brachlets pubescent. Leaves odd-pinnate; leaflets elliptic-oblong, 0.8-2 × 0.5-1.2 cm, pubescent, base obtuse, apex obtuse, margin entire. Flowers axillary, in racemes, pink-rose. Pods cylindrical, 4 × 1 cm, glabrous; seeds 8-10.

Fls & Frs. : December - March. *Vern.* : Karakandi.

Occasional in forests. Narsapur RF, **TP & CP** 14040.

Indigofera cordifolia Heyne ex Roth, Nov. Pl. Spec. 357, 1821; FBI 2:93; Gamble 309 (218).

Prostrate herbs; branchlets densely white-tomentose. Leaves ovate, 0.5-1 × 0.4-0.6 cm, tomentose, base cordate, apex acute, margin entire. Flowers axillary, in racemes, red. Pods linear-oblong, 2.5 × 0.3 cm, pubescent; seeds 2-4.

Fls. & Frs. : Round the year. *Vern.* : Budithaku.

Common weed of cultivated fields. Choutkoor. **TP & CP** 11932.

Indigofera glandulosa Roxb. ex Willd., Sp. Pl. 3:1227. 1800; FBI 2:94; Gamble 309 (218).

Undershrubs, about 50 cm tall; branchlets pubescent. Leaves 3-foliolate; leaflets oblanceolate, 0.4-2 × 0.2-0.6 cm, sparsely hairy, gland-dotted below, base attenuate, apex obtuse, margin entire. Flowers axillary, in subcapitate clusters, red. Pods cylindrical, 0.5 × 0.1 cm, hairy; seed 1.

Fls. & Frs. : March - September.

Occasional in waste lands. Narsapur forest nursery, **BR & CP** 11221.

Indigofera hirsuta L., Sp. Pl. 751. 1753; FBI 2:98; Gamble 312 (221).

Undershrubs, about 1 m tall; branchlets hirsute. Leaves odd-pinnate; leaflets elliptic-oblong, 1.5-3.5 × 1-1.5 cm, densely pubescent, base obtuse, apex apiculate, margin entire. Flowers axillary, in compact racemes, brickred. Pods linear, tetragonous, 1.8 × 0.3 cm, appressed-hirsute, mucronate; seeds 5.

Fls. & Frs. : March - September.

Common weed of cultivated fields. Poosanpalle, **TP & CP** 11051.

Indigofera linifolia (L.f.) Retz. Obs. Bot. 4:29. 1786; FBI 2:92; Gamble 309 (218). *Hedysarum linifolium* L.f., Suppl. Pl. 331. 1781.

Diffuse herbs, about 40 cm tall; branchlets silvery-canescent. Leaves linear, 0.8-1.5 × 0.2-0.4 cm, appressed pubescent, base cuneate, apex mucronate, margin entire. Flowers axillary, solitary or in racemes, bright red. Pods globose, 0.2 cm across, appressed silvery - pubescent; seed 1.

Fls. & Frs. : September - December. *Vern.* : Gudlaku.

Occasional in forests and moist areas. Pillutla RF, **BR & CP** 11258; Patancheru, **BR & CP** 11311.

Indigofera linnaei Ali in Bot. Notiser 111:549. 1958. *Hedysarum prostratum* L., Mant. Pl. 102. 1767. *Indigofera enneaphylla* L., Mant. Pl. 2:272. 1771. *nom. superfl.*; FBI 2:94; Gamble 309 (218).

Prostrate herbs; branchlets white-hirsute. Leaves odd-pinnate; leaflets oblanceolate, 0.2-1 × 0.2 × 0.4 cm, densely hairy, base cuneate, apex obtuse, retuse, margin entire. Flowers axillary, in racemes, red. Pods ellipsoid, 0.4 × 0.2 cm, pubescent; seeds 2.

Fls. & Frs. : July - December.

Common weed of cultivated fields and road sides. Toopran, **BR & CP** 11278; Mombojipally, **BR & CP** 11639.

Indigofera nummularifolia (L.) Livera ex Alston in Trimen, Handb. Fl. Ceylon 6 (suppl.): 72. 1931, *Hedysarum nummularifoium* L., Sp. Pl. 1051. 1753. *Indigofera echinata* Willd., Sp. Pl. 3: 1222. 1800; FBI 2: 92; Gamble 308 (218).

Prostrate herbs; branchlets appressed-pubescent, Leaves obovate, 0.5-2 × 0.5-1.5 cm, base obtuse, apex obtuse-apiculate. margin entire. Flowers axillary, in racemes, pink. Pods reniform, 0.5 × 0.2 cm, prickled; seed 1.

Fls. & Frs. : July - December.

Occasional in dry deciduous forests. Pandivagu, **BR & CP** 11209; Pocharam tank, **TP & CP** 12051; Rampur, **TP & CP** 14070.

Indigofera tinctoria L., Sp. Pl. 751. 1753; FBI 2:99; Gamble 312 (220); Khan in FFH 123.

Undershrubs, about 80 cm tall; branchlets pubescent. Leaves odd-pinnate; leaflets elliptic-obovate, 0.5-1.8 × 0.2-0.5 cm, glabrous above, hairy below, base cuneate, apex obtuse, margin mucronate. Flowers axillary, in racemes, red. Pods linear, 2.5 × 0.2 cm, pubescent; seeds 8-10.

Fls. & Frs. : February - September. *Vern.* : *Neeli.*

Common weed of cultivated fields. Thornal, **CP** 11708.

Indigofera trifoliata L., Cent. Pl. 2:29. 1756; FBI 2:96; Gamble 310 (219).

Undershrubs, about 60 cm tall; branchlets grey pubescent. Leaves 3-foliolate; leaflets oblanceolate, 1-1.5 × 0.6-1 cm, densely hairy, base cuneate, apex obtuse, margin entire. Flowers axillary, in racemes, flame-coloured. Pods linear, 1.3 × 0.2 cm, pubescent; seeds 3-6.

Fls. & Frs. : August - November.

Common weed of waste lands. Poosanpalle, **TP & CP** 11955.

Indigofera trita L.f., Suppl. Pl. 335. 1781; FBI 2: 96; Gamble 310 (219)

Undershrubs, about 1 m tall; branchlets appressed-pubescent. Leaves 3-foliolate; leaflets obovate, 1-2.5 × 0.8-1.5 cm, appressed-pubescent, base cuneate, apex emarginate, margin entire. Flowers axillary, in racemes, pink. Pods linear, 3 × 0.2 cm, pubescent; seeds 5-8.

Fls & Frs. : March - September.

Common in forests and waste lands. Narsapur RF, **TP & CP** 14039.

MELILOTUS P. Miller

Melilotus alba Desr. in Lam., Encycl. 4:63. 1789; FBI 2:89; Gamble 303 (215).

Erect herbs, about 50 cm tall; branchlets glabrous. Leaves 3-foliolate; leaflets obovate, 1-2 × 0.4-0.3 cm, glabrous, base cuneate, apex acute, margin serrate. Flowers axillary, in racemes, white. Pods oblong, 1 × 0.4 cm, glabrous; seeds 2.

Fls. & Frs. : January - April.

Occasional weed of cultivated fields. NSF, **BR & CP** 11662.

MUCUNA Adanson *nom. cons.*

Mucuna pruriens (L.) DC., Prodr. 2:405. 1825; FBI 2:189; Khan in FFH 127. *Dolichos pruriens* L. in Stickman Herb. Amb. 23. 1754. *Mucuna prurita Hook.* in Journ. Bot. Kew Gard. Misc. 2:348 t. 13. 1830. *nom. superfl.*; Gamble 356(251).

Woody climbers; branchlets pubescent. Leaves 3-foliolate, leaflets ovate-rhomboid, 5-8.5 × 4-6 cm, pubescent, base truncate, apex acute-mucronate, margin entire. Flowers axillary, in racemes, purple. Pods oblong, 4.5 × 1 cm, clothed with dense brown bristles, ends curved; seeds 5-6.

Fls. & Frs. : August - December. Vern. : Duldamma, Neela.

Common on hedges, Pathur, **TP & CP** 12104.

NEONOTONIA (Wight & Arnott) Lackey

Neonotonia wightii (Wight & Arn.) Lackey in Phytologia 37:210.1977. *Glycine javanica* L. Sp. Pl. 754. 1753; FBI 2:183; Gamble 351 (248). *Notonia wightii* Wight & Arn. Prodr. 208. 1834. *Johnia wightii* Wight & Arn.. Prodr. 449. 1834.

Climbing herbs., branchlets hispid. Leaves 3-foliolate, leaflets equal in size, 5-8 × 5-6 cm, ovate, acute, margin entire, base obtuse. Flowers in axillary racemes. Pod 2.5 × 0.4 cm, flat, pubescent, seeds 4-5.

Fls. & Frs. : September - December

Rare in forests. Domadugu, **BR & CP** 14199.

OUGEINIA Bentham

Ougeinia oojeinensis (Roxb.) Hochr. in Ann. Cons. Jard. Bot. Geneve 13 & 14:51.1909. *Dalbergia oojeinensis* Roxb. Fl. Ind. 3:220. 1832. *Ougeinia dalbergioides* Benth. in Miq., Pl. Jungh. 216. 1852; FBI 2:161; Gamble 340 (240).

Trees, about 14 m tall; branchlets glabrous. Leaves 3-foliolate; leaflets broadly ovate, 9-12 × 7-10 cm, coriaceous, glabrous above, pubescent below, base subcordate, apex obtuse or slightly notched, margin undulate. Flowers axillary, in dense fascicles, rose. Pods linear, 6 × 1 cm, 2-5-jointed; seeds 2-6.

Fls. & Frs. : February - September. *Vern.* : Dargu.

Occasional in dry deciduous forest. Gangapur RF, **BR & CP** 11426.

PARACALYX S.I. Ali

Paracalyx scariosus (Roxb.) Ali in Univ. Stud. (Karachi) 5(3):95. 1968. *Cylista scariosa* Roxb., Pl. Corom.t.92.1798; FBI 2:219; Gamble 371 (262).

Twining shrubs; branchlets densely pubescent. Leaves 3-foliolate; leaflets ovate-rhomboid, 1-4.5 × 1-3 cm, pubescent above, tomentose below, base rounded-obtuse, apex acute, margin entire. Flowers axillary, in racemes, yellow. Pods oblong, 1 × 0.2 cm, pubescent, concealed in persistent calyx; seed 1.

Fls. & Frs. : September - March.

Occasional in forests. Gangapur RF, **TP & CP 12100.**

PONGAMIA Ventenat *nom. cons*

Pongamia pinnata (L.) Pierre, Fl. Forest, Cohinch. sub. t. 385 1899. *Cytisus pinnatus* L., Sp. Pl. 741 1753. *Pongamia glabra* Vent., Jard. Mam. t. 28. 1803, *nom. illegit;* FBI 2:240; Gamble 385 (272); Khan in FFH 137.

Trees, about 10 m tall; branchlets glabrescent. Leaves odd-pinnate; leaflets ovate-elliptic, 5-10 × 3.5-5.5 cm, base truncate, apex acuminate, margin entire. Flowers axillary, in panicles, purplish-white. Pods obliquely oblong, 4 × 2.5 cm, glabrous; seed 1.

Fls. & Frs. : February - December. *Vern. :* Kanuga.

Common avenue tree, often as an escape. Narsapur RF, **CP** 11572.

PSEUDARTHRIA Wight & Arnott

Pseudarthria viscida (L.) Wight & Arn., Prodr. 209. 1834; FBI 2: 154: Gamble 334 (236). *Hedysarum viscidum* L., Sp. Pl. 747. 1753.

Undershrubs; branchlets densely pubescent, leaves 3-foliolate; leaflets ovate-rhomboid, pubescent, median leaflets : 2.5-7 × 2-4.5 cm, lateral leaflets : 1.5-4.5 × 1-3.2 cm, base rounded-truncte, apex acute-acuminate, margin entire. Flowers axillary or terminal, in lax racemes, pink, Pods linear, 1.5 × 0.5 cm, viscid; seeds 4-5.

Fls. & Frs. : September - January.

Common in dry deciduous forests and waste lands. Narsapur RF, **TP & CP** 11982, 14048 Gangapur RF, **TP & CP** 12070.

PSORALEA Linnaeus

Psoralea corylifolia L., Sp. Pl. 764. 1753; FBI 2:103; Gamble 315 (222).

Herbs, about 60 cm tall; branchlets glabrous. Leaves ovate-elliptic, 3-6 × 2.5-5 cm, pubescent, base rounded, apex acute, margin dentate. Flowers on peduncles, in clusters, bluish-purple. Pods oblong, 0.5 × 0.2 cm, glabrous; seed 1.

Fls. & Frs. : December - April.

Occasional in cultivated fields and waste lands. Manjira Barrage, **BR & CP** 14088.

RHYNCHOSIA Loureiro *nom. cons.*

1. Twining herbs or undershrubs :
    2. Leaflets small, apex obtuse **R. minima**
    2. Leaflets moderately large, apex acuminate **R. bracteata**
1. Erect shrubs :
    3. Calyx accrescent, lobes oblong **R. rufescens**
    3. Calyx not accrescent, lobes lanceolate **R. suaveolens**

Rhynchosia bractaeta Benth. in Hook. f, FBI 2:225, 1878; Gamble 374 (263).

Twining undershrubs; branchlets striate, velvety-pubescent. Leaves 3-foliolate, terminal leaflets rhomboid-orbicular, laterals obliquely obovate, 4-7.5 × 1-3 cm, pubescent, apex acuminate, margin entire. Flowers axillary, in branched racemes. Pods mucronate, 2.5 × 1 cm, turgid; seeds 2.

Fls. & *Frs.* : July - December.

Rare in waste lands. Annaram, **GNR** 4629.

Rhynchosia minima (L.) DC., Prodr. 2:385. 1825; FBI 2:223; Gamble 375 (264). *Dolichos minimus L.*, Sp. Pl. 726. 1753.

Spreading herbs; branchlets pubescent. Leaves 3-foliolate. Leaflets obovate-rhomboid, 0.8-1.4 × 0.7-2 cm, pubescent, gland dotted below, base obtuse, apex truncate-rounded, margin entire. Flowers axillary, in racemes, yellow. Pods oblong, 1 × 0.3 cm. pubescent, seeds 2.

Fls. & Frs. : June - November.

Common in hedges and waste lands. Ranjhole, **CP** 11678; Poosanpally, **TP & CP** 11963.

Rhynchosia rufescens (Willd.) DC. Prodr. 2:387. 1825; FBI 2. 220; Gamble 373 (263). *Glycine rufescens* Willd. in Ges. Naturf. Freunde Berlin Neue Schriften 4 : 222. 1803.

Erect shrubs; branchlets trailing, viscid-pubescent. Leaves 3-foliolate, terminal leaflets ovate-elliptic, lateral ovate-rhomboid, 2-5 × 1.5-2.5 cm, pubescent, base obtuse-acute, apex acute, margin entire. Flowers axillary, in lax racemes, yellow. Pods ovate, 1.2 × 0.8 cm, pubescent; seed 1.

Fls. & Frs. : Round the year.

Occasional along road sides in forests. Kondapur, **GNR** 4564.

Rhynchosia suaveolens (L.) DC., Prodr. 2:387. 1825; FBI 2:221; Gamble 374 (264). *Glycine suaveolens* L.f., Suppl. Pl. 326. 1781.

Undershrubs, about 75 cm tall; branchlets viscid. Leaves 3-foliolate; leaflets ovate, 3-7 × 2-4.5 cm, glabrous, base trauncate, apex acuminate, margin entire. Flowers axillary, solitary or in pairs, yellow. Pods oblong, 1.5 × 0.8 cm, pubescent; seeds 2.

Fls. & Frs : July - January. *Vern.* : Karra-vulava.

Occasional in forests. Gummadidala hills, **BR & CP** 14126.

ROTHIA Persoon *nom. cons.*

Rothia indica (L.) Druce in Bot. Exch. Club Soc. Brit. Isles 3:423. 1914. *Trigonella indica L.*, Sp. Pl. 778. 1753. *Rothia trifoliata* (Pers.) DC., Prodr. 2:382. 1825; FBI 2:63; Gamble 279 (199). *Dilwynia trifoliata* Pers., Syn. Pl. 2:302. 1807.

Prostrate herbs; branchlets hirsute. Leaves 3-foliolate; leaflets oblanceolate, 0.8-2.5 × 0.4-0.6 cm, hispid, base attenuate, apex acute, margin entire. Flowers axillary, solitary or in pairs, yellow. Pods linear, 6 × 0.2 cm, hairy; seeds many.

Fls. & Frs. : July - December.

Occasional weed of waste lands and forests. Gummadidala, **GNR** 4468.

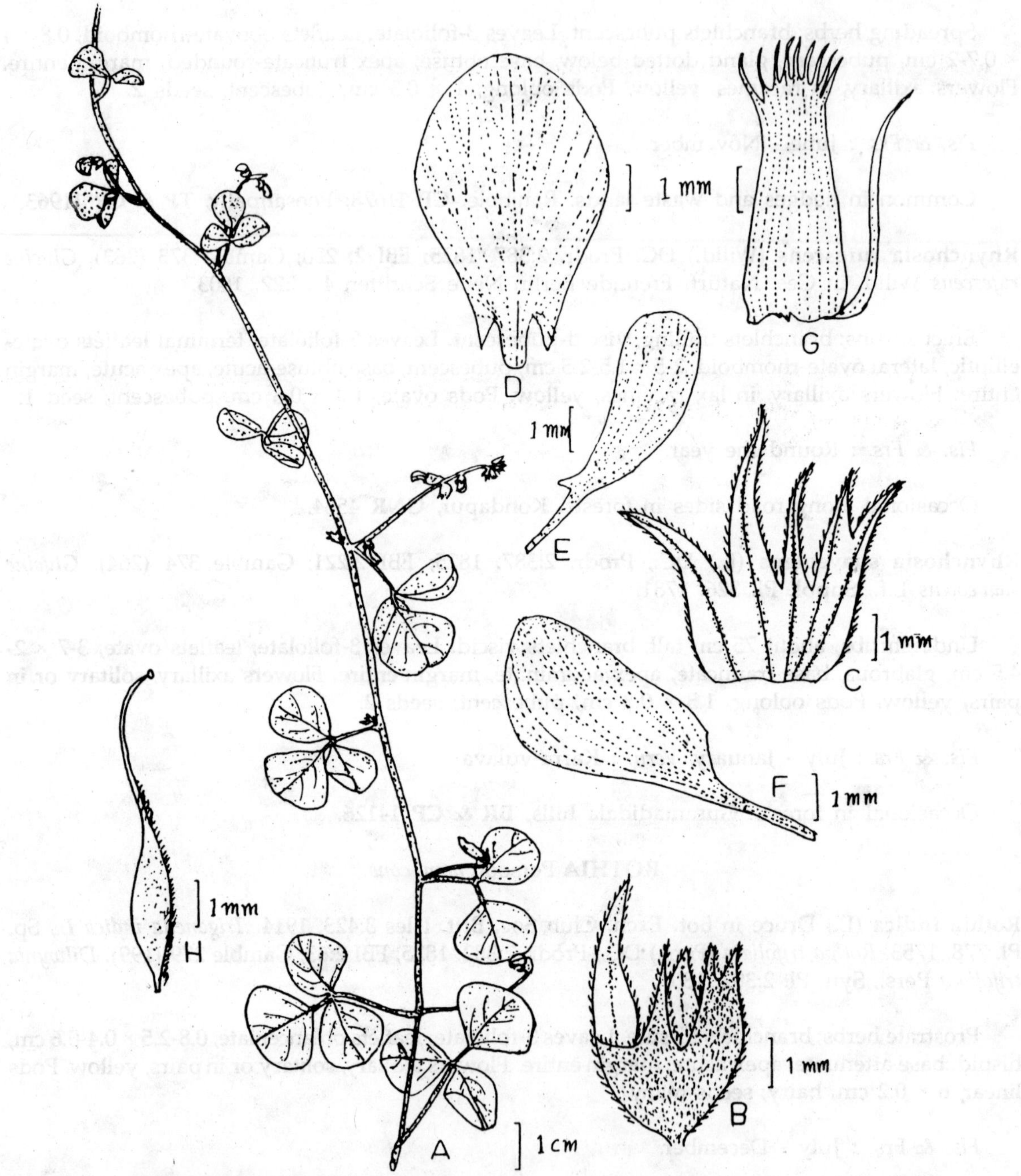

Fig. 6.11 : *Rhynchosia minima* **(L.) DC. A. Twig, B. Calyx, C. calyx split open, D. Standrad, E. Wing, F. Keel, G. Staminal colum split open, H. Pistil**

SESBANIA Scopoli *nom. cons.*

1. Leaf rachis armed, leaflets 40-50 pairs **S. bispinosa**
1. Leaf rachis unarmed, leaflets 15-20 pairs **S. sesban**

Sesbania bispinosa (Jacq.) W.F. Wight. in U.S.D.A. Bur. Pl. Industr. Bull. 137. 15. 1909. *Aeschynomene bispinosa* Jacq., Ic.Pl. Rar. 3: 13t. 564. 1793. *Sesbania aculeata* Poir. in Lam., Encycl. 7: 128. 1806; FBI 2: 114; Gamble 323 (228).

Armed shrubs, about 3 m tall; branchlets with short prickles, glabrous. Leaves even-pinnate; leaflets oblong, 0.5-1 × 0.1-0.2 cm, glabrescent, base cuneate, apex mucronate, margin entire. Flowers axillary, in drooping racemes, yellow. Pods linear, 20 × 0.2 cm, glabrous; seeds 30-40.

Fls. & *Frs.* : September - February.

Common in wet places and fields. Narsingi, **BR & CP** 11544; Manjira barrage **BR & CP** 14086.

Sesbania sesban (L.) Merr. in Philipp. Journ. Sci. 7:235. 1912. *Aeschynomene sesban* L., Sp. Pl. 714. 1753. *Sesbania aegyptiaca* Poir. in Lam., Encycl. 7:128. 1806; FBI 2:114; Gamble 323 (228); Khan in FFH 123.

Shrubs, about 4 m tall; branchlets glabrous. Leaves even-pinnate; leaflets oblong, 1-2 × 0.2-0.4 cm, glabrous, base cuneate, apex apiculate, margin entire. Flowers axillary, in drooping racemes, yellow. Pods linear, 30 × 0.2 cm, glabrous; seeds 20-30.

Fls. & *Frs.* : November - March. *Vern.* : Suiminta.

Cultivated, often as an escape. Gummadidala, **BR & CP** 11488.

STYLOSANTHES Swartz

Stylosanthes fruticosa (Retz.) Alston in Trimen Handb. Fl. Ceylon 6:77. 1931. *Arachis fruticosa* Retz., Obs. 5:26. 1789. *Stylosanthes mucronata* Willd. Sp. Pl. 3:1166. 1802; FBI 2:148; Gamble 326 (230).

Undershrubs, about 75 cm tall; branchlets appressed-tomentose. Leaves 3-foliolate; leaflets oblong-lanceolate, 1-3 × 0.3 - 0.5 cm, glabrous, base subacute, apex stiff-mucronate, margin ciliate, entire; Flowers axillary and terminal. in heads, yellow. Pods oblong, 0.4 cm, concealed by bracts; seeds 2-3.

Fls. & *Frs.* : November - February.

Common weed of waste lands and forests. Sangareddy tank, **BR & CP** 11624; Burugupally, **CP** 11765.

TEPHROSIA Persoon *nom. cons.*

1. Pods pubescent; standard petal < 1 cm in length :

2. Undershrubus; flowers in racemes **T. purpurea**
2. Procumbent herbs; flowers in clusters **T. pumila**
1. Pods woolly-tomentose; standard petal > 1 cm in length **T. villosa**

Tephrosia pumila (Lam.) Pers. Syn. Pl. 2:330. 1807. *Galega pumila* Lam., Encycl. 2:599. 1786. *Tephrosia purpurea* (L.) Pers. var. *pumila* (Lam.) Baker in FBI 2:113. 1876. *T.procumbens* (Buch. - Ham.) Benth. Gen. Index to Trans. Linn. Soc. London 101. 1866; Gamble 320 (226). *Galega procumbens* Buch. - Ham. in Trans. Linn. Soc. London 13:547. 1822.

Procumbent herbs; branchlets pubescent. Leaves odd-pinnate; leaflets 5-9. obovate-oblanceolate, 0.7-1.2 × 0.3-0.5 cm, base cuneate, apex truncate-mucronate, margin entire. Flowers axillary, solitary or 2-3 in clusters, yellowish-white. Pods oblong, 3.5 × 0.4 cm, slightly curved, pubescent; seeds 5-7.

Fls. & Frs. : June - November.

Occasional in waste lands and fields. Choutkoor, **TP & CP** 11935.

Tephrosia purpurea (L.) Pers. Syn. Pl. 2:329. 1807; FBI 2:112; Gamble 320 (226). *Cracca purpurea* L., Sp. Pl. 752. 1753. *Tephrosia hamiltonii* Drumm. in Gamble Fl. Pres. Madras 1:320 (223). 1918.

Undershrubs; about 1 m tall; branchlets glabrous. Leaves odd-pinnate; leaflets 1-17, obovate-oblanceolate, 1-2 × 0.5-1 cm, base cuneate, apex rounded-mucronate, margin entire. Flowers axillary and terminal, in racemes, purple. Pods oblong, 3 × 0.3 cm, slightly curved, pubescent; seeds 5-7.

Fls. & Frs. : Round the year. *Vern.* :Vempali.

Common in waste lands and fields. Nathyanipally, **BR & CP** 11253.

Tephrosia villosa (L.) Pers., Syn. Pl. 2:329. 1807; FBI 2:113. *Cracca villosa* L., Sp. Pl. 752. 1753. *Tephrosia hirta* (Buch. - Ham.) Benth., Gen. Index to Trans. Linn. Soc. London 101. 1866; Gamble 318 (225). *Galega hirta* Buch. - Ham. in Trans. Linn. Soc. London 13:546. 1822.

Undershrubs, about 1 m tall; branchlets silky-villous. Leaves odd-pinnate; leaflets obovate, 0.7-1.8 × 0.3-0.5 cm, silky, base cuneate, apex truncate-mucronate, margin entire. Flowers axillary and terminal, in racemes, white-lilac. Pods oblong, 3 × 0.3 cm, curved, woolly-tomentose; seeds 5-8.

Fls. & Frs. : July - December. *Vern.* : Nugu vempali.

Common in waste lands and in cultivated fields. Choutkoor, **TP & CP** 11936; Poosanpalle, **TP & CP** 11961.

Note : Most of the Indian floristic workers cited Gamble (1918) as authentic author for *Tephrosia procumbens* and *T. hirta*, synonyms of *T. pumila* and *T. villosa* respectively. But Bentham in Gen. Index to Trans. Linn. Soc. London (1855) already erected these species.

TERAMNUS P. Brown

Teramnus labialis (L.f.) Spreng. in L. Syst. Veg. (ed. 16) 3:235. 1826; FBI 2:184; Gamble 352 (249). *Glycine labialis* L.f., Suppl. Pl.325. 1781.

Twining herbs; branchlets pubescent. Leaves 3-foliolate; leaflets ovate-obovate 0.9-3 × 0.5-1.3 cm, glabrous above, pubescent below, base cuneate, apex obtuse-acute, margin entire. Flowers axillary, in racemes, pink. Pods linear, 3 × 0.3 cm, compressed, septate; seeds 4-8.

Fls. & Frs. : February - October.

Common in forests and field hedges. NSF, **BR & CP** 11665; Pathur, **TP & CP** 12112.

URARIA Desvaux

Uraria picta (Jacq.) Desv. ex DC., Prodr. 2:324. 1825; FBI 2:155; Gamble 336 (237). *Hedysarum pictum* Jacq., Collectania 2:262. 1789 & Ic. Pl. Rar. t. 567. 1793.

Undershrubs, about 1 m tall; branchlets pubescent. Leaves 5-7-foliolate; leaflets linear-oblong, 3-14 × 0.5-1.5 cm, glabrous, base cordate, apex acute, margin entire. Flowers terminal, in compact cylindric racemes, purple. Pods lomentum of 3-6 joints; seeds 3-6.

Fls & Frs. : July - December.

Rare in dry deciduous forests. Sathaihpalle, **GNR** 4545.

VIGNA Savi *nom. cons.*

Vigna trilobata (L.) Verdc. in Taxon 17: 172. 1968. *Dolichos trilobatus* L., Mant. Pl. 101. 1767. *Phaseolus trilobus* auct. non L.: FBI 2: 201; Gamble 362 (256).

Prostrate herbs; branchlets glabrous. Leaves 3-foliolate; leaflets palmately 3-lobed, obovate, 1.7-4 × 0.5-3.5 cm, glabrescent, base obtuse, apex acute, margin entire. Flowers axillary, in slender racemes, yellow. Pods cylindric, 3.5 × 0.2 cm, sparsely pubescent; seeds 7-10.

Fls. & Frs. : March - November.

Common weed of cultivated fields and waste lands. Pathur, **TP & CP** 12166; Sivampet, **TP & CP** 14012.

ZORNIA J.F. Gmelin

Zornia gibbosa Spanog. in Linnaea 15: 192. 1841. *Zornia diphylla* auct. non L. : FBI 2: 147; Gamble 325 (229).

Diffuse herbs; branchlets pubescent. Leaves 2-foliolate; leaflets lanceolate, 1.5-2 × 0.3-0.5 cm, base obtuse, apex acute, margin entire. Flowers axillary or terminal, in interrupted racemes, yellow. Pods moniliform, 1 × 0.2 cm, with glochidiate prickles; seeds 3-5.

Fls. & Frs. : August - November.

Occasional in forests and waste lands. Sangareddy tank, **BR & CP** 11298.

Arachis hypogaea L. is cultivated for its oil seeds. *Cajanus cajan* Millsp, *Cicer arietinum* L. *Vigna mungo* (L.) Hepp. and *V. radiata* (L.) Hepp. are common pulse crops. *Cyamopsis tetragonoloba* (L.) Taub., *Lablab purpureus* (L.) Sweet are cultivated as fruit vegetables.

Gliricidia sepium (Jacq.) Kunth. is commonly planted along road sides as hedge.

CAESALPINIACEAE

1. Armed stragglers **Caesalpinia**
1. Herbs, shrubs or trees:
   2. Leaves all bipinnate:
      3. Trees; branchlets reddish-brown **Peltophorum**
      3. Shrubs; branchlets not as above **Parkinsonia**
   2. Leaves simple or 1-pinnate :
      4. Leaves simple, dividied to the middle **Bauhinia**
      4. Leaves 1-pinnate :
         5. Petals 3; stamens 3, connate **Tamarindus**
         5. Petals 5; stamens 5 or 10, free **Cassia**

BAUHINIA Linnaeus

Bauhinia racemosa Lam., Encycl. 1:390. 1795; FBI 2:276; Gamble 406 (288); Khan in FFH 147.

Trees, about 10 m tall; branchlets densely tomentose. Leaves orbicular, 2.5-5.5 × 2.6-5.3 cm, glabrous, base cordate rounded, apex mucronate, margin entire. Flowers axillary or terminal, in panicles, white. Pods oblong, 17 × 2.5 cm, turgid; seeds 8-10.

Fls. & Frs. : March - November. *Vern.* : Are.

Common in dry deciduous forests. Narsapur RF, **CP** 11585 and **TP & CP** 14066; Pocharam, **BR & CP** 11603.

CAESALPINIA Linnaeus

1. Unarmed shrubs; pods smooth **C. pulcherrima**
1. Armed stragglers; pods prickled **C. bonduc**

Caesalpinia bonduc (L.) Roxb., Fl. Ind. 2:362. 1832. *emend.* Dandy & Exell in Journ. Bot. 76; 175. 1938. *Guilandinia bonduc* L., Sp. Pl. 381. 1753. *Caesalpinia crista* L., Sp. Pl. 380. 1753; Gamble 393 (278) *C. bonducella* Flem. in Asiat. Res. 11:159. 1810 *nom. illegit.* : FBI 2:254; Khan in FFH 139.

Armed stragglers, about 4 m tall; branchlets densely pubescent; prickles recurved. Leaves bipinnate; pinnae 6-7 pairs; leaflets ovate-elliptic, 1.5-2.5 × 0.9-1.3 cm, glabrous above, pubescent below, base truncate, apex obtuse mucronate, margin entire. Flowers axillary, in racemes, yellow. Pods oblong, 6 × 5 cm, prickled; seeds 1-2.

Fls. & Frs. : August - February. *Vern.* : Gachakaya.

Occasional in field hedges. Choutkoor, **TP & CP** 11926.

Caesalpinia pulcherrima (L.) Sw., Obs. Bot. 166. 1791; FBI 2:255; Gamble 394 (279); Khan in FFH 140. *Poinciana pulcherrima* L., Sp. Pl. 380. 1753.

Shrubs, about 1.5 m tall; branchlets glabrous. Leaves bipinnate; pinnae 8 pairs; leaflets oblong - elliptic, 1-2 × 0.5-0.8 cm, base obliquely rounded, apex emarginate, margin entire; small prickles in the axils of pinnae and on the peduncle. Flowers axillary, in corymbose cymes, scarlet-orange. Pods oblong, 8 × 1.5 cm, compressed, glabrous; seeds 5-8.

Fls. & Frs. : Round the year. *Vern.* : Pedda thangedu.

Common in field hedges. Medak, **CP** 11720.

CASSIA Linnaeus

1. Herbs :
   2. Prostrate herbs :
      3. Stamens 5 — **C. pumila**
      3. Stamens 10 — **C. mimosoides**
   2. Erect herbs :
      4. Pinnae with 2 pairs of leaflets — **C. absus**
      4. Pinnae with 3 pairs of leaflets — **C. tora**
1. Shrubs or trees :
   5. Shrubs :
      6. Stipules foliaceous — **C. auriculata**
      6. Stipules not foliaceous :
         7. Pinnae with 5-6 pairs of leaflets — **C. occidentalis**
         7. Pinnae with 8-10 pairs of leaflets — **C. sophera**
   5. Trees :
      8. Leaflets < 2 cm across; pods compressed — **C. siamea**
      8. Leaflets > 3 cm across; pods cylindric — **C. fistula**

Cassia absus L., Sp. Pl. 376. 1753; FBI 2:265; Gamble 403 (285); Khan in FFH 145.

Erect herbs, about 50 cm tall; branchlets pubescent. Leaves to 5 cm; leaflets elliptic-obovate, 1-2.5 × 0.5-1.5 cm, glabrous above, pubescent below, base cuneate, apex obtuse-subacute, margin entire. Flowers axillary and terminal, in racemes, reddish-yellow. Pods oblong, 4 × 1 cm, compressed, strigose, shortly, stipitate; seeds 4-6.

Fls. & *Frs.* : July - December. *Vern.* : Chanupala vithulu.

Common in open forests and along road sides. Pathur. **TP & CP** 12128; Narsapur RF, **TP & CP** 14043.

Cassia auriculata L., Sp. Pl. 379. 1753; FBI 2:263; Gamble 402 (284); Khan in FFH 143.

Shrubs, about 3 m tall; branchlets densely pubescent. Leaves to 10 cm; leaflets oblong-elliptic, 1-2 × 0.5-1 cm, glabrous, base cuneate-rounded, apex mucronate, margin entire; stipules foliaceous. Flowers axillary or terminal, in corymbose racemes, yellow. Pods oblong, 10 × 1.5 cm, mucronate; seeds 6-10.

Fls. & *Frs.* : August - December. *Vern.* : Thangedu.

Frequent in waste lands. Nathyanipalle, **BR & CP** 11248.

Cassia fistula L., Sp. Pl. 377. 1753; FBI 2:261; Gamble 400 (283); Khan in FFH 143. *C. rhombifolia* Roxb. Fl. Ind. 3:334. 1832.

Trees, about 10 m tall; branchlets glabrous. Leaves to 25 cm; leaflets broadly ovate, 5-9 × 3-4.5 cm, glabrous, base obtuse, apex slightly notched, entire. Flowers terminal, in pendulous racemes, yellow. Pods cylindric, 50 × 2 cm, glabrous; seeds many.

Fls. & *Frs.* : March - November. *Vern.* : Rela.

Common in dry deciduous forests. Akkannapet, **BR & CP** 11553; Pocharam, **BR & CP** 11611.

Cassia mimosoides L., Sp. Pl. 379. 1753; FBI 2:266; Gamble 403 (285).

Prostrate herbs; branchlets densely pubescent. Leaves to 4 cm; leaflets oblong, 0.5-0.8 × 0.1-0.2 cm, pubescent, base obtuse, apex obtuse-acute, margin entire. Flowers axillary, solitary, yellow. Pods oblong, 3.5 × 0.2 cm, compressed; seeds 8-10.

Fls. & *Frs.* : Round the year.

Common in wet places. Narsapur forest nursery, **BR & CP** 11225; Narsapur RF, **TP & CP** 14050.

Cassia occidentalis L., Sp. Pl. 377. 1753; FBI 2:262; Gamble 401 (284); Khan in FFH 145.

Undershrubs, about 2 m tall; branchlets glabrescent. Leaves to 15 cm; leaflets ovate, 1.5-3.5 × 0.5-2 cm, glabrous, base obtuse, apex acute-mucronate, margin entire. Flowers axillary and terminal, in racemes, yellow. Pods oblong, 10 × 0.5 cm, compressed, glabrous, horned; seeds 15-20.

Fls. & *Frs.* : Round the year. *Vern.* : Thangedu.

Abundant in all places. Narsapur RF, **CP** 11577; Pathur, **TP & CP** 12115.

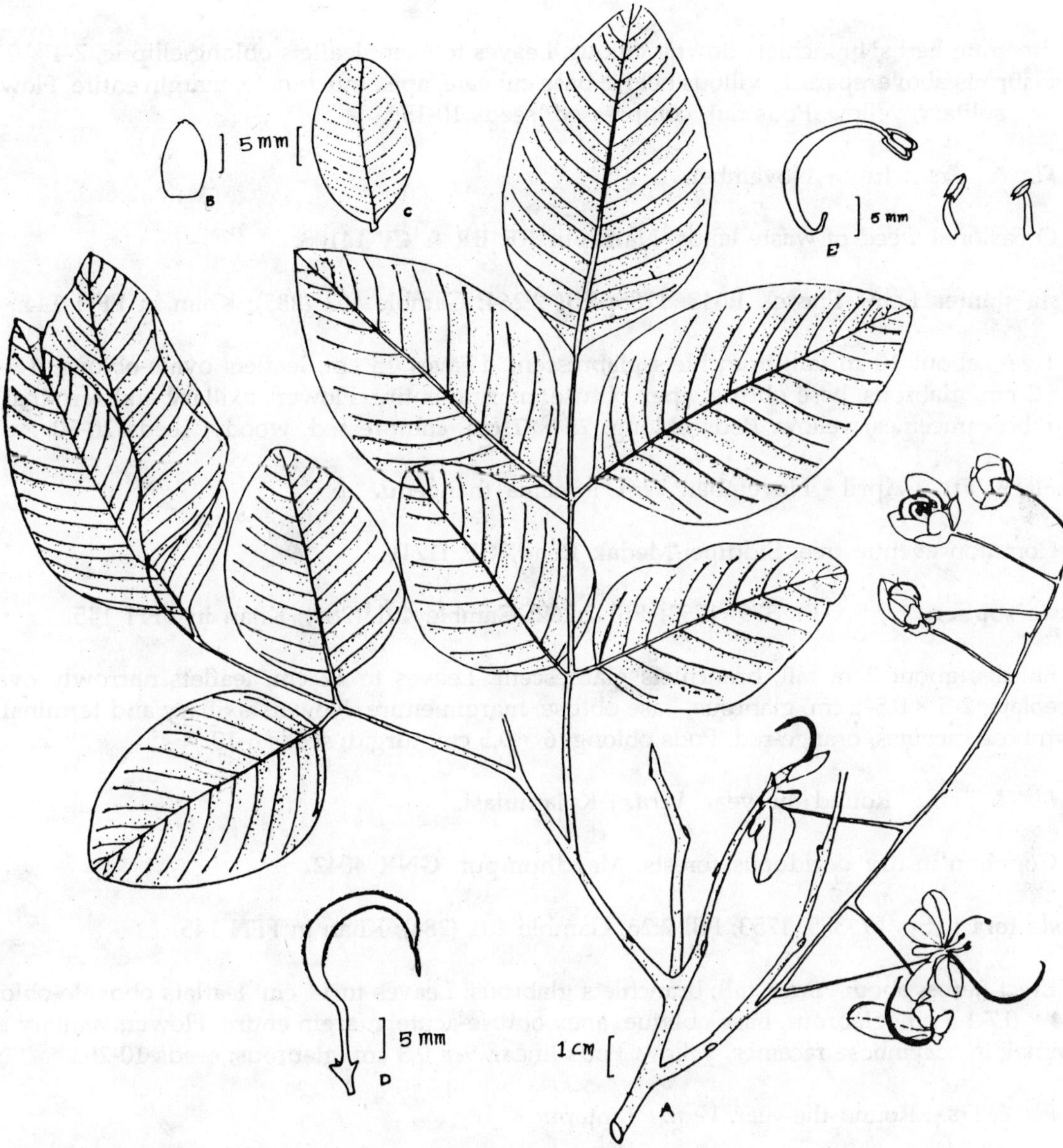

Fig. 6.12 : *Cassia fistula* L. A. Twig, B. Sepal, C. Petal, D. Pistil, E. Stamens

Cassia pumila Lam., Encycl. 1:651. 1785; FBI 2:266; Gamble 403 (285). *Senna prostrata* Roxb. Fl. Ind. 2:352. 1832.

Prostrate herbs; branchlets downy villous. Leaves to 6 cm; leaflets oblong-elliptic, 2-4 × 0.2-1 cm, glabrous above, sparsely villous below, base cuneate, apex mucronate, margin entire. Flowers axillary, solitary, yellow. Pods flat, compressed; seeds 10-15. .

Fls. & *Frs.* : July - November.

Occasional weed of waste lands. Narsapur RF, **BR & CP** 14108.

Cassia siamea Lam., Encycl. 1:648. 1785; FBI 2:264; Gamble 402 (285); Khan in FFH 144.

Trees, about 10 m tall; branchlets glabrescent. Leaves 15 cm, leaflets ovate-oblong, 1.5-3.5 × 0.8-2 cm, glabrous, base obtuse, apex obtuse, margin entire. Flowers axillary and terminal, in corymbose racemes, yellow. Pods oblong, 25 × 2 cm, compressed, woody; seeds 20-30.

Fls. & *Frs.* : April - November. *Vern.* : Seema thangedu.

Common avenue tree. Siddipet-Medak Road, **CP** 11711.

Cassia sophera L., Sp. Pl. 379. 1753; FBI 2:262; Gamble 401 (284); Khan in FFH 145.

Shrubs, about 2 m tall; branchlets glabrescent. Leaves to 12 cm; leaflets narrowly ovate-lanceolate, 2-5 × 0.5-1 cm, glabrous, base obtuse, margin entire. Flowers axillary and terminal, in corymbose racemes, orange-red. Pods oblong, 6 × 0.5 cm, turgid; seeds 6-10.

Fls. & *Frs.* : Round the year. *Vern.* : Kolathulasi.

Common in dry deciduous forests. Magdhumpur, **GNR** 4042.

Cassia tora L.,Sp. Pl. 375. 1753; FBI 2:26; Gamble 401 (284); Khan in FFH 145.

Erect herbs, about 75 cm tall; branchlets glabrous. Leaves to 14 cm; leaflets obovate-oblong, 1.5-4 × 0.7-1.5 cm, glabrous, base oblique, apex obtuse-acute, margin entire. Flowers axillary and terminal, in corymbose racemes, yellow. Pods linear, 9 × 0.3 cm, glabrous; seeds 10-20.

Fls. & *Frs.* : Round the year. *Vern.* : Tantepi.

Common in waste lands and along road sides. Ramayampet, **BR & CP** 11507.

PARKINSONIA Linnaeus

Parkinsonia aculeata L., Sp. Pl. 375. 1753; FBI 2:260; Gamble 397 (281); Khan in FFH 142.

Armed shrubs, about 3.5 m tall; branchlets glabrous.Leaves bipinnate; pinnae 2-3 pairs, flattened; leaflets oblanceolate, 0.2-0.4 × 0.3 cm, base cuneate, apex obtuse, margin entire. Flowers axillary and terminal, in racemes, yellow. Pods torulose, 10 × 0.6 cm, glabrous; seeds 4-6.

Fls. & Frs. : July - February. *Vern.* : Seemathumma.

Occasional in hedges. Hugelli, **CP** 11680.

PELTOPHORUM (Vogel) Bentham *nom. cons.*

Peltophorum pterocarpum (DC.) Baker ex K. Heyne, Nutt. Pl. Ned.-Ind. (ed.2) 2:755. 1927. *Inga pterocarpa* DC., Prodr. 2:441. 1825.

Trees, about 20 m tall; branchlets reddish-brown, tomentose. Leaves bipinnate; pinnae 8-10 pairs; leaflets oblong, 0.6-1.5 × 0.3-0.6 cm, subcoriaceous, base unequal, apex obtuse, margin entire. Flowers terminal, in panicled racemes, yellow. Pods oblong, 10 × 2 cm, narrowed at both ends; seeds 1-4.

Fls. & Frs. : May - October.

Common along road sides. Siddipet - Hyderabad road, **CP** 11380.

TAMARINDUS Linnaeus

Tamarindus indica L., Sp. Pl. 34. 1753; FBI 2:273; Gamble 409 (290); Khan in FFH 146.

Trees, about 15 m tall; branchlets warty, tomentose. Leaves bipinnate; leaflets oblong, 1.5-1.7 × 0.4-0.6 cm, base oblique, apex obtuse, margin entire. Flowers terminal, in racemes, pinkish-yellow. Pods oblong, 10 × 1 cm, sub-compressed, turgid; seeds 3-8.

Fls. & Frs. : April - December. *Vern.* : Chinta.

Common along the roads and in waste lands. Ramayampet. **BR & CP** 11511.

Cassia roxburghii DC., *Delonix elata* (L.) Gamble and *D. regia* (Hook.) Raf. are planted as avenue trees.

MIMOSACEAE

1. Plants unarmed :
    2. Stamens indefinite — **Albizzia**
    2. Stamens definite — **Leucaena**
1. Plants armed :
    3. Inflorescence bicoloured — **Dichrostachys**
    3. Inflorescence concolorous :
        4. Flowers tetramerous; Pods with bristles — **Mimosa**
        4. Flowers pentamerous; pods not with bristles :
            5. Pinnae pairs 3-many — **Acacia**

5. Pinnae pairs 2 :
   6. Leaflets 2; pods twisted **Pithecellobium**
   6. Leaflets many; pods straight **Prosopis**

ACACIA P. Miller

1. Straggling shrubs; stipular spines absent :
   2. Leaflets overlapping; flowers pedicellate **A. pennata**
   2. Leaflets not overlapping; flowers sessile **A. torta**
1. Erect trees or shrubs; stipular spines present :
   3. Flowers in globose heads :
      4. Pods moniliform **A. nilotica**
      4. Pods not moniliform :
         5. Flower heads on axillary peduncles **A. farnesiana**
         5. Flower heads in panicles **A. leucophloea**
   3. Flowers in elongate spikes **A. chundra**

Acacia chundra (Roxb. ex Rottler) Willd., Sp. Pl. 4:1078. 1806. *Mimosa chundra* Roxb. ex Rottler in Ges. Naturf. Freunde Berlin Neue Schriften 4:207. 1803. *Acacia sundra* DC., Prodr. 2:458. 1825; FBI 2:295; Gamble 428(303). *A. catechu* Willd. var. *sundra* Prain in Journ. Asiat. Soc. Beng. Pt. 2.66:508-510. 1898; Khan in FFH 159 (*excl.* var.).

Trees, about 8 m tall; branchlets glabrous; Leaves bipinnate; pinnae 14-20 pairs; leaflets oblong, 0.2-0.6 × 0.1-0.2 cm, glaucous, base oblique, apex acute, margin entire. Flowers axillary, in spikes, yellowish-white. Pods oblong, 5.5 × 1.5 cm, compressed, glabrous, shortly stipitate; seeds 6-8.

Fls. & Frs. : March - October. *Vern.* : Sundra.

Common in dry deciduous forests. Burugapally, **BR & CP** 11616; Pochammaralu, **CP** 11757.

Acacia farnesiana (L.) Willd., Sp. Pl. 4:1083. 1806; FBI 2:292; Gamble 425 (301); Khan in FFH 156. *Mimosa farnesiana* L., Sp. Pl. 521. 1753. *Vachellia farnesiana* (L.) Wight & Arn., Prodr. 272. 1834.

Armed shrubs, about 4 m tall; branchlets warty. Leaves bipinnate; pinnae 3-5 pairs; leaflets oblong, 0.2-0.4 × 0.1 cm. glabrous, base oblique, apex acute, margin entire. Flowers axillary, in globose heads, yellow. Pods oblong, 5 × 0.5 cm, turgid, apically horned; seeds 4-6.

Fls. & Frs. : December - April. *Vern.* : Murki-tumma.

Common in waste lands. Pochammaralu, **BR & CP** 11500.

Acacia leucophloea (Roxb.) Willd. Sp. Pl. 4:1083, 1806; FBI2:294. Gamble 427 (302); Khan in FFH 158. *Mimosa leucophloea Roxb.*, Pl. Corom. t. 150. 1800.

Trees, aboout 15 m tall; branchlets striate, puberulous. Leaves bipinnate; pinnae 4-6 pairs;

leaflets elliptic, 0.2-0.3 × 0.1 cm, glabrous above, pubescent below, base oblique, apex acute, margin entire. Flowers axillary and terminal, in globose heads, pale yellow. Pods linear, rusty tomentose, 10 × 1 cm; seeds many.

Fls. & Frs. : July - December. *Vern.* : Tela tumma.

Occasional in waste lands and dry deciduous forests. Ponnala, **CP** 11393.

Acacia nilotica (L) Willd. ex Del., Descr. Egypt. Hist. Nat. 3. 79. 1813. ssp.*indica* (Benth.) Brenan in Kew Bull 12:84. 1957. *Mimosa nilotica* L., Sp. Pl. 521. 1753. *Acacia arabica* (Lam.) Willd. var. *indica* Benth. in Hook., London Journ. Bot. 1:500. 1842. *A. arabica* auct.non Willd. 1806; FBI 2:293; Gamble 425 (301); Khan in FFH 156. *Mimosa arabica* Lam., Encycl. 1: 19. 1793.

Trees, about 8 m tall; branchlets pubescent. Leaves bipinnate; pinnae 4-5 pairs; leaflets elliptic, 0.3-0.4 × 0.1 cm, base oblique, apex rounded, margin entire. Flowers axillary, in globose heads,yellow. Pods moniliform, 16 × 1.5 cm, grenish- tomentose, apex horned; seeds 10-12.

Fls. & Frs. : December - June. *Vern.* : Nallatumma.

Common in Wastelands and field hedges.Mombojipally, **BR & CP** 11633.

Acacia pennata (L.) Willd. Sp. Pl. 4:1090. 1806; FBI 2:297; Gamble 429 (309). *Mimosa pennata* L. Sp.Pl. 522. 1753.

Straggling shrubs; branchlets yellowish-tomentose; internodes with scattered prickles. Leaves bipinnate; pinnae 8-12 pairs; leaflets elliptic, 0.3-0.5 × 0.1 cm, glabrous, base truncate, apex acute, margin entire. Flowers axillary or terminal, in globose heads, cream. Pods strap-shaped, flat, 14 × 3 cm, glabrous; seeds 8-10.

Fls. & Frs. :August- December.

Occasional in forests. Gangapur RF, **BR & CP** 11444.

Acacia torta (Roxb.) Craib in Bull. Misc. Inform. 410. 1915; Gamble 428 (304). *Mimosa torta* Roxb. Fl. Ind. 2:566. 1832.

Armed stragglers; branchlets rusty-pubescent, Leaves bipinnate; pinnae 10-12 pairs; leaflets oblong, 0.4 - 05 × 0.1-0.2 cm, glabrous, base cordate-truncate, apex subacute, margin entire. Flowers terminal in globose heads or in panicles, yellow. Pods oblong, 14 × 1.5 cm, woody, compressed; seeds 6-7.

Fls. & Frs. : March- January. *Vern.* : Konda-korinta.

Common in dry deciduous forests. Burugupally, **BR & CP** 11615; Pochammaralu, **CP** 11758.

ALBIZIA Durazzini

1. Leaflets large, usually < 1 cm in length **A. odoratissima**
1. Leaflets small, usually > 1 cm in length:
    2. Stipules conspicuous; leaflets acute at apex **A. chinensis**
    2. Stipules minute; leaflets rounded at apex **A. amara**

Albizia amara (Roxb.) Boiv., Encycl. 19. Seicle 2:34. 1838; FBI 2:301; Gamble 432 (306); Khan in FFH 163. *Mimosa amara* Roxb., PI. Corom. t. 122. 1799.

Trees, about 10 m tall; branchlets pubescent. Leaves bipinnate; pinnae 4-10 pairs; leaflets oblong, 0.5- 1 × 0.1 cm, base oblique, apex acute, margin entire. Flowers axillary, in heads, yellowish-white. Pods oblong, 20 × 3 cm, glabrous; seeds 8-10.

Fls. & *Frs.* : March- November. *Vern.* : Nalla renga.

Common in forests and villages along road sides. Narsingi, **BR & CP** 11533; Akkannapet, **BR & CP** 11583;

Albizia chinensis (Osbeck) Mer. in Amer.J. Bot. 3:575. 1916. *Mimosa chinensis* Osbeck, Dagb Ostind. Resa 2 : 33. 1757. *M. marginata* Lam., Encycl. 1: 12 1783. *Albizia stipulata* (Roxb.) Boiv., Encycl. 19. Seicle 2: 33. 1838; FBI 2: 300. *A. marginata* (Lam.) Merr. in Philipp. J. Sci. 5:23. 1910; Gamble 433 (307).

Trees, about 20 m tall; branchlets densely pubescent. Leaves bipinnate; pinnaae 8-13 pairs; leaflets narrow; elliptic, 0.5-0.6 × 0.1-0.6 cm, pubescent, base oblique, apex acute, margin ciliate. Flowers axillary and terminal, in globose heads, white. Pods flat, compressed, 25 × 4 cm; seeds 5.

Fls. & *Frs.* : April- December. *Vern.* : Konda chindugu, Nalla Regi.

Occasional in dry deciduous forests. Narsapur, **GNR** 4070.

Albizia odoratissima (L.f.). Benth. in Hook., London Journ. Bot. 3:88. 1844; FBI 2:299; Gamble 431 (306); Khan in FFH 162. *Mimosa odoratisima* L.f., Suppl. Pl. 437. 1781.

Trees, about 15m tall; branchlets appressed-pubescent. Leaves bipinnate, pinnae 3-5 pairs; leaflets oblong-elliptic, 0.7-2.5 × 0.4-1cm, base oblique, apex obtuse oblique, margin entire. Flowers in globose heads, white. Pods oblong, compressed, 15 × 2 cm, glabrescent; seeds 8-10.

Fls. & *Frs.*: March-November. *Vern.*: Chindugu, Chinta-elagu.

Common in dry deciduous forests. Akkannapet, **BR & CP** 11551; Pocharam RF, **BR & CP** 11602; Narsapur, RF. **TP & CP** 14063.

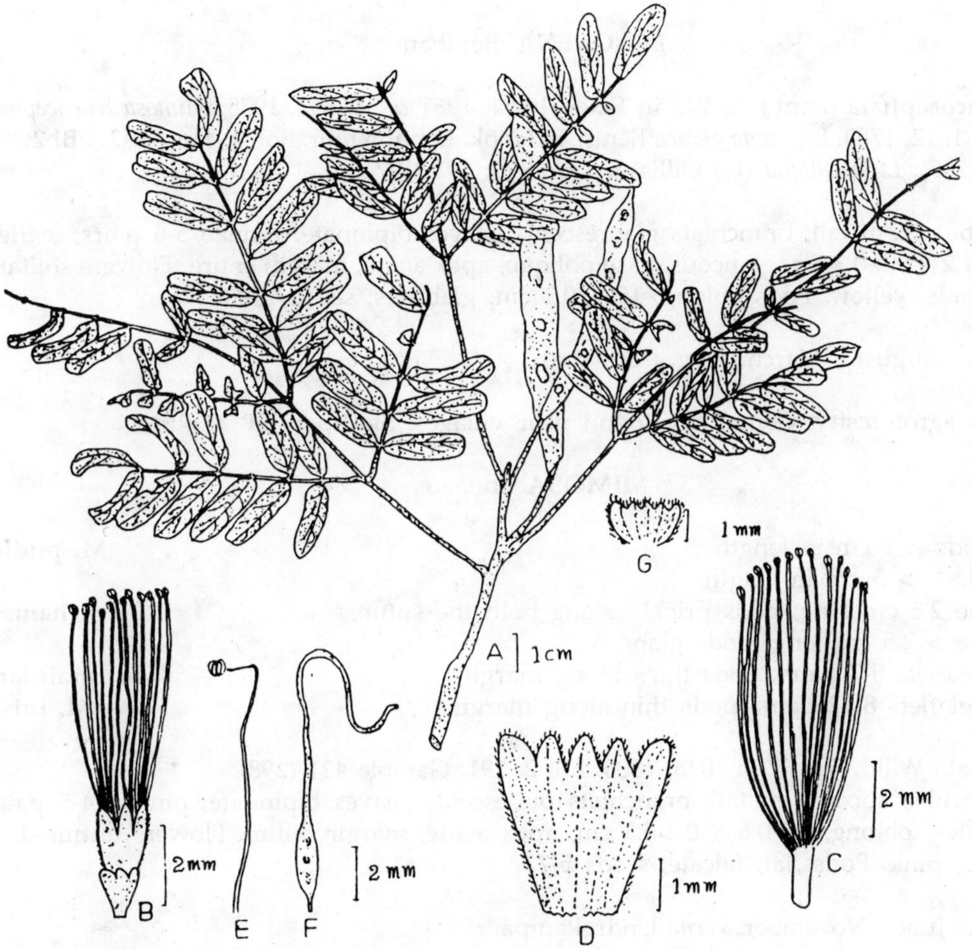

Fig. 6.13 : *Albizia odoratissima* (L.f.) Benth. A. Twig, B. Flower, C. Staminal column, D. Corolla split open, E. Stamen, F. Pistil

DICHROSTACHYS (DC.) Wight & Arntt

Dichrostachys cinerea (L.) Wight & Arn., Prodr. 271. 1834; FBI 2: 288; Gamble 419 (297); Khan in FFH 154. *Mimosa cinerea* L., Sp. Pl. 520. 1753.

Shrubs, about 2 m tall, branchlets pubescent. Leaves bipinnate; leaflets elliptic, 0.2 × 0.1 cm, base obtuse, apex acute, margin entire. Flowers axillary, in cylindrical spikes, yellow. Pods linear, 4 × 0.2 cm, twisted; seeds 6-10.

Fls. & Frs. : April - November. *Vern.* : Velturu.
Common in forests. Narsapur RF, **TP & CP** 14052.

LEUCAENA Bentham

Leucaena leucocephala (Lam.) de Wit in Taxon 10:54. 1961 & 24: 352. 1975. *Mimosa leucocephala* Lam., Encycl. 1: 12. 1783. *Leucaena glauca* Benth. in Hook. London Jour. Bot. 4: 416. 1842; FBI 2:290; Gamble 419 (297). *L. latisiliqua* (L.) Gillis in Taxon 23: 191. 1974.

Trees, about 15 m tall; branchlets glabrescent. Leaves bipinnate; pinnae 5-6 pairs; leaflets lanceolate, 1-1.2 × 0.2-0.3 cm, glaucous, base oblique, apex acute, margin entire. Flowers solitary, in globose heads, yellow. Pods oblong, 15 × 0.2 cm, glabrous; seeds 15-20.

Fls. & Fls.: August - March. *Vern.* : Subabul.

Grown in agroforestry. Commonly found near villages. Siddipet, **CP** 11375.

MIMOSA linnaeus

1. Herbs; pods < 2 cm in length **M. pudica**
1. Shrubs; pods > 5 cm in length:
   2. Pinnae 2.5 cm long; pods prickly along both the sutures **M. hamata**
   2. Pinnae > 2.5 cm long; pods glabrous:
      3. Leaflets 5-7 pairs; pods thick along margins **M. prainiana**
      3. Lelaflets 8-12 pairs; pods thin along margins **M. intsia**

Mimosa hamata Willd. Sp. Pl. 4: 1033, 1806; FBl 2: 291; Gamble 421 (298).

Prickly shrubs, about 2 m tall; branchlets pubescent. Leaves bipinnate; pinnae 4-5 pairs; leaflets obovate- oblong, 0.4-0.6 × 0.2-0.3 cm, apex acute, margin entire. Flowers terminal, in globose heads, pink. Pods flat, falcate; seeds 5-7.

Fls. & Frs. : June - November. *Vern.*: Undrakampa.

Ocasional in dry deciduous forests. Pathur RF, **TP & CP** 12125.

Mimosa intsia L., Sp. Pl. 522. 1753. *Mimosa rubicaulis* Lam., Encyl. 1: 20. 1784; FBl 2: 291; Gamble 421 (279); Khan in FFH 15.

Straggling shrubs; branchlets pubescent; prickles short, curved. Leaves bipinnate; leaflets obliquely oblobng, 0..2 - 0.5 × 0.1-0.2 cm, sparsely hairy, base truncate, apex acute, margin entire. Flowers terminal, in fasciculate heads, pinkish-yellow. Pods falcate, 6 × 1.5 cm, glabrous; seeds 6-10.

Fls. & Frs. : July - December.

Common in forests along road sides. Narsapur, **GNR** 3936.

Mimosa prainiana Gamble, Fl. Pres. Madras 421 (299). 1919.

Armed shrubs, about 3 m tall; branchlets pubescent. Leaves bipinnate; pinnae 3-5 pairs; leaflets oblong, 0.2-0.3 × 0.1-0.2 cm, pubescent, base oblique, apex obtuse-mucronate, margin entire. Flowers axillary, in globose heads, pink. Pods oblong, 6 × 1.5 cm, prickly along the sutures; seeds 6-8.

Fls. & Frs. : August - December.

Occasional in dry deciduous forests. Sardhana, **BR & CP** 11622 and **CP** 11772.

Mimosa pudica L., SP., Pl. 518. 1753; FBl 2:291; Gamble 421 (298); Khan in FFH 155.

Ascendiing prickled herbs, about 75 cm tall; branchlets glabrescent. Leaves bipinnate; pinnae 1-2 pairs; leaflets oblong, 0.2-0.4 × 0.2 cm, hispid, base obtuse, apex acute, margin entire. Flowers axillary, in globose heads, pink. Pods oblong, 1.5 × 0.5 cm, bristly along margins, horned; seeds 2-5.

Fls. & Frs. : Round the year. *Vern.* : Attipathi.

Occasional in moist places. Medak, **CP** 11727.

PITHECELLOBIUM C.F.P. Martius *nom. cons.*

Pithecellobium dulce (Roxb.) Benth. in Hook. London Journ. Bot. 3:199. 1844; FBl 2:302; Gamble 434 (308); Khan in FFH 163 (*Pithecolobium*). *Mimosa dulcis* Roxb., Pl. Corom. t. 99. 1798.

Armed trees, about 15 m tall; branchlets densely tomentose. Leaves bipinnate; leaflets obliquely oblong, 1.5-2.5 × 0.3-1 cm, glabrous, base obtuse, apex obtuse, margin entire. Flowers axillary and terminal, in panicled heads, white. Pods spirally twisted, 5 × 1 cm; seeds 2-6.

Fls. & Frs. : January - June. *Vern.* : Pul-chinta.

Common in villages and fields hedges. Zaheerabad, **CP** 11598.

PROSOPIS Linnaeus

Prosopis chilensis (Molina) Stunz in U.S.D.A. Bur. Pl. Industr. Invent. Seeds 31: 85. 1914. *Ceratonia chilensis* Molina, Sagg. Chil. 172. 1782. *Prosopis juliflora* (Sw.) DC. Prodr. 2: 447. 1825. *Mimosa juliflora* Sw., Prodr. 85. 1788; Khan in FFH 153.

Armed trees, about 5 m tall; branchlets glabrous. Leaves bipinnate; pinnae 1-2 pairs; leaflets oblong. 0.8 - 1.5 × 0.1 -0.2 cm, glabrous, base obtuse, apex obtuse, margin entire. Flowers axillary, in spikes, yellow. Pods linear-oblong, 10 × 0.7 cm, slightly curved, septate, glabrous; seeds 8-12.

Fls. & Frs. : Round the year. *Vern.* : Thumma Kampa.

Common in waste lands. Mambojipally, **BR & CP** 11451; Patancheru, **CP** 11692.

Samanea saman (Jacq.) Merr. is planted along road sides, as a shade tree.

DROSERACEAE

DROSERA Linnaeus

1. Leaves rosulate, stipulate **D. burmannii**
1. Leaves cauline, exstipulate **D. indica**

Drosera burmannii Vahl, Symb. Bot. 3:50. 1794; FBl 2: 424; Gamble 452 (320).

Acaulescent herbs, about 5 cm tall. Leaves obovate-orbicular, 0.5-0.8 × 0.3-0.5 cm, reddish-green, glandular tentacles elongate, base attenuate, apex; obtuse, margin entire. Flowers terminal, in racemes, white. Capsules 5-valved; seeds 5-6.

Fls. & Frs. : August - February.

Rare in moist places. Pocharam tank, **TP & CP** 12026.

Drosera indica L. Sp. Pl. 282. 1753; FBl 2: 424; Gamble 452 (320).

Caulescent herbs, about 8 cm tall; branchlets hairy. Leaves : lower ones deflexed, 1-2.5 cm, upper ones linear, 1-2.5 cm, glandular, tentacled above, glabrous below. Flowers leaf-opposed, in racemes, rose. Capsules 3-valved; seeds many.

Fls. & Frs. : November - March.

Rare in moist places in forests. Nagulabanda, **CP** 11355; Pocharam tank, **TP & CP** 12025 and 12028.

Rosa spp. (ROSACEAE) are grown as ornamentals.

COMBRETACEAE

1. Trees:
   2. Flowers in heads **Anogeissus**
   2. Flowers in panicles or spikes **Terminalia**
1. Climbing shrubs **Combretum**

ANOGEISSIUS Wallich ex Guillemin & Perrottet

Anogeissus latifolia (Roxb., ex DC.) Wall. ex Beddome, Fl. Sylv. S. Ind. t. 15. 1869; FBl 2: 450; Gamble 466 (330); Khan in FFH 171. *Conocarpus latifolia* Roxb. ex DC., Prodr. 3:17. 1828.

Trees, about 20 m tall; branchlets grey-pubescent. Leaves oblong- obovate, 10-12.5 × 4.5-6 cm, glabrous, base subcordate, apex obtuse-retuse, margin entire. Flowers axillary, in cymose heads, yellow. Drupes compressed, 1 × 1.2 cm, glabrous, 2-winged.

Fls. & Frs. : May - November. *Vern*.: Tiruman.

Common in dry deciduous forests. Sardhana, **BR & CP** 11621.

COMBRETUM Loefling *nom. cons.*

Combretum albidum G. Don in Trans. Linn. Soc. London 15:429. 1827. *C. ovalifolium* Roxb.,Fl. Ind. 2:226. 1832; FBl 2:458; Gamble 469 (332); Khan in FFH 173.

Climbing shrubs; branchlets glabrous. Leaves ovate-oblong, 9-11 × 5-6.5 cm, glabrous, base cuneate-obtuse, apex acuminate, margin entire. Flowers in panicled spikes, greenish-yellow. Drupes globose, 2 cm across; glabrous, 4-winged.

Fls. & Frs. : February - May.

Common in dry deciduous forests. Gangapur RF, **BR & CP** 11432.

TERMINALIA Linnaeus *nom. cons.*

Note: The flowers of *Terminalia* are remarkably uniform through-out the genus and scarcely ever provide any taxonomic useful characters and great reliance must therefore be placed on leaf, bark and fruit characters (Flora of Tropical East Africa 74. 1973).

1. Fruits winged:
    2. Fruits shortly winged or angled **T. arjuna**
    2. Fruits always winged, not-angled:
        3. Leaves tomentose; drupes glabrous **T. alata**
        3. Leaves coriaceouss; drupes velvety **T. coriacea**
1. Fruits not winged:
    4. Leaves broadly elliptic; drupes tomentose **T. bellirica**
    4. Leaves ovate-drupes glabrous **T. chebula**

Terminalia alata Heyne ex Roth, Nov. Pl. Spec. 379. 1821. T. *tomentosa* (Roxb. ex DC.) Wight & Arn. Prodr. 314. 1834; FBi 2: 447; Gamble 465 (329); Khan in FFH 169. *Pentaptera tomentosa* Roxb. ex DC., Prodr. 3; 14. 1828.

Trees, about 25 m tall; branchlets tomentose. Leaves elliptic-oblong, 2-8 × 1.5-4 cm, tomentose, base subcordate, apex obtuse, margin crenate. Flowers axillary and terminal, in panicled racemes, yellow. Drupes oblong, 4 × 3 cm, glabrous.

Fls. & Frs. : May - December. *Vern.* : Nallamaddi.

Common in dry deciduous forests. Narsapur RF, **CP** 11573 and **TP & CP** 11984.

Terminalia arjuna (Roxb. ex DC.) Wight & Arn., Prodr. 314. 1834; FBl 2: 447; Gamble 465 (329); Khan in FFH 168. *Pentaptera arjuna* Roxb. ex DC, Prodr. 3:14. 1828.

Trees, about 30 m tall; branchlets pubescent. Leaves oblong-obovate, 8-15 × 5-7 cm, coriaceous, glabrous, base subcordate-obtuse, apex obtuse, margin crenate-serrate. Flowers terminal, in panicled spikes, dull yellow. Drupes ovate, 4 × 2.5 cm, glabrous.

Frs. & Frs. : April - December. *Vern.* : Tellamaddi.

Common in dry deciduous forests on the banks of streams. Narsapur, *KMS* 7981 (CAL); Narsapur RF, **TP & CP** 14047.

Terminalia bellirica (Gaertn.) Roxb., Pl. Corom. t. 198. 1805. (*bellerica*); FBI 2: 445; Gamble 463 (328); Khan in FFH 166. *Myrobalanus bellirica* Gaertn., Fruct. 2:90 t. 97.1791. (*berllirina*).

Trees, about 30 m tall; branchlets tomentose. Leaves broadly elliptic, 8-20 × 6-14 cm, coriaceous, glabrous, base cuneate-attenuate or obtuse, apex rotund, emarginate, margin entire-subcrenulate. Flowers axillary, in spikes, pale-white. Drupes ovate, 4 × 3.5 cm, tomentose.

Fls. & Frs. : February - September. *Vern.* : Thani.

Occasional in dry deciduous forests. Narsapur RF, **CP** 11586 and **TP & CP** 14035.

Terminalia chebula Retz. Obs. Bot. 5: 31. 1788; FBI 2: 446 p.p.; Gamble 464 (328); Khan in FFH 167. *Myrobalanus chebula* (Retz.) Gaertn. Fruct. 2:91. t. 97. 1791.

Trees, about 10 m tall; branchlets glabrescent. Leaves elliptic-obovate, 7-14 × 3.5-6 cm, glabrous above, woolly below, base cordate, apex obtuse, margin entire. Flowers axillary, in spikes, cream. Drupes oblong, 3 × 2 cm, horned.

Fls. & Frs. : June - November. *Vern.* : Karaka.

Occasional in forests. Narsapur, **GNR** 4379.

Terminalia coriacea (Roxb.) Wight & Arn., Prodr. 315. 1834; Gamble 465 (329). *Pentaptera coriacea* Roxb. Fl. Ind. 2: 438. *1824. Terminalia tomentosa* (Roxb. ex DC.) Wight & Arn. var. *coriacea* (Roxb.) C.B Clarke in FBI 2: 448. 1878.

Trees, about 20 m tall; branchlets velvety-pubesscent. Leaves elliptic-oblong, 7-16 × 3-8 cm, glabrous above, glaucous below, base cordate, apex obtuse emarginate, margin entire. Flowers terminal, in panicles, pale yellow. Drupes globose, 4.5 cm across, velvety.

Fls. & Frs. : July - December.

Occasional in forests. Narsapur RF, **TP & CP** 14015.

MYRTACEAE

SYZYGIUM Gaertner *nom. cons.*

Syzygium cumini (L) Skeels in U.S.D.A. Bur. Pl. Industr. Bull. 248. 2.1912. *Myrtus cumini* L., Sp. Pl. 471. 1753. *Eugenia jambolana* Lam., Encycl. 3: 198. 1789; FBI 2:499; Khan in FFH 176. *Syzygium jambolanum* (Lam.) DC., Prodr. 3: 259. 1828; Gamble 481(340).

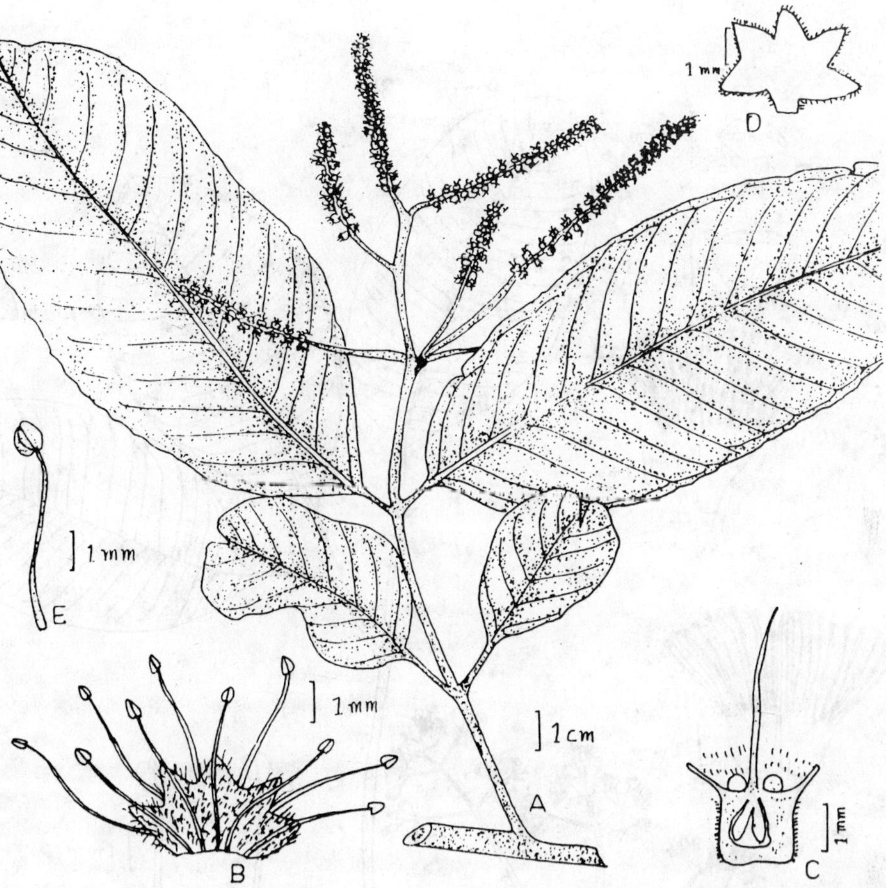

Fig. 6.14 : *Terminalia coriacea* (Roxb.) Wt. & Arn. A. Twig, B. Calyx stamens, C. Pistil, D. Calyx, E. Stamen

Trees, about 20 m tall; branchlets glabrous. Leaves elliptic-ovate, 8-10 × 4-5.5 cm, coriaceous, base cuneate, apex acuminate, margin entire. Flowers axillary or terminal, in trichotomous cymes, white. Berries oblong, 3 × 2 cm; seed 1.

Fls. & *Frs.* : March - July. *Vern.* : Neredu.

Common along road sides and in villages. Akkannapet, **BR & CP** 11563; Zaheerabad-Hyderabad Road, **CP** 11681.

Eucalyptus spp. are planted by Social forestry department. *Psidium guajava L.* is cultivated for its edible fruits.

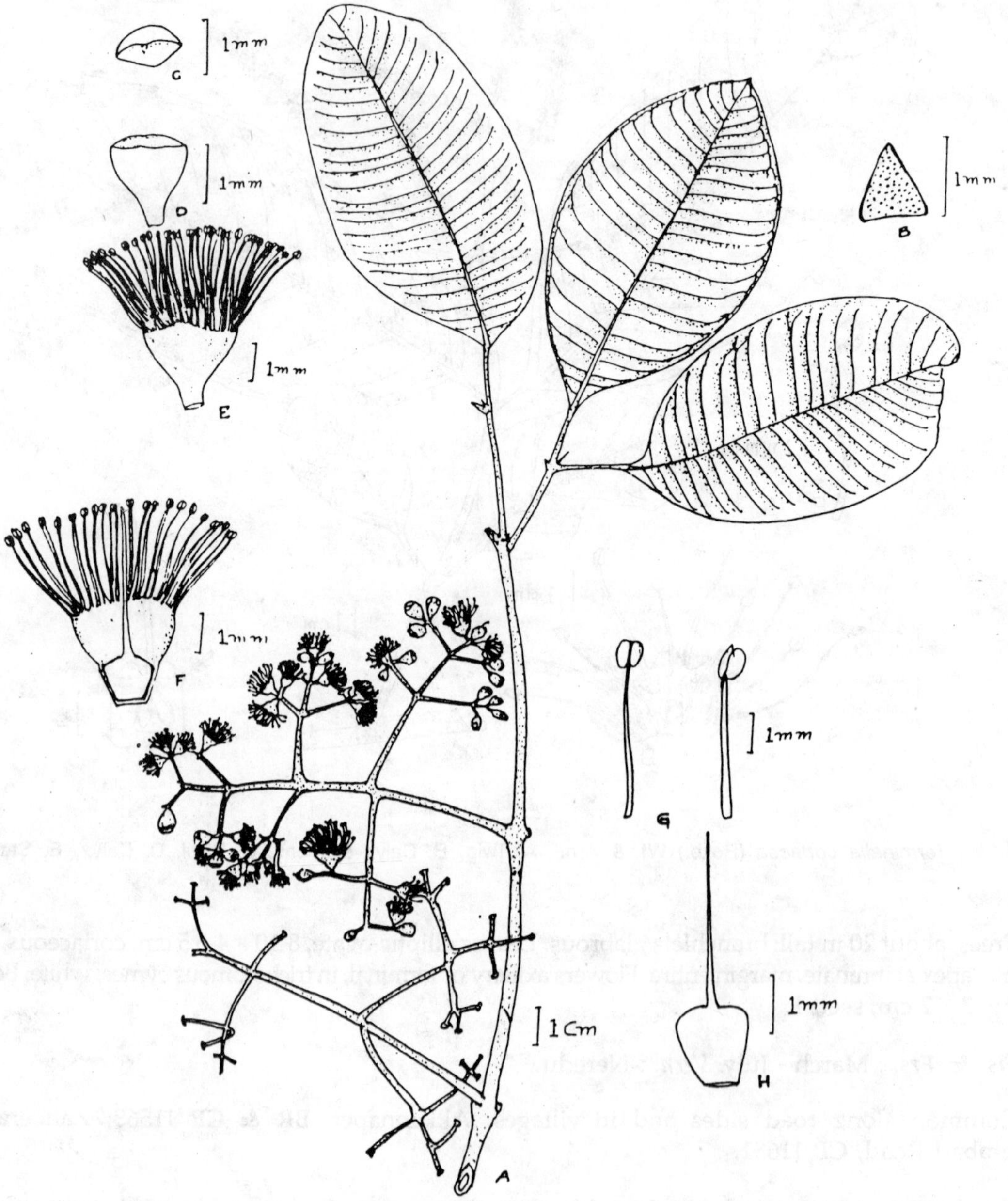

Fig. 6.15 : *Syzygium cumini* (L.) Sk. A. Twig, B. Bract, C. Calyptra, D. Calyx, E. Calyx with stamens, F. Calyx with ovary and stamens, G. Stamens, H. Pistil

BARRINGTONIACEAE

CAREYA Roxburgh *nom. cons.*

Careya arborea Roxb. Pl. Corom. t. 218. 1811; FBI 2: 511; Gamble 488 (345); Khan in FFH 178.

Trees, about 15 m tall; branchlets pubescent. Leaves obovate-oblong, 7-16 × 2.5-7 cm, thin-coriaceous, base cuneate-attenuate, apex obtuse-subacute, margin obscurely serrulate. Flowers terminal, in dense spikes, white. Berries globose, 3.5 cm across, crowned by persistent calyx; seeds many.

Fls. & Frs. : February - April.

Occasional in dry deciduous forests. Pocharam RF, **BR & CP** 11605 .

Note: *Barringtonia* J.R. & J.G.A. Forster and *Careya* Roxb. have been conserved in forming a distinct family *Barringtoniaceae* (ICBN, 1966).

MELASTOMATACEAE

OSBECKIA Linnaeus

Osbeckia zeylanica L.f. Suppl. Pl. 215. 1781; FBI 2:516; Gamble 494 (349).

Herbs, about 16 cm tall; branchlets quadrangular, striate. Leaves oblong-lanceolate, 2-2.5 × 0.5-1 cm, hirsute, base subacute-cuneate, apex acute, margin entire-serrulate. Flowers axillary, solitary or 3-5 in clusters, pinkish-purple. Capsules ovate, 0.6 × 0.4 cm; seeds many.

Fls & Frs. : November - March.

Occasional in wet grass lands, near streams. Pocharam tank, **TP & CP** 12080.

LYTHRACEAE

1. Herbs :
   2. Flowers pedicelled, in axillary cymes **Ammannia**
   2. Flowers sessile, in leaf axils or in spikes **Rotala**
1. Shrubs or trees:
   3. Shrubs :
      4. Flowers in terminal panicles, white **Lawsonia**
      4. Flowers in axillary clusters, brick-red **Woodfordia**
   3. Trees **Lagerstroemia**

AMMANNIA Linnaeus

1. Flowers in clusters, petals absent:
   2. Leaf base attenuate **A.baccifera ssp baccifera**

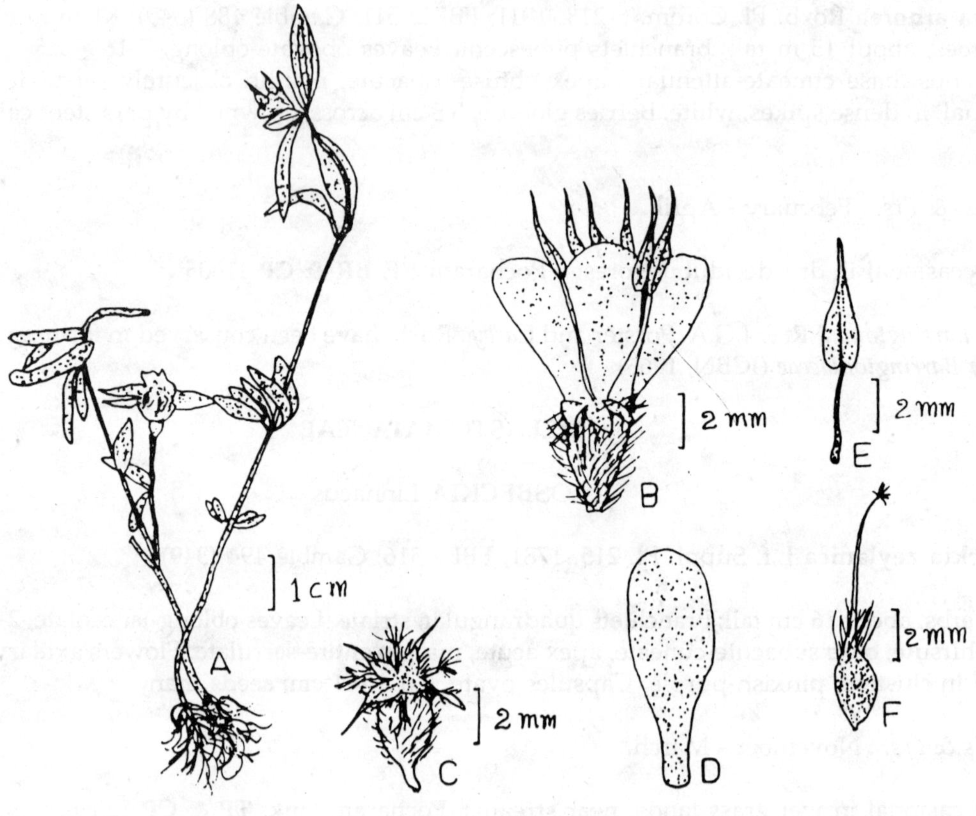

Fig. 6.16 : *Osbeckia zeylanica L.f.* A. Twig, B. Flower, C. Calyx, D. Petal, E. Stamens, F. Pistil

2. Leaf base cordate **A. baccifera ssp aegyptica**
1. Flowers in simple cymes or solitary, petals present **A. multiflora**

Ammannia baccifera L., Sp. Pl. 120. 1753; FBI 2:569; Gamble 510 (360) spp. **baccifera**.

Erect herbs, about 75 cm tall; branchlets glabrous. Leaves lanceolate-oblong, 1-3 × 0.2-0.7 cm, glabrous, base attenuate, apex acute, margin entire. Flowers at nodes, in clusters, reddish-brown. Capsules pyriform, 0.1 cm across, glabrous; seeds many.

Fls. & Frs. : Round the year.

Common in moist places. Narsapur forest nursery, **BR & CP** 11220; Pochammaralu, **BR & CP** 11492; Patancheru, **CP** 11691; Sardhana, *CP* 11776.

Ammannia baccifera L. spp. **aegyptiaca** (Willd.) Koehne in Engl. Bot. Jahrd. Syst. 1: 260. 1880; FBI 2:569; Gamble 510 (360). *A. salicifolia* sensu C.B. Clarke in FBI 2:569. 1879, non Blume 1856.

Flowers in dense clusters at nodes. Capsules 0.2 cm across.

Fls. & Frs. : Round the year.

Common in water-logged areas. Patancheru, **BR & CP** 11304 & 11305.

Ammannia multiflora Roxb., Fl. Ind. 1:447. 1820; FBI 2: 570; Gamble 509 (360).

Erect herbs, about 30 cm tall; branchlets quadrangular, glabrous. Leaves linear-oblong, 0.5-4 × 0.1-0.6 cm, glabrous, base attenuate, apex acute, margin entire. Flowers axillary, solitary or in simple cymes, red. Capsules globose, 0.1 cm across; seeds many.

Fls. & Frs. : Round the year.

Occasional in moist fields. Gummadidala, **BR & CP** 4116.

LAGERSTROEMIA Linnaeus

Lagerstroemia parviflora Roxb., Pl. Corom. t. 66. 1796; FBI 2:575; Gamble 512 (362); Khan in FFH 183.

Trees, about 12 m tall; branchlets glabrous. Leaves elliptic-obovate, 4.5-10.5 × 2.5-9.5 cm, glabrous, base cuneate, apex obtuse, margin entire. Flowers axillary and terminal, in panicles, creamish-white. Capsules ellipsold, 3 × 1.5 cm, glabrous; seeds many, winged.

Fls. & Frs. : June - December. *Vern.* : Chennangi.

Common in dry deciduous forests. Nathyanipalle, **BR & CP** 11251; Thornal, **CP** 11705.

LAWSONIA Linnaeus

Lawsonia inermis L., Sp. Pl. 349. 1753; Gamble 514 (363); Khan in FFH 182. *L. alba* Lam., Encyl. 3:106. 1789; FBI 2:573.

Shrubs, about 4 m tall; branchlets spine - tipped. Leaves ovate-obovate, 1-2.5 × 0.5-1.2 cm glabrous, base cuneate, apex acute, margin entire. Flowers terminal, in compact panicles, white. Capsules globose, 0.8 cm across; seds many.

Fls. & Frs. : February - May. *Vern.* : Gorinta, Maidaaku.

Common in field hedges and villages. Pochammaralu, **BR & CP** 11495.

ROTALA Linnaeus

1. Leaves linear-lanceolate, > 0.5 cm in length **R. densiflora**

1. Leaves ovate, < 0.5 cm in length **R. serpyllifolia**

Rotala densiflora (Roth. ex Roemer & Schultes) Koehne in Bot. Jahrb. Syst. 1:164. 1880; Gamble 508 (359). *Ammannia densiflora* Roemer & Schultes in L. Syst. Veg. 3:304. 1818. *A. pentandra* Roxb. Fl. Ind. 1:448. 1820; FBI 2:568 *p.p.*

Erect herbs, about 40 cm tall; branchlets glabrous. Leaves linear-lanceolate, 0.5-2 × 0.1-2 cm, base subcordate, apex acute, margin entire. Flowers axillary, solitary or in spikes, reddish- purple. Capsules globose, 0.1 cm across; seeds many.

Fls. & Frs. : June - November.

Common in moist localities. Usirikapalle, **GNR** 4307.

Rotala serpyllifolia (Roth) Bremek. in Acta Bot. Neerl. 3:149. 1954. *Micranthus serpyllifolia* Roth, Nov. Pl. Spec. 282. 1821. *Ammannia tenuis* (Wight) C.B Clarke in FBI 2:567. 1876. *Ameletia tenuis* Wight, lc. t. 257. 1840.

Spreading herbs, about 15 cm tall; branchlets reddish- brown, glabrous. Leaves ovate, 0.2-0.5 × 0.1-0.4 cm, glabrous, base sub-cordate, apex acute, margin entire. Flowers axillary and terminal, in spikes, pink. Capsules ellipsoid, 0.2 × 0.1 cm, glabrous; seeds many.

Fls. & Frs. : June - September.

Rare in moist places. Pochammaralu, **BR & CP** 11491.

WOODFORDIA Salisbury

Woodfordia fruticosa (L.) Kurz in Journ. Asiat. Soc. Beng. 40:56. 1871; Gamble 511 (361); Khan in FFH 182. *Lythrum fruticosum* L. Syst. Nat. ed. 10: 1045. 1759. *Woodfordia floribunda* Salisb., Parad. London t. 42. 1818; FBI 3: 653.

Shrubs, about 3 m tall; branchlets pubescent. Leaves lanceolate, 2-5 × 0.5-1.7 cm, pubescent above, softly velvety below, base cordate, apex acuminate, margin entire. Flowers axillary, in clustered cymes, brick red. Capsules ellipsoid, 1 × 0.5 cm; seeds many.

Fls. & Frs. : February - May. *Vern* : Pagadam.

Occasional along road sides and open forests. Pochammaralu, **BR & CP** 11475.
Punica granatum L. (PUNICACEAE) is grown for its edible fruits.

ONAGRACEAE

LUDWIGIA Linnaeus

1. Plants aquatic **L. adscendens**
1. Plants terrestrial :

2. Branchlets densely tomentose; stamens 8. **L. octovalvis**
2. Branchlets glabrous; stamens 4 **L. perennis**

Ludwigia adscendens (L.) H. Hara in Journ. Jap. Bot. 28:291. 1953. *Jussiaea adscendens* L., Mant. Pl. 69. 1767. *J. repens* L., Sp. Pl. 388. 1753. non Forster; FBI 2: 587; Gamble 516 (365).

Floating herbs; branchlets ribbed, clusters of spongy aerophores at nodes. Leaves obovate-oblanceolate, 1-4 × 0.6-1.5 cm, chartaceous, glabrous, base attenuate, apex obtuse, margin entire. Flowers axillary, solitary, creamish-yellow; Capsules linear, 2.5 × 0.5 cm, terete; seeds many.

Fls. & *Frs.* : Round the year.

Rare in stagnant waters. Medak, **BR & CP** 11567.

Ludwigia octovalvis (Jacq.) Raven in Kew Bull. 15: 476. 1962. *Oenothera octovalvis* Jacq., Enum. Syst. Pl. 19. 1760. *Jussiaea suffruticosa* L., Sp. Pl. 388. 1753; FBI 2:587; Gamble 516 (365).

Undershrubs, about 1.5 m tall; branchlets densely tomentose. Leaves lanceolate, 2-7 × 0.5-1.5 cm, base attenuate, apex acuminate, margin entire. Flowers axillary, solitary, yellow. Capsules oblong, 4 × 0.3 cm, terete; seeds many.

Fls. & *Frs.* : June - December.

Occasional in forests and in wet places. Gangapur RF, **TP & CP** 12092.

Ludwigia perennis L., SP. Pl. 119. 1753. *L. parviflora* Roxb., Fl. Ind. 1:419. 1820; FBI 2:588; Gamble 517 (365).

Erect herbs, about 60 cm tall; branchlets glabrous. Leaves lanceolate, 1-3.5 × 0.3-0.8 cm, glabrous, base attenuate, apex acute, margin entire. Flowers axillary, solitary, yellow. Capsules oblong, 1.5 × 0.5 cm, glabrous; seeds many.

Fls. & *Frs.* : June - November.

Common in paddy fields and moist places. Narsapur RF, **BR & CP** 11276; NSF quarters, **BR & CP** 11646; Sardhana, **CP** 11775.

TURNERACEAE

TURNERA Linnaeus

Turnera ulmifolia L., Sp. Pl. 271. 1753. var. *angustifolia* (Miller) Willd. ex Urban, Monogr. Turn. 141. 1883; Gamble 523 (369). *T. angustifolia* Miller, Gard. Dict. ed. 8. n.2. 1768.

Erect herbs, about 60 cm tall; branchlets tomentose. Leaves lanceolate. 2.5-7.5 × 1-2.5 cm, pubescent, base cuneate, apex acuminate, margin serrate. Flowers axillary, solitary, yellow. Capsules globose, 0.6 cm across, pubescent; seeds many.

globose, 0.6 cm across, pubescent; seeds many.

Fls. & Frs. : June - November.

Common in waste lands. Siddipet-Hyderabad Road, **CP** 11712; Medak, **CP** 11722.

PASSIFLORACEAE

PASSIFLORA Linnaeus

Passiflora foetida L., Sp. Pl. 959. 1753; FBI 2:599; Gamble 524 (370).

Climbing herbs; branchlets hairy. Leaves palmately 3- lobed, 3-7 × 2-3.5 cm, densely hispid, base cordate, apex acute, margin entire; tendrils simple. Flowers axillary, solitary, yellow. Berries ovate, 2 × 1.5 cm, glabrous; seeds many.

Fls. & Frs. : June - December.

Common in hedges. Medak, **CP** 11733.

Carica papaya L. (CARICACEAE) is cultivated for its fruits.

CUCURBITACEAE

Note : Generic delimitations are based on Jeffrey (1980).

1. Tendrils simple :
    2. Calyx tube closed at base by scales **Momordica**
    2. Calyx tube without scales:
        3. Male flowers in clusters:
            4. Bracts ciliate **Ctenolepis**
            4. Bracts glabrous:
                5. Anthers straight **Mukia**
                5. Anthers conduplicate **Cucumis**
        3. Male flowers solitary **Coccinia**
1. Tendrils branched:
    6. Petals fimbriate; white **Trichosanthes**
    6 Petals entire; greenish-yellow **Diplocyclos**

COCCINIA Wight & Arnott

Coccinia grandis (L.) Voigt., Hort. Subrub. Calc. 59. 1845; Chakravarthy in Fasc. Fl. Ind. 11:24. 1982. *Bryonia grandis* L., Mant. Pl. 1:126. 1767. *Coccinia indica* Wight & Arn., Prodr. 1:347. 1834. *nom. illegit:* Gamble 537 (379). *Cephalandra indica* Naud. in Ann. Sci. Nat. Bot. (Ser.5) 5:16 1859; FBI 2:621. *Coccinia cordifolia* auct. non (L.): Cogn. in A. & C. DC., Monogr. Phan 3:529. 1881.

Climbing herbs; branchlets glabrous. Leaves 5-angled or lobed, 4-5 × 4.5-7 cm, glabrous, landular, base cordate, apex obtuse, margin denticulate. Male and female flowers solitary, white. Berries

Fls. & Frs. : Round the year. *Vern.* : Kaki-donda.

Common in cultivated fields and waste lands. Nathyanipally, **BR & CP** 11249; Marrikunta, **CP** 11389.

CTENOLEPIS J. Hooker

Ctenolepis garcinii (Burm.f.) C.B. Clarke in FBI 2:629. 1879; Jeffrey in Kew Bull. 34 : 793. 1980. *Sicyos garcinii* Burm.f., Fl. Ind. 211. 1768. *Blastania garcinii* (Burm.f.) Cogn. in A. & C.DC., Monogr. Phan. 3:629. 1881; Gamble 540 (381); Chakravarthy in Fasc. Fl. Ind. 11:13. 1982.

Climbing herbs, branchlets sub-filiform, glabrous. Leaves deeply 5-lobed, 2-4.5 × 2.5-5 cm; lobes unequal, obovate, apex acute, margin denticulate. Male flowers 3-4 in clusters, yellowish-white; female flowers solitary, dullwhite. Berries globose, reddish, 2 × 1 cm; seeds 1-2.

Fls. & Frs. : August - March.

Occasional in field hedges. Choutkoor, **TP & CP** 11927.

CUCUMIS Linnaeus

Cucumis melo L., Sp. Pl. 1011. 1753; Jeffrey in Kew Bull. 34: 793. 1980; FBI 2: 620; Gamble 535 (378); Khan in FFH 188. *Bryonia callosa* Rottler in Ges. Naturf. Freunde Berlin Neue Schriften 4:210. 1803 (*collosa*). *Cucumis callosus* (Rottler) Cogn. in Engl., Pflanzeur. 88: 129. 1924; Chakravarthy in Fasc. Fl. Ind. 11:31. 1982. *Cucumis melo* L. var. *agrestis* Naud. in Ann. Sci. Nat. Bot. (Ser.4.) 11:73. 1859.

Trailing or climbing herbs; branchlets stiff-hispid. Leaves suborbicular, 5-angled or lobed, 4-8.5 x 8-9.5 cm, hispid, base cordate, apex obtuse, margin denticulate. Male flowers axillary, in clusters, female flowers solitary, yellow. Berries ellipsoid or oblong, 2.5 x 0.7 cm; seeds many.

Fls. & Frs. : July - December.

Common in waste lands. Lingareddypally. **CP** 11716.

DIPLOCYCLOS (Endlicher) Post & Kuntze

Diplocyclos palmatus (L.) Jeffrey in Kew Bull. 15:352. 1962; Chakravarthy in Fasc. Fl. Ind. 11: 48. 1982. *Bryonia palmata* L. Sp. Pl. 1012. 1753. *B. laciniosa* L. Sp. Pl. 1013. *p.p.*, excl. typus; FBI 2:622. *Bryonopsis laciniosa* sensu Naud. in Ann. Sci. Nat. Bot. Ser. 4. 12:141. 1859; Gamble 534 (377).

Climbing herbs; branchlets angular. Leaves orbicular, deeply 5-lobed, 8-12 × 4-6 cm, scabrous; lobes elliptic-oblong, base cordate, margin denticulate. Male flowers axillary, 3-6 in clusters greenish-yellow; female flowers solitary or in clusters, yellowish-green. Berries globose, 1.5 × 1 cm, green; seeds many.

Fls. & Frs. : April - December.

Common in hedges and on bushes. Pathur, **TP & CP** 12122.

MOMORDICA Linnaeus

1. Plants monoecious; berries oblong, tubercled **M. charantia**
1. Plants dioecious; berries ovate, softly spinous **M. dioica**

Momordica charantia L., Sp. Pl. 1009. 1753; FBI 2:616; Gamble 532 (380); Chakravarthy in Fasc. Fl. Ind. 11: 84. 1982.

Trailing or climbing herbs; branchlets villous. Leaves orbicular, deeply 5-7-lobed, 2-5 × 1-3 cm, sparsely villous, base cordate, apex acuminate-mucronate, margin denticulate. Male and female flowers solitary. yellow. Berries oblong, 6 × 2.5 cm, muricate; seeds many.

Fls. & Frs. : July - December. *Vern.* : Kakara.

Commonly cultivated, occasional in waste lands. Mombojipally, **BR & CP** 11642; Medak, **CP** 11723.

Momordica dioica Roxb. ex Willd., Sp. Pl. ed. 4. 665. 1805; FBI 2: 617; Gamble 532 (375); Chakravarthy in Fasc. Fl. Ind. 11:94. 1982.

Climbing herbs; branchlets glabrous. Leaves broadly ovate, 3-lobed, 3-8.5 × 3-5.5 cm, glabrous, base cordate, apex mucronate, margin denticulate. Flowers solitary, yellow. Berries ovate, 3 × 3 cm; seeds 6-7.

Fls. & Frs. : Round the year. *Vern.* : Adavi kakara.

Occasional in cultivated fields. Narsapur, **GNR** 4414.

MUKIA Arnott

1. Leaves sessile, densely villous below **M. leiosperma**
1. Leaves petiolate, shortly hispid below **M. maderaspatana**

Mukia leiosperma (Wight & Arn.) Wight in Ann. Mag. Nat. Hist. Ser. 8: 208. 1842; FBI 2: 623. *Bryonia leiosperma* Wight & Arn., Prodr. 1: 345. 1834. *Melothria leiosperma* (Wight & Arn.) Cogn. in DC. Mong. Phan. 3: 622. 1881; Gamble 539 (380); Chakravarthy in Fasc. Fl. Ind. 11:81. 1982.

Climber; branchlets apically villous, basally hispid. Leaves broadly ovate, 5-angled, 5-8.5 x 4-6 cm, scabrid above, densely villous below, base cordate, apex acute, margin denticulate. Male and female flowers in fascicles, yellow. Berries globose, 1 x 0.5 cm, seeds 5-6.

Fls. & Frs. : July - September.

Occasional on hedges and on shrubs. Choutkoor, **TP & CP** 11908.

Mukia maderaspatana (L.) M. Roemer. Fam.2. Peponif. 47. 1846. *Cucumis maderaspatanus* L., Sp. Pl. 1072. 1753. *Bryonia scabrella* L.f., Suppl. Pl. 424. 1781. *Mukia scabrella* (L.f.) Arn. in Hook. Journ. Bot. 3:276. 1841; FBI 2:623. *Melothria maderaspatana* (L.f.) Cogn. in A. & C. DC. Monogr. Phan. 3:623. 1881; Gamble 539 (381); Chakravarthy in Fasc. Fl. Ind. 11: 83. 1982.

Climbing herbs; branchlets glabrous. Leaves ovate, 3 - lobed. 4-10 × 3-6 cm, scabrid above, hispid below, base cordate, apex acuminate, margin denticulate. Male and female flowers in clusters, yellow. Berries globose, 1 cm across; seeds 6-8.

Fls. & Frs. : August - December.

Occasional in hedges. Nagulabanda, **CP** 11335.

TRICHOSANTHES Linnaeus

Trichosanthes tricuspidata Lour. Fl. Cochinch. 589. 1790; Jeffrey in Kew Bull. 34: 797. 1980. *Modecca bracteata* Lam., Encycl. 4:210. 1796. *Trichosanthes palmata* Roxb., Fl. Ind.3: 704. 1832; FBI 2: 606; Gamble 530 (374). *Trichosanthes bracteata* (Lam.) Voigt. Hort. Suburb. Calc. 58. 1845; Chakravarthy in Fasc. Fl. Ind. 11: 109. 1982.

Climbing herbs; branchlets angled, sub-glabrous. Leaves palmately 3-5-lobed, 4-6.5 × 4.5-6 cm; lobes ovate-elliptic, scabrid, base cordate, apex acuminate, margin denticulate. Male flowers in racemes, female flowers solitary, white. Berries ovate, 9 × 6 cm, red when ripe; seeds many.

Fls. & Frs. : May - September.

Rare in forests. Medak, **CP** 11736.

Benincasa hispida (Thunb.) Cogn., *Citrullus vulgaris* Schrad, *Coccinia grandis* (L.) Voigt, *Cucumis melo* L., *Cucurbita pepo* DC., *Lagenaria siceraria* (Molina) Standley, *Luffa acutangula* (L.) Roxb., are cultivated for their fruits.

CACTACEAE

OPUNTIA P. Miller

1. Spines 1-3 per areole; berries obovate **O. vulgaris**
1. Spines 3-7 per areole; berries globose **O. stricta**

Opuntia stricta (Haw.) Haw., Syn. Pl. Succ. 191. 1812. var. **dillenii** (Ker-Gawl.) L. Benson in Cact. & Succ. Journ. Amer. 41. 126. 1969. *Cactus strictus* Haw., Misc. 188. 1803. *Cactus dillenii* Ker-Gawl., Edward's Bot. Reg. 3.255. 1818. *Opuntia dillenii* (Ker-Gawl.) Haw., Suppl. Syn. Pl. Succ. 79. 1819; FBI 2: 657; Gamble 548 (387); Khan in FFH 189.

Shrubs, about 2 m tall; stems jointed, 5-12 × 2-5 cm, fleshy, glaucous-green, tubercled; tubercles spiny, hirsute. Leaves O. Flowers lateral, solitary, yellow or reddish- purple. Berries globose, 4 × 2 cm, fleshy, glochidiate.

Fls. & Frs. : August - April.

Occasional in hedges and waste lands. Nagulabanda, CP 11328; Sivanoor, **BR & CP** 11548.

Opuntia vulgaris Mill. Gard. Dict. ed. 8.n. 1.1768. *Opuntia monacantha* (Willd.) Haw. Suppl. Pl. Succ. 81. 1819; Gamble 548 (387). *Cactus monacanthos* Willd., Enum. Pl. Suppl. 33. 1814. *Opuntia dillenii* auct non. Haw. : Wight & Arn., Prodr. 363. 1834: *p.p.*; FBI 2: 657 *p.p.*

Undershrubs, about 2 m tall; stems jointed; joints fleshy, flattened. Leaves O.Flowers lateral, solitary, pinkish-yellow. Berries obovate, 6 x 3.5 cm, fleshy; seeds many.

Fls. & Frs. : Round the year.

Rare in waste lands. Domadugu, **GNR** 4617.

AIZOACEAE

Note : A. Linn Bogle (1970) split the family Aizoaceae into two separate families Aizoaceae and Molluginaceae based on free perianth segments and presence of anthocyanin pigments in the latter, which are absent in the former. Bogle concept is follwed in the present work.

TRIANTHEMA Linnaeus

Trianthema portulacastrum L., Sp. Pl. 223. 1753; Gamble 550 (389). *T. monogyna* L., Mant. Pl. 1: 69. 1767; FBI 2: 660.

Prostrate herbs; branchlets subglabrous. Leaves orbicular-obovate, 0.8-2.2 × 0.5-2 cm, glabrous, base cuneate, apex obtuse-apiculate, margin entire. Flowers solitary, white. Capsules oblong, 0.5 × 0.2 cm, glabrous; seeds many.

Fls. & Frs. : June - January.

Common weed of cultivated fields and in moist places. Toopran, **BR & CP** 11281.

MOLLUGINACEAE

1. Fruit syncarpous:
   2. Leaves all cauline; flowers in fascicles **Glinus**
   2. Leaves radical and cauline; flowers in cymes **Mollugo**
1. Fruit apocarpous **Gisekia**

GISEKIA Linnaeus

Gisekia pharnaceoides L. Mant. Pl. 562. 1771; FBI 2: 664; Gamble 553 (391).

Diffuse herbs; branchlets glabrous. Leaves oblong-obovate, 1-2 x 0.5-1 cm, pubescent, base cuneate,

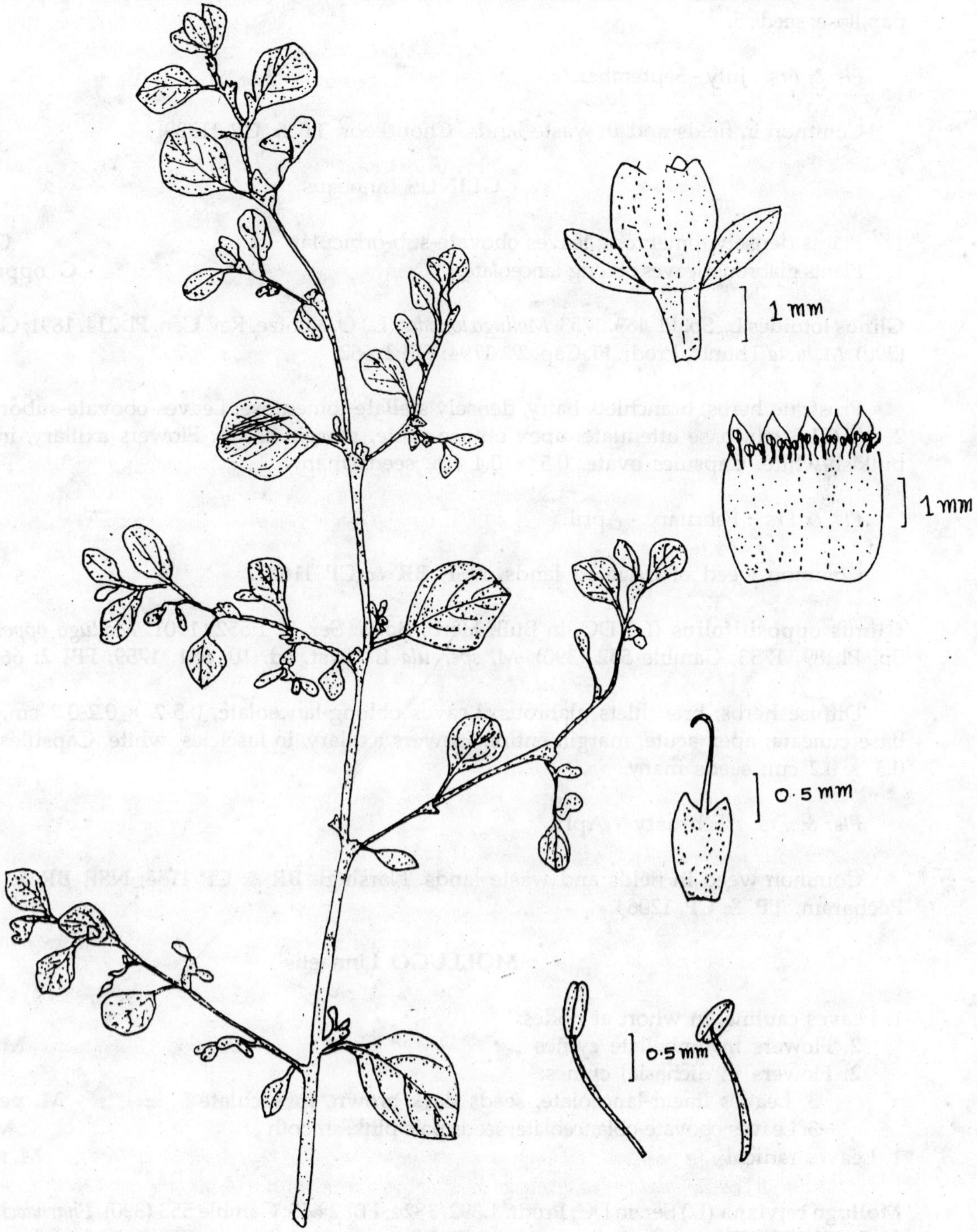

Fig. 6.17 : *Trianthema portulacastrum* L. A. Twig, B. Calyx, C. Corolla with stamens, D. Pistil, E. Stamens

apex acute-obtuse, margin entire. Flowers axillary, in umbellate cymes, whitish-pink. Capsules 5-celled, papillose; seeds 5.

Fls. & Frs. : July - September.

Common in fields and in waste lands. Choutkoor, **TP & CP** 11939.

GLINUS Linneaus

1. Plants densely tomentose; leaves obovate-sub-orbicular **G. lotoides**
1. Plants glabrous; leaves oblong-lanceolate **G. oppositifolius**

Glinus lotoides L., Sp. Pl. 463. 1753. *Mollugo lotoides* (L.) O. Kuntze, Rev. Gen. Pl. 214. 1891; Gamble; 552 (390). *M. hirta* Thunb., Prodr. Fl. Cap. 24. 1794; FBI 2: 662.

Prostrate herbs; branchlets hairy, densely stellate-tomentose. Leaves obovate-suborbicular, 1-2 × 0.6-1.2 cm, base attenuate, apex obtuse-acute, margin entire. Flowers axillary, in fascicles, pinkish-white. Capsules ovate, 0.5 × 0.4 cm; seeds many.

Fls. & Frs. : February - April.

Common weed of irrigated lands. NSF, **BR & CP** 11672.

Glinus oppositifolius (L.) DC. in Bull. Herb. Boiss. Ser. 2. 1:552. 1901. *Mollugo oppositifolia* L., Sp. Pl. 89. 1753; Gamble 552 (390). *M. spergula* L., Syst. ed. 10: 881. 1759; FBI 2: 662.

Diffuse herbs; branchlets glabrous. Leaves oblong-lanceolate, 0.5-2 × 0.2-0.3 cm, glabrous, base cuneate, apex acute, margin entire. Flowers axillary, in fascicles, white. Capsules ellipsoid, 0.3 × 0.2 cm; seeds many.

Fls. & Frs. : February - April.

Common weed in fields and waste lands. Narsingi, **BR & CP** 1154; NSF, **BR & CP** 11671; Pocharam, **TP & CP** 12064.

MOLLUGO Linnaeus

1. Leaves cauline, in whort at nodes:
    2. Flowers in umbellate cymes **M. cerviana**
    2. Flowers in dichasial cymes:
        3. Leaves linear-lanceolate; seeds dark brown, tuberculate **M. pentaphylla**
        3. Leaves obovate-oblanceolate; seeds pale pink, smooth **M. disticha**
1. Leaves radical **M. nudicaulis**

Mollugo cerviana (L.) Ser. in DC. Prodr. 1:392, 1824; FBI 2:663; Gamble 553 (390). *Pharnaceum cerviana* L., Sp. Pl. 272. 1753.

Erect herbs, about 12 cm tall; branchlets glabrous. Leaves 5-7 in whorl, linear, 0.5-1 × 0.1-0.2 cm, glabrous, base attenuate, apex obtuse, margin entire. Flowers terminal, in umbellate cymes, green. Capsules ellipsoid, 0.2 × 0.1 cm; seeds many.

Fls. & Frs. : January - April.

Occasional in cultivated fields. Domadugu, **GNR** 4697.

Mollugo disticha (L.) Ser. in DC. Prodr. 1:392. 1824; FBI 2:663; Gamble 553 (390). *Pharnaceum distichum* L., Mant. Pl. 221. 1771.

Erect herbs, about 20 cm tall; branchlets pubescent. Leaves 3-5 in whorl, oblanceolate-elliptic, 0.5-2 × 0.2-0.3 cm, pubescent, base cuneate, apex acute, margin entire. Flowers terminal, in racemes, white. Capsules ovate, 0.3 × 0.2 cm; seeds many.

Fls. & Frs. : January - April.

Occasional in moist places. Domadugu, **GNR** 4633.

Mollugo nudicaulis Lam., Encycl. 4:234. 1797; FBI 2:664; Gamble 553 (390).

Erect herbs, about 25 cm tall; branchlets glabrous. Leaves radical, rosette at base, spathulate, 2.5-5 × 0.5-1.2 cm, base attenuate, apex acute, margin entire. Flowers in branched peduncles, green. Capsules globose, 0.1 cm across; seeds many.

Fls. & Frs. : February - April.

Common in cultivated fields. Poosanpalle, **TP & CP** 11953.

Mollugo pentaphylla L. Sp. Pl. 89. 1753; Gamble 553 (390). *M. stricta* sensu C.B. Clarke in FBI 2: 663. 1879. *p.p.* non L. 1762.

Erect herbs, about 25 cm tall; branchlets glabrous. Leaves 3-5 in whorl, linear-lanceolate, 0.5-2.5 × 0.2-0.4 cm, glabrous, base attenuate, apex aucte, margin entire. Flowers in dichasial cymes, green. Capsules globose, 0.1 cm across; seeds many.

Fls. & Frs.: February - April.

Common in waste lands and cultivated fields. Sangareddy tank, **BR & CP** 11293; Nagulabanda, **CP** 11326; NSF , **BR & CP** 11655; Gangapur, CP 11787.

APIACEAE

CENTELLA Linnaeus

Centella asiatica (L.) Urban in C. Martius, Fl. Bras. 11(1):287.t.1. 1879; Gamble 556 (392). *Hydrocotyle asiatica* L., Sp. Pl. 234. 1753; FBI 2:669.

asiatica L., Sp. Pl. 234. 1753; FBI 2:669.

Prostrate herbs; stem with long stolons, rooting at nodes. Leaves reniform, 1.5-2.5 × 2.5-4.1 cm, glabrous, base cordate, apex round, margin crenate-dentate. Flowers in umbels, purple. Fruits ovate, 0.2 × 0.1 cm, mericarps laterally compressed; seeds 2.

Fls. & Frs.: June - October. *Vern.* : Saraswathiaku.

Occasional in moist places. Medak, **BR & CP** 11466.

Coriandrum stivum L. is cultivated for its leaves and fruits which are used as flavouring agents.

ALANGIACEAE

ALANGIUM Lamarck *nom. cons.*

Alangium salvifolium (L.f.) Wang. in Engl., Pflanzenr. 41.4. (2206) 9. 1910; Gamble 572 (404). *Grewia salvifolia* L.f., Suppl. Pl. 409. 1781. *Alangium lamarckii* Thw., Enum, Pl. Zeyl. 133. 1859; FBI 2: 741.

Trees, about 5 m tall; branchlets grey-pubescent, spinescent. Leaves oblong-ovate or lanceolate, 2.5-6.2 × 1-1.5 cm, glabrous, base obtuse-subacute, apex subacute, margin entire. Flowers axillary, in cymes, white. Berries globose, 2 × 1 cm, crowned by calyx-lobes; seed 1.

Fls. & Frs.: February - June. *Vern.* Udugu, Urgu.

Common in dry deciduous forests. Gangapur RF, **BR & CP** 11435 and **TP & CP** 12067.

RUBIACEAE

1. Stipules intrapetiolar **Gardenia**
1. Stipules interpetiolar:
    2. Stipules fringed or pectinate:
        3. Prostrate; fruits indehiscent **Dentella**
        3. Erect or diffuse; fruits dehiscent:
            4. Ovules more than 2, usually numerous in each locule; flowers in cyme **Hedyotis**
            4. Ovules solitary in each locule; flowers in fascicled cyme **Borreria**
    2. Stipules entire, rarely bifid :
        5. Stipules foliaceous **Mitragyna**
        5. Stipules triangular - ovate or lanceolate, rarely sheathing:
            6. Fruit a syncarpium; calyx connate below **Morinda**
            6. Fruit a drupe, capsule or berry; calyx not connate below :
                7. Style as long as corolla-tube, stigma bifid **Ixora**
                7. Style much longer than corolla-tube, stigma entire:
                    8. Flowers 4-merous:
                        9. Corolla-tube glabrous within; berry globose **Pavetta**
                        9. Corolla-tube villous within; drupe ellipsoid-globose **Canthium**

8. Flowers 5-merous:
10. Calyx cupular or disciform; berry with pulp, many-seeded **Catunaregam**
10. Calyx turbinate; berry without pulp, few-seeded **Benkara**

BENKARA Tirvengadum

Benkara malabarica (Lam.) Tirvengadum in Taxon 32:440. 1983. *Randia malabarica* Lam., Encycl. 3:325. 1789; FBI 3:111; Gamble 616 (435); Khan in FFH 198.

Armed shrubs, about 2 m tall; branchlets glabrous. Leaves oblong-elliptic; 1-2.5 × 0.6-1 cm, glabrous, base cuneate, apex obtuse, margin entire. Flowers axillary, in cymes, white. Berries globose, 0.8 cm across, glabrous; seed 6-8.

Fls. & Frs.: March - October. *Vern.* : Pedda-manga.

Occasional in scrub jungles and villages. Nathyanipally, **BR & CP** 11252.

BORRERIA G.F.W. Meyer *nom. cons.*

Note : Although the genus *Borreria* G.F.W. Meyer has been conserved, some Indian workers followed the treatment of Verdcourt (Fl. Trop. East Africa 1:341. 1976), who considered that *Spermacoce* L. should be the appropriate genus instead of *Borreria* G.F.W. Meyer. This involves several nomenclatural changes for the Indian taxa under *Spermacoce* L. in the present study, Meyer's treatment of the genus *Borreria* G.F.W. Meyer is followed.

1. Erect slender herbs; leaves whorled at nodes **B. pusilla**
1. Diffuse or prostrate herbs; leaves opposite at nodes **B. articularis**

Borreria articularis (L.f.) F. Will. In Bull. Herb. Boiss. (Ser.2.) 5:956. 1905. *Spermacoce articularis* L.f., Suppl. Pl. 119. 1781. *S. hispida* L., Sp. Pl. 102. 1753; FBI 3:200. *Borreria hispida* (L.) K. Schum. in Engl. Nat. Pflanzenfam. 4(4) : 144. 1891. non. K. Schum. 1888; Gamble 654 (461).

Ascending herbs, about 40 cm tall; branchlets 4-angled, hispid. Leaves elliptic-ovate, 1-3.5 × 0.5-1.1 cm, scarbid, base attenuate, apex acute, margin entire. Flowers axillary, within the stipular cup, white-pale violet. Capsules ellipsoid, 0.5 × 0.4 cm, hairy; seeds 2.

Fls. & Frs.: July - December. *Vern.* : Madanaku.

Common in all places. Pandivagu, **BR & CP** 11204; Pillutla RF, **BR & CP** 11255; Marrikunta, **CP** 11394.

Borreria pusilla (Wall.) DC., Prodr. 4:543. 1830. *Spermacoce pusilla* Wall. in Roxb., Fl. Ind. 1:379. 1820. *S. stricta* auct. non. L.F: FBI 3:200. *Borreria stricta* auct. non K. Schum.: Gamble 654 (461).

Erect herbs, about 50 cm tall; branchlets 4-angled, minutely prickled. Leaves linear-lanceolate or oblong, 1-2.5 × 0.2-0.5 cm, scabrid, base attenuate, apex acute, margin entire. Flowers axillary and terminal, in dense

globose heads, white. Capsules subglobose, 0.2 cm across, hairy.
Fls. & Frs.: July - November.

Common in forests. Narsapur RF, **TP & CP** 11967.

CANTHIUM Lamarck

Canthium parviflorum Lam., Encycl. 1:602. 1785; FBI 3:136; Khan in FFH 202. *Plectronia parviflora* (Lam.) Beddome, Fl. Sylv. S. Ind. 124. 1872; Gamble 625 (441).

Armed shrubs, about 4 m tall; branchlets glabrous. Leaves elliptic-ovate, 1.5-3.5 × 0.8-2.5 cm, glabrous, base cuneate-obtuse, apex obtuse, margin entire. Flowers axillary, in cymes, greenish-white. Dupes ellipsoid, 1 cm across, with 2 one-celled pyrenes.

Fls. & Frs.: March - May. *Vern.* : Balusu.

Common in field hedges and in scrub jungles. Rampur, **TP & CP** 14069.

CATUNAREGAM Wolf

Note: Tirvengadum (1978a) reestablished the genus *Catunaregam* for the old world *Randia sensu lato*. He describes that *Randia* is an exclusively American genus characterised by its one-celled ovary with parietal placentation. He transferred the South Indian *Xeromphis (Randia)* to *Catunaregam*.

Catunaregam spinosa (Thunb.) Tirvengadum in Bull. Mus. Hist. Nat. (Paris) Ser. 3.35: 13. 1978. *Gardenia spinosa* Thunb., Diss. Gard 7:16 1780. *Randia dumetorum* (Retz.) Poir. in Lam., Encyl. 2:829. 1811; FBI 3:110; Khan in FFH 198. var. *floribunda* (DC) Gamble, Fl. Pres. Madras 616. 1921 & 434. 1957. *Gardenia dumetorum* Retz., Obs. Bot. 2:14 1781. *Randia floribunda* DC., Prodr. 4:386. 1830. *R. brandisii* Gamble, Fl. Pres. Madras 616. 1921 & 434. 1957.

Armed shrubs, about 4m tall; branchlets glabrous. Leaves clustered at nodes, obovate, 1-4.5 × 1-2.5 cm, pubescent, base attenuate, apex obtuse, margin entire. Flowers axillary, 1-few in clusters yellowish-white. Berries globose, 2 cm across, glabrous; seeds many.

Fls. & Frs.: March - October. *Vern.* : Manga.

Common in forests and waste lands. Sardhana x roads, **BR & CP** 11620.

DENTELLA J.R. Forster et J.G.A. Forster

Dentella repens (L.) Forster & Forster f., Char. Gen. Pl. 26. t. 13. 1776; FBI 3:42; Gamble 590 (416).

Prostrate herbs; branchlets glabrous. Leaves elliptic- oblong 0.1-0.4 × 0.1-0.2 cm, sparsely hairy, base attenuate, apex acute, margin entire. Flowers at the forks of branches, solitary, white. Capsules globose, 0.3

Fls. & Frs. : June - September.

Occasional in moist places. Pochammaralu, **BR & CP** 11490.

GARDENIA Ellis *nom. cons.*

1. Calyx lobes triangular, acute; stamens included **G. gummifera**
1. Calyx lobes lanceolate, acuminate; stamens excluded **G. latifolia**

Gardenia gummifera L.f., Suppl. Pl. 164. 1781; FBI 3:116; Gamble 618 (436); Khan in FFH 200.

Trees, about 5m tall; branchlets glabrous. Leaves obovate-oblong, 1-5.2 × 0.5-2.7 cm, glabrous, base rounded, apex acute-obtuse, margin entire. Flowers axillary or terminal, solitary, deep yellow. Berries oblong 2.5 × 2 cm; seeds rugose.

Fls. & Frs.: March - September. *Vern.* : Chit-mit.

Common in dry deciduous forests. Narsapur RF, **TP & CP** 11979.

Gardenia latifolia Aiton, Hort. Kew. 1:294. 1789; FBI 3:116; Gamble 619 (436); Khan in FFH 200.

Trees, about 10 m tall; branchlets puberulous. Leaves broadly elliptic, coriaceous, 12-23 × 7-14 cm, glabrous, base attenuate, apex acute, margin entire. Flowers axillary, solitary, white. Berries globose, 4 cm across; seeds many.

Fls. & Frs. : Round the year.

Occasional in dry deciduous forests. Narsapur RF, **CP** 11571 and **TP & CP** 11978.

HALDINIA Ridsdale

Haldinia cordifolia (Roxb.) Ridsdale in Blumea 24: 361. 1978. *Nauclea cordifolia* Roxb., Fl. Ind. 1: 514. 1832. *Adina cordifolia* (Roxb.) Hook.f. ex Brandis, Forest Fl. N.W. India t. 33. 1874; FBI 3: 24; Gamble 584 (412); Khan in FFH 193.

Trees, about 20 m tall; branchlets pubescent. Leaves orbicular, cordate, 10-17 × 12-20 cm, glabrous above, pubescent below, base cordate, apex abruptly acuminate, margin entire. Flowers axillary, in globose heads, cream. Capsules oblong, 0.4 × 0.2 cm, cocci 2.

Fls. & Frs. : April - October. *Vern.* : Bandaru.

Occasional in dry deciduous forests. Madikonakutta, **KMS** 7644.

Note : Earlier floristic workers treated this taxon under *Adina cordifolia* (*loc. cit.*) But Ridsdale (*loc. cit.*) in his revision of tribe Naucleeae *s.s.* feels that Indian taxon does not belong to the genus *Adina* Salisbury, and has raised the monotypic new genus *Haldinia* to accommodate it.

HEDYOTIS Linnaeus

Note : In the present study the concept of W.H. Lewis (1961) and Backer & Bakhuizen f. (1965) is followed. They treated the genus *Oldenlandia* L. under *Hedyotis* L.

1. Corolla tube long and slender **H. aspera**
1. Corolla tube short:
    2. Peduncles many-flowered; flowers blue, pink or purple:
        3. Ascending herbs; stamens exserted **H. puberula**
        3. Diffuse herbs; stamens inserted **H. affinis**
    2. Peduncles 1-4-flowered:flowers white:
        4. Erect herbs **H.herbacea**
        4. Diffuse herbs:
            5. Leaves elliptic-lanceolate; base or calyx-teeth touching in fruits **H. pumila**
            5. Leaves linear; base or calyx-teeth not touching in fruits **H. corymbosa**

Hedyotis affinis Roemer & Schultes in L. Syst. Veg. 3: 194. 1818. *Oldenlandia affinis* (Roemer & Schultes) DC., Prodr. 4: 428. 1830. *O. dichotoma* (Roth) Hook. f. in FBI 3:67. 1810; Gamble 601 (424). *Hedyotis dichotoma* Roth, Nov. Pl. Spec. 93. 1821.

Diffuse herbs, about 60 cm tall; branchlets glabrous. Leaves linear - lanceolate, 1-3 × 0.1-0.2 cm, sparsely hairy, base decurrent, apex acute, margin entire. Flowers axillary, 2-3 on filiform peduncles, blue. Capsules subglobose, 0.1 cm across, glabrous; seeds many.

Fls. & Frs. : July - January.

Common in dry deciduous forests. Narsapur RF, **TP & CP** 11974.

Hedyotis aspera Heyne ex Roth, Nov. Pl. Sp. 94. 1821. *Oldenlandia aspera* (Heyne ex Roth) DC., Prodr. 4: 428. 1830; FBI 3:68; Gamble 602 (424).

Erect herbs, about 30 cm tall; branchlets scabrid. Leaves lanceolate, 2-4.5 × 0.2-0.6 cm, pubescent, base attenuate, apex acute, margin undulate-entire. Flowers terminal, in lax panicled cymes, white-light green. Capsules globose, loculicidal. 0.3 cm across, white-dotted.

Fls. & Frs. : July - December.

Occasional in waste lands. Choutkoor, **TP & CP** 11916.

Hedyotis corymbosa (L.) Lam., Encycl. 1:272. 1792. *Oldenlandia corymbosa* L., Sp. Pl. 119. 1753; FBI 3: 64; Gamble 600 (423).

Diffuse herbs; branchlets pubescent. Leaves linear-lanceolate, 1-2 × 0.1-0.2 cm, glabrous, base decurrent, apex acute, margin entire. Flowers axillary, solitary or in cymes, white. Capsules globose, 0.2 cm across; seeds many.

globose, 0.2 cm across; seeds many.

Fls. & *Frs.* : July - March.

Common in cultivated fields. Choutkoor, **TP** & **CP** 11944.

Hedyotis herbacea L., Sp. Pl. 102. 1753. *Oldeniandia heynii* R.Br. ex G. Don, Gen. Hist. 3: 531. 1834; FBI 3: 65. *O. herbacea* (L.) Roxb., Ind. 1: 445. 1820; Gamble 601 (424).

Erect herbs, about 50 cm tall; branchlets glabrous. Leaves linear-lanceolate, 1.5-3 × 0.1-0.2 cm, glabrous, base decurrent, apex acute, margin entire. Flowers axillary, solitary or 1-2 at nodes, white. Capsules subglobose, 0.2 cm, across, glabrous; seeds many.

Fls. & *Frs.* : August - January.

Occasional in plains and hills. Narayanapur RF, **BR** & **CP** 11274.

Hedyotis puberula (G.Don) Arn., Pugill. Pl. Ind. Orient. 342. 1836. *Oldenlandia puberula* G. Don. Gen. Hist. 3: 530. 1834. *O. umbellata* L., Sp. Pl. 119. 1753; FBI 3: 66; Gamble 601 (424).

Ascending herbs, about 25 cm tall; branchlets pubescent. Leaves linear - lanceolate, 0.5-1.2 × 0.1-0.2 cm, scabrous, base decurrent, apex acute, margin entire. Flowers axillary, 3-10 in peduncled umbels, pinkish-white. Capsules globose, 0.2 cm across, glabrous; seeds many.

Fls. & *Frs.* : July - January.

Common weed of waste lands and road sides. Nagulabanda, **CP** 11325; Medak, **BR** & **CP** 11468.

Hedyotis pumila L.f. Suppl. 119. 1781. *Oldenlandia pumila* (L.f.) DC., Prodr. 4: 425. 1830. *O. crystallina* Roxb., Fl. Ind. 1: 443. 1820; FBI 3: 65; Gamble 602 (424).

Prostrate herbs. branchlets scabrid. Leaves elliptic - lanceolate, 0.5 - 1.6 × 0.1-0.6 cm, base tapering, apex subacute, margin entire. Flowers axillary or terminal, solitary or in pairs, white. Capsules obovate, 1 × 0.2 cm, glabrous; seeds many.

Fls. & *Frs.* : July - Ocotober.

Rare in waste lands and fields. Narsapur, **GNR** 4201.

IXORA Linnaeus

Ixora arborea Roxb. ex Smith in Rees, Cyclop. 19: 5. 1811. *I. pavetta* Andrews., Bot. Repos. t. 78. 1799. *I. parviflora* Vahl, Symb. Bot. 3: 11 t. 52. 1794; FBI 3: 142; Gamble 631 (445); Khan in FFH 203.

Trees, about 5 m tall; branchlets glabrous. Leaves obovate-elliptic, 6-10.5 × 3-5.2 cm, coriaceous, glabrous, base cuneate-rounded, apex obtuse, margin entire-undulate. Flowers axillary, in trichotomously branched cymes, white; corolla tube orange. Drupes subglobose, 0.6

Fls. & Frs. : March - September. *Vern.* : Pona.

Common in dry deciduous forests and hedges. Akkannapet, **BR & CP** 11562; Burugupally, **BR & CP** 11618.

MITRAGYNA P.W. Korthals *nom. cons.*

Mitragyna parvifolia (Roxb.) Korth., Observ. Naud. Ind. 19.1839; Gamble 585 (413). *Nauclea parvifolia* Roxb., Pl. Corom. t. 52. 1796. *Stephygyne parvifolia* (Roxb.) Korth., Verhn. Nat. Gesch. Ned. Bezitt. 161. 1842; FBI 3:25; Khan in FFH 194.

Trees, about 20 m tall; branchlets glabrous. Leaves elliptic-oblong or obovate, 5-11 × 3-8 cm, glabrous, base cuneate, apex obtuse, margin entire. Flowers axillary, in heads, creamish-white. Capsules aggregated into heads, heads 1 cm across, ribbed; seeds many.

Fls. & Frs. : March - September. *Vern.* : Battaganapu.

Common in dry deciduous forests. Narsapur RF, **CP** 11582; Sardhana, **CP** 11771.

MORINDA Linnaeus

1. Trees; flowers in globose heads **M. pubescens**
1. Shrubs; flowers in umbellate heads **M. umbellata**

Morinda pubescens Smith in Rees, Cyclop. 24:3. 1813. var. *pubescens*. *M. tinctoria* Roxb., Fl. Ind. 2: 197. 1824 & 1.543. 1832. non Noronha 1790; FBI 3:156; Gamble 651 (459); Khan in FFH 205. *M. tinctoria* Roxb. var. *tomentosa* (Heyne ex Roth) Hook.f. in FBI 3:156. 1880; Gamble 651 (459). *M. tomentosa* Heyne ex Roth. Nov. Pl. Spec. 147. 1821.

Trees, about 6 m tall; branchlets tomentose. Leaves obovate-elliptic, 9-14 × 4-6 cm, tomentose, base attenuate round, apex acuminate, margin undulate. Flowers axillary, in globose heads, white. Fruit syncarpium, 2 cm across.

Fls. & Frs. : May - November. *Vern.* : Togara, Mogilli.

Common in forests. Narsapur RF, **CP** 11578.

Morinda umbellata L. Sp. Pl. 176. 1753; FBI 3: 157; Gamble 652 (460).

Climbing shrubs, about 5m tall; branchlets glabrous. Leaves oblong-elliptic, 9-14.5 × 3-6.5 cm, coriaceous, glabrous, base cuneate, apex acute, margin entire. Flowers terminal, in umbellate heads, white. Fruit syncarpium, 1 cm across; seeds 4.

Fls. & Frs. : March - September.

Occasional in forests. Narsapur RF, **TP & CP** 14033.

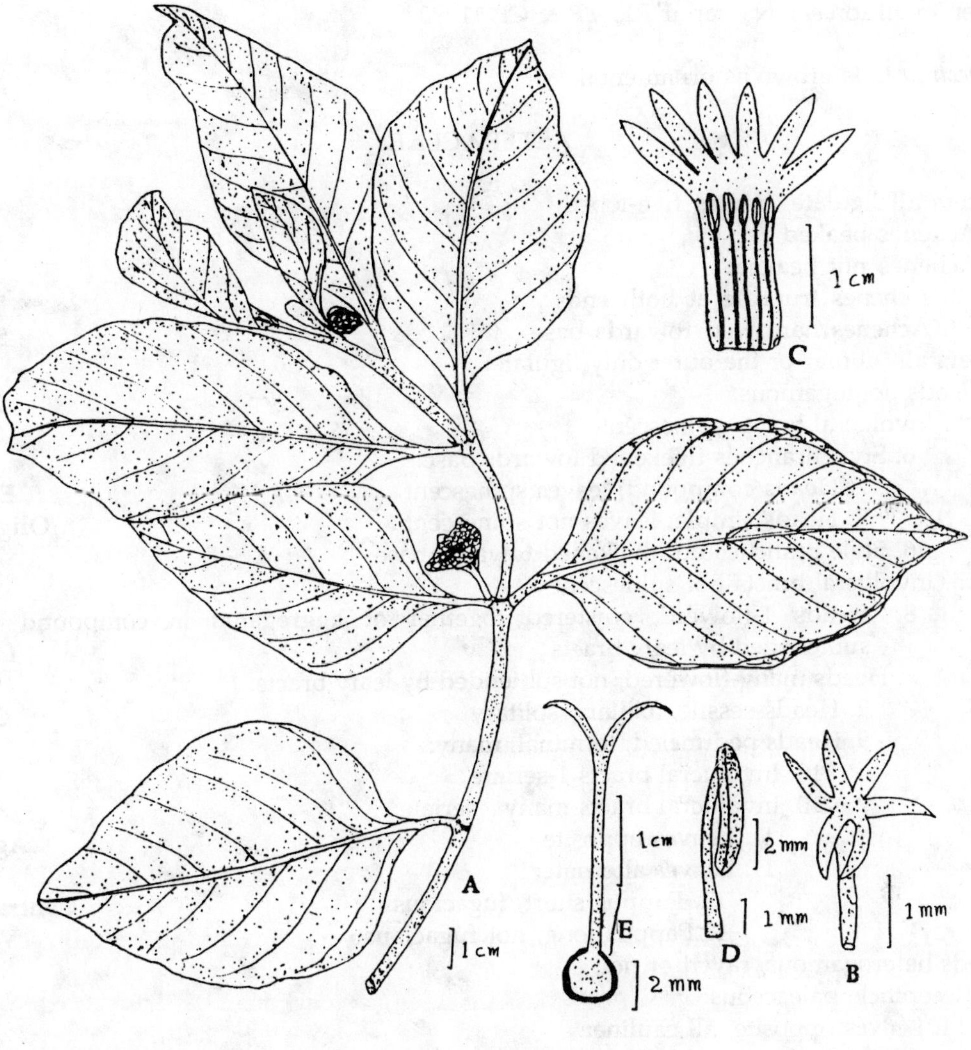

Fig. 6.18 : *Morinda pubescens* Smith A. Twig, B. Corolla, C. Corolla with stamens, D. Stamens, E. Pistil

PAVETTA Linnaeus

Pavetta tomentosa Roxb. ex Smith in Rees. Cyclop. 26:n.2.1819; Gamble 1879 (1301). *P. indica* L. var. *tomentosa* (Smith) Hook. f. Fl. Brit. India 3; 150. 1880; Gamble 683 (446).

Trees, about 3 m tall; branchlets tomentose. Leaves elliptic-oblong, 8-12 × 4.5-7 cm, softly pubescent above, tomentose below, base acute-rounded, apex acute, margin entire. Flowers terminal, in trichotomously branched corymbose cymes, white. Fruit a drupaceous berry, globose.

Fl.s & Frs. : May - February.

Common in all forests. Narsapur RF, **TP & CP** 11990.

Ixora coccinea L. is grown as ornamental.

ASTERACEAE

1. Flowers all ligulate, ligules five-lobed:
    2. Achenes beaked **Lactuca**
    2. Achenes not beaked:
        3. Achenes truncate at both ends **Launaea**
        3. Achenes narrowed towards base **Sonchus**
1. Flowers all tubular or the outer only ligulate:
    4. Heads homogamous:
        5. Involucral bracts spinescent:
            6. Style branches thickened towards base:
                7. Heads compound; leaves spinescent **Echinops**
                7. Heads simple; leaves not spinescent **Oligochaeta**
            6. Style branches not thickened towards base **Dicoma**
        5. Involucral bracts not spinescent :
            8. Heads 1-flowerd, clustered together or aggregated in compound fashion, subteneded by leafy bracts **Lagascea**
            8. Heads many-flowered, not subtended by leafy bracts:
                9. Heads sessile, axillary, solitary **Caesulia**
                9. Heads peduncled, terminal, many:
                    10. Involucral bracts 1-seriate **Emilia**
                    10. Involucral bracts many - seriate:
                        11. Leaves opposite **Ageratum**
                        11. Leaves alternate:
                            12. Pappus short, fugacious **Centratherum**
                            12.Pappus long, not fugacious **Vernonia**
    4. Heads heterogamous, rayed or not:
        13. Receptacle paleaceous:
            14. Leaves opposite, all cauline:
                15. Inner involucral bracts partly or completely enclosing the achenes of ray florets **Acanthospermum**
                15. Inner involucral bracts do not enclose the achenes of ray florets:
                    16. Pappus of awns:
                        17. Awns plumose **Tridax**
                        17. Awns retrorsely barbellate **Bidens**
                    16. Pappus scales, bristles or absent:
                        18. Receptacle conical **Spilanthes**
                        18. Receptacle flat or slightly convex:
                            19. Ligules white:
                                20. Plants of aquatic condition; stems usually decumbent; involucral bracts foliaceous **Eclipta**

20. Plants of terrestrial condition; stems erect; involucral bracts otherwise **Blainvillea**

19. Ligules yellow:
   21. Bisexual florets completely enclosed by palea **Sclerocarpus**
   21. Bisexual florets free from palea **Wedelia**
      23. Leaves deeply dissected, achenes densely hairy along margins **Glossocardia**
      23. Leaves not much dissected as above, achens glabrous **Chrysanthellum**

14. Leaves alternate, radical and cauline or cauline only:
   22. Plants prostrate
   22. Plants erect:
      24. Heads unisexual **Xanthium**
      24. Heads bisexual **Parthenium**

13. Receptacle epaleaceous:
   25. Leaves opposite **Flaveria**
   25. Leaves alternate:
      26. Heads rayed:
         27. Pappus two seriate, outer reduced into a cup **Pulicaria**
         27. Pappus one seriate **Vicoa**
      26. Heads not rayed:
         28. Heads aggregated into ovoid or spherical clusters:
            29. Stems winged **Sphaeranthus**
            29. Stems not winged **Blepharispermum**
         27. Heads not aggregated into clusters:
            30. Leaves pinnately dissected **Grangea**
            30. Leaves simple:
               31. Leaves and inflorescences are covered by white woolly tomentum **Gnaphalium**
               31. Leaves and inflorescences are variously pubescent **Blumea**

ACANTHOSPERMUM Schrank

Acanthospermum hispidum DC., Prodr. 5:522. 1836; Gamble 704 (495).

Erect herbs, about 60 cm tall; branchlets hairy. Leaves ovate, 2-7 × 1.5-3.5 cm, hairy, base cuneate, apex acute, margin coarsely serrate. Heads axillary, solitary, heterogamous, yellow. Achenes of ray florets compressed; pappus absent.

Fls. & Frs. : August - November. *Vern.* : Guntakalagara.

Common weed of waste lands and fields. Pilutla RF, **BR & CP** 11271; Nagulabanda. CP 11354.

AGERATUM Linnaeus

Ageratum conyzoides L., Sp. Pl. 839. 1753; FBI 3:243; Gamble 677 (476).

Erect herbs, about 75 cm tall; branchlets pubescent. Leaves ovate, 2-6 × 1-4 cm, puberulous, base rounded - obtuse, apex acute, margin bluntly serrate. Heads terminal, in corymbs, homogamous, white. Achenes scrabrous; pappus awned.

Fls. & Frs. : October - April.

Common weed of cultivated fields and moist localities. Narsapur, **BR & CP** 11232; Narsingi, **BR & CP** 11536.

BIDENS Linnaeus

1. Leaves with 3-5 deeply dentate or pinnatifid leaflets **B. biternata**
1. Leaves ternate, leaflets simple with serrate margins **B. pilosa**

Bidens biternata (Lour.) Merr. & Scherff in Bot. Gaz. 88: 293. 1929. *Coreopsis biternata* Lour. Fl. Cochinch. 508. 1790. *Bidens wallichii* DC., Prodr. 5: 598. 1836. *B. pilosa auct.* non L., 1753 : FBI 3:309; Gamble 709 (499).

Erect herbs, about 1.5 m tall; branchlets glabrous. Leaves 3-7-foliolate; leaflets ovate-lanceolate, 1.5-7 × 1-2.5 cm, scabrous, base attenuate, apex acute, margin dentate. Heads axillary or terminal, solitary or in pairs, heterogamous, yellow. Achenes linear; pappus awned.

Fls. & Frs. : August - December.

Common in moist and shady places. Narsapur forest nursery. **BR & CP** 11222; Nagulabanda, **CP** 11351; Ranjhole, **CP** 11676.

Bidens pilosa L., Sp. Pl. 832. 1753; FBI 3:309; Gamble 709 (499).

Erect herbs, about 75 cm tall; branchlets striate. Leaves 3- foliolate; terminal leaflet obovate, lanceolate, 4-10 × 1.8-4.5 cm, scabrous, base rounded-acute, apex acuminate, margin serrate; lateral ones ovate. heads terminal, solitary or in pairs, hetero-gamous, yellow. Achenes linear black; pappus bristly.

Fls. & Frs. : October - April.

Occasional in waste lands and cultivated fields. Way to Pegarikutta, **KMS** 6641.

BLAINVILLEA Cassini

Blainvillea acmella (L.) Philipson in Blumea 6:350. 1950. *Verbesina acmella* L., Sp. Pl. 901. 1753. *Eclipta latifolia* L.f. Suppl. Pl. 378. 1781. *Blainvillea latifolia* (L.f.) DC. in Wight. Contrib. Bot. 17: 1834; FBI 3: 305. *B. rhombiodea* Cass. in Dict. Sci. Nat. 29:494. 1823: Gamble 706 (496).

Erect herbs, about 1.5 cm tall; branchlets grooved, scabrid. Leaves ovate, 2.5-6 × 1-4 cm, scabrous, base obtuse, apex acute, margin serrate. Heads axillary and terminal, solitary, heterogamous, white. Achenes laterally compressed. pappus bristled.

Fls. & Frs. : July - December.

Occasional weed in waste places and cultivated fields. Bonthapalle, **GNR** 4478; Narsapur RF, **KMS** 6692.

BLEPHARISPERMUM Wight

Blepharispermum subsessile DC. in Wight, Contrib. Bot. Ind. 11.1834; FBI 3: 26; Gamble 692(487); R.R. Rao et al., Fl. Ind. Enum. Ast. 14. 1988.

Erect herbs, about 60 cm tall; stems striated, glabrous. Leaves elliptic, 2.5-8.5 × 1-4 cm, glabrous, base cuneate or obtusely attenuate, apex acute, margin entire. Heads terminal, in globose clusters, heterogamous, white. Achenes obovate, trigonous, black; pappus of unequal bristles.

Fls. & Frs. : November - February.

Rare in open forests, Kulcharam RF, **TP & CP** 14079; Narsapur RF, **KMS** 6770.

BLUMEA A.P. de Condolle *nom. cons.*

1. Plants with globose woolly buds at the base of stem and on roots, bisexual florets multicellular hairy except on tube **B. oxyodonta**
1. Plants with out globose woolly buds, bisexual florets multicelullar hairy on lobes and tubes **B. obliqua**

Blumea obliqua (L.) Druce Rep.Bot.Exch.Club Brit. Isles 4: 609: 1916. *Erigeron obliqum* L. Mant. 2:573.1771. *Blumea amplectans* DC. in Wight Contrib. Bot. Ind. 13. 1834; FBI 3: 260; Gamble 686 (482).

Erect herb, to 40 cm, stem striated. Leaves oblong, 2-8 × 0.5-2.5 cm, base semiamplexicaul, margin dentate, apex acute. Heads terminal, solitary or two, pinkish. heterogamous, Achenes oblong.

Fls. & Frs. : October - February.

Common near wet places of black soils. Poosanpalle, **CP** 11958.

Blumea oxydonata DC., in Wight, Contrib, Bot. Ind. 15: 1834; FBI 3: 266; Gamble 686 (482).

Erect herbs, about 15 cm tall; branchlets pubescent. Leaves oblong-obovate, 2-5 × 1-2 cm, scabrous above, softly pubescent below, base decurrent, apex obtusely mucronate, margin dentate. Heads axillary and terminal, soitary or in clusters, heterogamous, yellow. Achenes angled, pubescent; pappus white.

Fls. & Frs. : January - April.

Occasional on road sides. Kondapur lake. **GNR** 4575.

CAESULIA Roxburgh

Caesulia axillaris Roxb. Pl. Corom. 1:64. t. 93. 1795; FBI 3:291; Gamble 704 (494).

Suberect herbs, about 60 cm tall; branchlets glabrous. Leaves lanceolate, 4-8 × 0.5-1.2 cm, glabrous, base attenuate, apex acute, margin distantly serrate. Heads axillary, solitary, homogamous, white. Achenes obovate, winged; pappus sparsely hairy.

Fls. & Frs. : Round the year.

Common in cultivated fields and muddy areas. Poosanpalle, **TP & CP** 11952.

CENTRATHERUM Cassini

Centratherum anthelminticum (L.) O. Kuntze, Rev. Gen. Pl. 1:320. 1891; Gamble 667 (469). *Conyza anthelmintica* L. Sp. Pl. 1753. *Vernonia anthelmintica* (L.) Willd. Sp. Pl. 3:1634. 1880; FBI 3:326.

Erect herbs, about 80 cm tall; branchlets hirsute. Leaves ovate-lanceolate, 4-6 × 2-2.5 cm, pubescent, apex acute, margin serrate. Heads terminal, in panicles, homogamous, purple. Achenes obtuse, 8-ribbed; pappus paleaceous.

Fls. & Frs. : October - April.

Occasional in waste lands and forests. Way to Rusthempet, **KMS** 6783.

CHRYSANTHELLUM L.C. Richard

Chyrsanthellum americanum (L.) Vatke, Abh.Nat. Ver. Brem. 9:122.1887. *Artemesia americana* L. Sp. Pl. 895. 1753. *Chrysanthellum indicum* DC. Prodr. 5:631. 1836; FBI 3:310. 1861.

Erect or diffuse herb, 10-20 cm tall, stem 4-ribbed. Leaves bipinnately compound, segments oblong, entire, sharply apiculate. Heads solitary, axillary or terminal, yellow, heterogamous. Achenes dimorphic, marginal ones oblong, those towards centre obovate or elliptic.

Fls. & Frs. : September - December.

Rare in wet places. Pagarikutta. *KMS 6632*.

DICOMA Cassini

Dicoma tomentosa Casess; Bull. Sci. Soc. Philom. Paris. 12. 1818; FBI 3:387; Gamble 728 (512).

Erect herbs, about 50 cm tall; branchlets densely pilose. Leaves linear-lanceolate, 2-4.5 × 0.2-0.5 cm, tomentose above, white woolly below, base cuneate, apex apiculate, margin minutely dentate. Heads axillary, solitary, homogamous, white. Achenes enlarged, densely villous; pappus barbellate.

Fls. & Frs. : November - March.

Occasional in fields and dry habitats. Choutkoor, **TP & CP** 11942.

ECHINOPS Linnaeus

Echinops echinatus Roxb., Fl. Ind. 3:447. 1832; FBI 3:358; Gamble 724 (509).

Erect herbs, about 40 cm tall; branchlets hairy. Leaves lanceolate, 2-5 × 0.5-1.5 cm, pinnatisect, pubescent, base amplexicaul, lobes spine-tipped. Heads terminal, densely aggregated into clusters, homogamous, white. Achenes elongate, villous; pappus barbellate.

Fls. & Frs. : November - April. *Vern.* : Brahmadandi.

Common weed of waste lands and fields. Mombojipally, **BR & CP** 11452.

ECLIPTA Linnaeus *nom. cons.*

Eclipta prostrata (L.) L., Mant. Pl. 286. 1771. *Verbesina prostrata* L., Sp. Pl. 902. 1753. *Eclipta alba* (L.) Hassk., Pl. Jav. Rar. 528. 1848; FBI 3: 304; Gamble 705 (496). *Verbesina alba* L., Sp. Pl. 902. 1753.

Creeping herbs, branches ascending, about 60 cm tall; branchlets hirsute. Leaves lanceolate, 1.2-4 × 0.7-1.2 cm, hirsute, base cuneate, apex acute, margin irregularly serrate-dentate. Heads axillary or terminal, solitary, heterogamous, white. Achenes of ray florets triquetrous, of disc florets compressed; pappus biseriate.

Fls. & Frs. : Round the year.

Common in moist waste lands and cultivated fields. Marrikunta, **CP** 11388; Patancheru, **CP** 11690.

EMILIA Cassini

Emilia sonchifolia (L.) DC. in Wight, Contrib. Bot. 24. 1834; FBI 3: 336; Gamble 716 (503).

Suberect herbs, about 60 cm tall; branchlets, glabrous. Leaves ovate and lanceolate, 2.5-10 × 1-4.5 cm, glabrescent, base amplexicaul, apex acute, margin coarsely crenate. Heads terminal, in lax corymbs, homogamous, pink. Achenes angled, 5-ribbed, scabrid; pappus copious.

Fls. & Frs. : August - November.

Common in cultivated fields and waste lands. Nagulabanda, **CP** 11356.

FLAVERIA A.L. Jussieu

Flaveria trinervia (Spreng.) C. Mohr, U.S.D.A. Contrib. Nat. Herb. 6: 810. 1910. *Oedera trinervia* Spreng., Bot. Gard. Halle 63. 1800. *Flaveria australasica* Hook. in T.L. Mitch. Journ, Exep. Trop. Austral. 118. 1848; Gamble 711 (501).

Erect herbs, about 50 cm tall; branchlets glabrous. Leaves obovate-lanceolate, 2-5 × 0.5-1.5 cm, chartaceous, base cuneate, apex acute, margin dentate. Heads terminal or at the forks of dichotomous branches, solitary, heterogamous, yellow. Achenes ribbed; pappus absent.

Fls. & Frs. : September - March.

Occasional in waste lands. Achampet, **GNR** 9760.

GLOSSOCARDIA Cassini

Glossocardia bosvallea (L.f.) DC. in Wight, Contrib. Bot. 19. 1934; Gamble 709 (499). *Verbesina bosvallea* L.f. Suppl. Pl. 379. 1781. *Glossocardia linearifolia* Cass. in Dict. Sci. Nat. 19:62. 1821; FBI 3:308.

Prostrate herbs; stem glabrous. Leaves pinnatisect, glabrous, pinnules 1-2 cm long, filiform, base decurrent, apex apiculate. Heads axillary, solitary, heterogamous, golden yellow. Achenes compressed, margins ciliate; pappus 2-awned.

Fls. & Frs. : August - November.

Common weed in cultivated fields and waste lands. Thornal, **CP** 11400; Choutkoor, **TP & CP** 11933.

GNAPHALIUM Linnaeus

Gnaphalium polycaulon pers. Syn. Pl. 2: 421. 1807. *G. indicum auct.* non L. : FBI 3:289; Gamble 698 (491).

Erect slender herbs, about 15 cm tall; branchlets white-woolly. Leaves lanceolate-spathulate, 2-3.5 × 0.2-0.6 cm, pilose, base cuneate, apex acute, margin entire. Heads terminal, in clusters, heterogamous, silverywhite. Achenes oblong. minutely scaly; pappus hairy.

Fls. & Frs. : September - March.

Occasional weed of paddy fields, waste places and along river banks. Narsapur tank, **TP & CP** 12004.

GRANGEA Adanson

Grangea maderaspatana (L.) Poir. in Lam., Encycl. Suppl. 2:825. 1812; FBI 3: 247; Gamble 580 (478). *Artemesia maderaspatana* L., Sp. Pl. 849. 1753.

Prostrate herbs; branchlets pubescent. Leaves pinnatifid, 2-5 × 1-2.5 cm, lobes oblong-lanceolate, pubescent, base decurrent, apex apiculate, margin entire. Heads axillary and terminal, solitary, heterogamous, yellow. Achenes oblong, truncate; pappus margin fimbriate.

Fls. & Frs. : September - April. *Vern.* : Nemalipadam.

Occasional in dried up tanks and moist fields. Narsapur, **GNR** 4081; Narsapur, RF, **KMS** 7965.

LACTUCA Linnaeus

Lactuca runcinata DC. in Wight, Contrib. Bot. 26. 1834; Gamble 731 (514). *Lactuca heyneana* DC., Prodr. 7: 140. 1838; FBI 3: 403.

Erect herbs, about 1 m tall; branchlets glabrous. Leaves radical, 6-14 × 2-10 cm, pinnatifid, base auricled, irregularly serrate. Heads in small clusters, homogamous, light yellow. Achenes ovate-oblong, compressed.

Fls. & *Frs.* : August - March.

Common weed in cultivated fields. Mombojipally, **BR & CP** 11631.

LAGASCEA Cavanilles *nom. cons.*

Note : In many Indian Floras this has been spelt as "*Lagasca*". This has been corrected by Willdenow, Enum. Pl. Hort. Berol. 941. 1809 as *Lagascea.*

Lagascea mollis Cav. in Anal. Cienc. Nat. 6:332. t. 44. 1803; FBI 3: 302; Gamble 703 (494).

Erect herbs, about 75 cm tall; branchlets pubescent. Leaves lanceolate, 1-3 × 0.5-1.5 cm, pubescent, base acute, apex acuminate, margin serrate. Heads terminal, 1-flowered, densely aggregated, hamogamous, white. Achenes 3-angled; pappus fimbriate.

Fls. & *Frs.* : August - March.

Common weed in cultivated fields and waste lands. Narsapur forest nursery, **BR & CP** 11217; Choutkoor, **TP & CP** 11903.

LAUNAEA Cassini

Launaea sarmentosa (Willd.) Schultz-Bip: ex O. Kuntze, Rev. Gen. Pl. 1: 350. 1891. *Prenanthes sarmentosa* Willd., Phytographia 10. t. 6.f.2 1794. & SP. Pl. 3:1540. 1803. *Launaea pinnatifida* Cass. in Ann. Sci. (Paris) 23: 85. 1831; FBI 3: 416; Gamble 733 (515).

Procumbent or prostrate herbs, about 1 m tall; stem with milky sap. Leaves in rosettes, lanceolate, pinnatifid, 3-8 × 0.5-2 cm, glabrescent, base cuneate, apex acute, margin denticulate. Heads solitary or 1-3 in clusters, homogamous, yellow. Achenes 3-5 ribbed, beaked above; pappus fimbriate.

Fls. & *Frs.* : September - March.

Occasional in moist places. Pochammaralu, **BR & CP** 11486.

OLIGOCHAETA (A.P. de Condolle) K.H.E. Koch

Oligochaeta ramosa (Roxb.) Wagentiz in Veroff. Gebot. Inst. ETH Stiftung Rubel Zurich. 37: 323. 1962. *Carduus ramosus* Roxb., Fl. Ind. 3: 407. 1832. *Volutarella ramosa* (Roxb.) Kitam. Pl. W. Pakistan

& Afghan. 159. 1964. *V. divaricata* Benth. et Hook.f. Gen. Pl. 2: 476. 1873; FBI 3: 383; Gamble 727 (511).

Erect herbs, about 60 cm tall; branchlets grooved, glabrous. Leaves lanceolate, pinnatifid., 1.5-3.5 × 0.2-0.6 cm, scabrous, base decurrent, apex acute, margin spine-tipped. Heads terminal, solitary, heterogamous, violet. Achenes oblong, ribbed; pappus filiform.

Fls. & Frs. : September - March.

Occasional in waste lands. Choutkoor, **TP & CP 11943.**

PARTHENIUM Linnaeus

Parthenium hysterophorus L., Sp. Pl. 988. 1753; Rolla S. Rao in Journ. Bombay Nat. Hist. Soc. 54: 219. 1956.

Erect herbs, about 1.5 m tall; branchlets pubescent. Leaves pinnatifid, lanceolate, 2-8 × 0.5-2 cm, pubescent, base decurrent, apex acute, margin entire. Heads terminal, in panicles, heterogamous, white. Achenes compressed; pappus awned.

Fls. & Frs. : Round the year. *Vern.* : Congresspoolu.

Abundant weed of waste lands, road sides and fields. Narsapur forest nursery, **BR & CP 11219.**

PULICARIA Gaertner

1. Ray-florets tubular **P. foliolosa**
1. Ray-florets ligulate:
   2. Peduncles thickened below the heads **P. wightiana**
   2. Peduncles not thickened below the heads **P. angustifolia**

Pulicaria angustifolia DC., Prodr. 5:479. 1836; FBI 3:299; Gamble 702 (493).

Erect herbs, about 40 cm tall; branchlets pubescent. Leaves obovate-oblong, 1.5-4 × 0.5-1 cm, pubescent, base cordate, margin serrulate. Heads terminal, solitary, heterogamous, yellow. Achenes terete, scarcely quadrangular; pappus hairy.

Fls. & Frs. : November - December.

Occasional in open forests. Sathihpalle, **GNR 4336.**

Pulicaria foliolosa DC., Prodr. 5. 480. 1836; FBI 3: 298; R.R. Rao et. al., Fl. Ind. Enum. Ast. 60. 1988.

Erect herbs, about 40 cm tall; branchlets terete. Leaves spathulate, 2-7 × 0.2-1.5 cm, base obtuse, apex apiculate, margin obscurely serrate. Heads terminal and axillary, solitary or few, heterogamous. Achenes ellipsoid, brown; pappus biseriate.

Fls. & Frs. : October - January.

Rare in moist places. Pochammaralu, **BR & CP** 11485.

Pulicaria wightiana (DC.) C.B. Clarke, Comp. Ind. 118. 1876; Gamble 702 (493). *Poloa wightiana* DC. in Arch. Bot. 2 : 514. 1833. *Callistephus wightianus* (DC.) DC., Prodr 5:275. 1836.

Erect herbs, about 50 cm tall; branchlets hairy. Leaves lanceolate, 1.5-4.5 × 0.4-0.8 cm, pubescent, base amplexicul, apex acute, margin serrate-dentate. Heads terminal, solitary, heterogamous, bright yellow. Achenes ribbed; pappus fimbriate.

Fls. & Frs. : September - March.

Occasional in forests and waste lands. Pandivagu, **BR & CP** 11207.

SCLEROCARPUS N.J. Jacquin

Sclerocarpus africanus Jacq., in Ic. Pl. Rar. 1:17. t. 176. 1780; FBI 3: 305, Gamble 706 (496).

Erect herbs; branchlets villous. Leaves ovate, strigosely hairy, apex acute, margin serrate. Heads axillary or terminal, heterogamous, yellow. Achenes ovate, pappus absent.

Fls. & Frs. : September - March.

Occasional in forests. Way to Peddhakutta, **KMS** 6760.

SONCHUS Linnaeus

1. Leaves with well-developed auricles; corolla tube and ligule equal in length **S. oleraceus**
1. Leaves with small auricles; corolla tube twice as long as ligule **S. wightianus**

Sonchus oleraceus L., Sp. Pl. 794. 1753; FBI 3: 414; Gamble 732 (514).

Erect herbs, about 90 cm tall; branchlets grooved, glabrous. Leaves obovate-lanceolate, 8-20 × 2-5 cm, base amplexicaul, apex acute-rounded, margin dentate. Heads terminal, solitary or in clusters, homogamous, yellow. Achenes obovate, ribbed; pappus muricate, silky.

Fls. & Frs. : December - April.

Occasional on road sides. Gummadiadala, **GNR** 4258

Sonchus wightianus DC., Prodr. 7:187. 1838. *S. arvensis* auct. non L : C.B. Clarke, Compos. Ind. 276. 1876; FBI 3: 414; Gamble 732 (515).

Erect herbs, about 80 cm tall; branchlets glabrescent. Leaves oblong-oblanceolate, 5-20 × 0.2-8 cm, glabrous, base attenuate, apex acute-acuminate, margin sharply dentate. Heads terminal, in corymbose cymes, homogamous, yellow. Achenes oblong, tetragonous, ribbed; pappus silky.

Fls. & Frs. : Round the year.

Occasional weed of road sides. Gummadidala, **GNR** 4251.

SPHAERANTHUS Linnaeus

Sphaeranthus indicus L., Sp. Pl. 927. 1753; FBI 3:275 p.p. Gamble 692 (487).

Ascending herbs, about 50 cm tall; branchlets pubescent, narrowly winged, dentate. Leaves ovate-obovate, 2-4 × 1-2.5 cm, pubescent, base decurrent, apex acute, margin sharply dentate. Heads axillary, solitary, heterogamous, purple. Achenes oblong, glabrous; pappus absent.

Fls. & Frs. : October - March. *Vern.* : Bodasaramu.

Common weed in fields and waste lands. Medak, **BR & CP** 11418.

SPILANTHES N.J. Jacquin

Spilanthes uliginosa Sw. Nov. Sen. Pl. Sci. Prod. Descr. Veg. Ind. Occ. 110. 1788; DC. Prodr. 5: 264. 1836; Moore, Proc. Amer. Acad. Art. Sci. 42(20) : 537. 1907. *S. iabadicensis* A.H. Moore, Proc. Amer. Acad. Art. Sci. 42 (20) : 524. 1907; Koster & Phillipson, Blumea 6 : 354. 1950.

Erect herbs, about 40 cm tall; stems striated, fistular. Leaves ovate, 3-6 × 1-3 cm, sparsely pilose, base cuneate-attenuate, apex obtuse, margin dentate-serrate. Heads terminal or axillary, solitary, heterogamous, yellow. Achenes dimorphic, marginal ones triquetrous, inner ones flat to biconvex, black; pappus absent.

Fls. & Frs. : October - January.

Rare in wet places. Gummadidala, **BR & CP** 14129.

TRIDAX Linnaeus

Tridax procumbens L., Sp. Pl. 900. 1753; FBI 3:311; Gamble 711 (500).

Straggling herbs, about 60 cm tall; branchlets hirsute. Leaves ovate, 2.5-5 × 1-2.5 cm, strigose, base acute, apex acute, margin coarsely serrate. Heads terminal, solitary, heterogamous, yellow. Achenes oblong; pappus of feathery bristles.

Fls. & Frs. : Round the year. *Vern.* Gaddichamanthi.

Common weed of all localities. Sangareddy tank, **BR & CP** 11288.

VERNONIA Schreber *nom. cons.*

Vernonia cinerea (L.) Less. in Linnaea 4: 291. 1829; FBI 3:233; Gamble 676 (476). *Conyza cinerea* L., Sp. Pl. 862. 1753.

Erect herbs, about 75 cm tall; branchlets pubescent. Leaves ovate, 1.5-6 × 1-2.5 cm, hairy, base decurrent, apex acute, margin crenate. Heads terminal, in corymbose panicles, homogamous, pink. Achenes 2-3-angled, densely hairy; pappus biseriate, white.

Fls. & Frs. : August - March. *Vern.* : Sahadevi.

Common in paddy fields, waste lands and road sides. Patancheru, **BR & CP** 11313.

VICOA Cassini

Vicoa indica (L.) DC. in Wight. Contrib. Bot. 10. 1834; Gamble 701 (493). *Inula indica* L., Sp. Pl. 2:1236. 1762. *Vicoa auriculata* Cass. in Ann. Sci. Nat. (Paris) Ser. 1.17: 418. 1829; FBI 3:297.

Erect herbs, about 1 m tall; branchlets striate, hispid. Leaves lanceolate, 2-6 × 0.5-1.5 cm, minutely scabrous, base amplexicaul, apex gradually acute, margin serrate. Heads terminal, solitary, heterogamous, yellow. Achenes obovate, terete, villous; pappus filiform in disc florets, absent in ray florets.

Fls. & Frs. : August - February.

Occasional in dry deciduous forests and fields. Choutkoor, **TP & CP** 11923.

WEDELIA N.J. Jacquin

Wedelia urticaefolia DC. in Wight, Contrib. Bot. Ind. 18. 1834; FBI 3:306; Gamble 707 (497).

Erect herbs, about 60 cm tall; branchlets hirsute. Leaves ovate, elliptic, 3-6 × 2.5-4 cm, scabrous, base truncate, apex acute, margin serrate. Heads axillary or terminal, solitary, heterogamous, yellow. Achenes angled, pubescent; pappus reduced.

Fls. & Frs. : Round the year.

Occasional in moist places. Gummadidala, **GNR** 4267

XANTHIUM Linneaus

Xanthium indicum J. Koenig in Roxb., Fl. Ind. 3:601. 1832. *X. strumarium* L., Sp. Pl. 987. 1753 *p.p.*; FBI 3: 303 Gamble 703 (494).

Erect herbs, about 85 cm tall; branchlets hispid. Leaves ovate, 3-lobed, 5-12 × 4.5-10 cm, scabrous, base acute-subcordate, apex acute, margin crenate. Heads monoecious, male in upper axils, female in lower axils, yellow. Achenes 2, enclosed in spinescent utricle; pappus absent.

Fls. & Frs. : August - March. *Vern.* : Maraluteega.

Common weed of cultivated fields and waste lands. Patancheru, **BR & CP** 11320.

Helianthus annuus L. is commonly cultivated for its seed oil. *Chrysanthemum indicum* L., *Cosmos* spp. *Tagetus erecta* L., are planted as ornamentals.

LOBELIACEAE

LOBELIA Linnaeus

Lobelia alsinoides Lam., Encycl. 3: 588. 1792. *L. trigona* Roxb., Fl. Ind. 2: 111. 1824; FBI 3: 423; Gamble 736(518).

Ascending herbs, about 30 cm tall; branchlets glabrous, winged. Leaves ovate, 0.6-1.5 × 0.5-1.5 cm, glabrous, base truncate, apex acute, margin crenate-dentate. Flowers axillary, solitary, violet. Capsules ellipsoid, 0.4 × 0.3 cm, glabrous; seeds many.

Fls. & Frs. : Round the year.

Occasional in cultivated fields and in forests. Narsapur RF, **BR & CP** 14113.

PLUMBAGINACEAE

PLUMBAGO Linnaeus

Plumbago zeylanica L., Sp. Pl. 151. 1753; FBI 3: 480; Gamble 744(523).

Undershrubs, about 1 m tall; branchlets glabrous, Leaves ovate, 2-6.5 × 0.5-3 cm, glabrous, base attenuate, apex acute, margin entire. Flowers in spiciform racemes, white. Capsules oblong, 1 × 0.4 cm; seed 1.

Fls. & Fls. : September - December.

Occasional in waste lands and scrub jungles. Medak, **BR & CP** 11408; Ghanapur, **CP** 11784.

SAPOTACEAE

1. Calyx-lobes 6-8; staminodes present **Mimusops**
1. Calyx-lobes 4; staminodes absent **Madhuca**

MADHUCA Hamilton ex J.F. Gmelin

Madhuca indica Gmel. in L., Syst. Nat. 2(1): 799. 1791. *M. longifolia* (Koenig) Macbr. var. *latifolia* A. Chev. in Rev. Int. Bot. Appl. Agric. Trop. 23: 149. 1943. *Bassia latifolia* Roxb., Pl. Corom. t. 19. 1795; FBI 3:544; Gamble 763 (536); Khan in FFH 211.

Trees, about 15 m tall; branchlets tomentose. Leaves broadly elliptic, 7-14 × 5-8.5 cm, glabrous, base rounded-truncate, apex abtuse, margin entire. Flowers axillary, in clusters, creamish-white. Berries ovate, fleshy, 3 × 2.3 cm, tomentose; seds 1-4.

Fls. & Frs. : March - September. *Vern.*: Ippa.

Common in forests. Gangapur RF, **BR & CP** 11427; Narsapur RF, **CP** 11587.

MIMUSOPS Linnaeus

Mimusops elengi L., Sp. Pl. 349. 1573; Roxb. Fl. Ind. 2: 236. 1832; FBI 3: 548; Gamble 765 (538); Khan in FFH 213.

Trees, about 20 m tall; branchlets rusty-pubescent. Leaves ovate-lanceolate or elliptic, 4-12 × 3-5.5 cm, coriaceous, base acute-attentuate, apex acute-acuminate, margin entire. Flowers axillary, solitary or in clusters, white. Berries ovate, fleshy; seed 1.

Fls. & Frs. : February - May. *Vern.*: Pogada.

Occasional in forests. Narsapur RF-North end, **KMS** 7958.

Manilkura zapota (L.) P.Royen is grown for its edible fruits.

EBENACEAE

DIOSPYROS Linnaeus

1. Leaves thin-coriaceous, < 5 cm in length **D. chloroxylon**
1. Leaves coriaceous, > 5 cm in length **D. melanoxylon**

Diospyros chloroxylon Roxb., Pl. Corom. t. 49. 1795; FBI 3:560; Gamble 776 (546); Khan in FFH 217.

Trees, about 5 m tall; branchlets tomentose. Leaves oblong-obovate, 1.5-5 × 0.5-3 cm, base truncate- rounded, apex acute-obtuse, margin entire. Flowers axillary, unisexual, white; male in clusters, female solitary. Berries globose, 1.2 cm across, glabrous; seeds 2-3.

Fls. & Frs. : Round the year. *Vern.* : Illinta.

Common in forests and in waste lands. Pillutla RF, **BR & CP** 11266; Gangapur RF, **BR & CP** 11437; Akkannapet, **BR & CP** 11566; Pochammaralu, **CP** 11755; Narsapur RF, **TP & CP** 11986 and 14071.

Diospyros melanoxylon Roxb., Pl. Corom t. 46. 1795; FBI 3:564; Gamble 772 (543); Khan in FFH 218. *D. tupru* Buch. - Ham. in Trans. Linn. Soc. London 15:111. 1827; FBI 3:563.

Trees, about 20 m tall; branchlets rusty tomentose. Leaves ovate-elliptic. 5-15 × 3-7 cm, tawny-tomentose above, pubescent below, base rounded, apex obtuse- acute, margin entire. Flowers axillary, unisexual, white; male in clusters, female solitary. Berries globose, 3 cm across, glabrous; seeds 3-8.

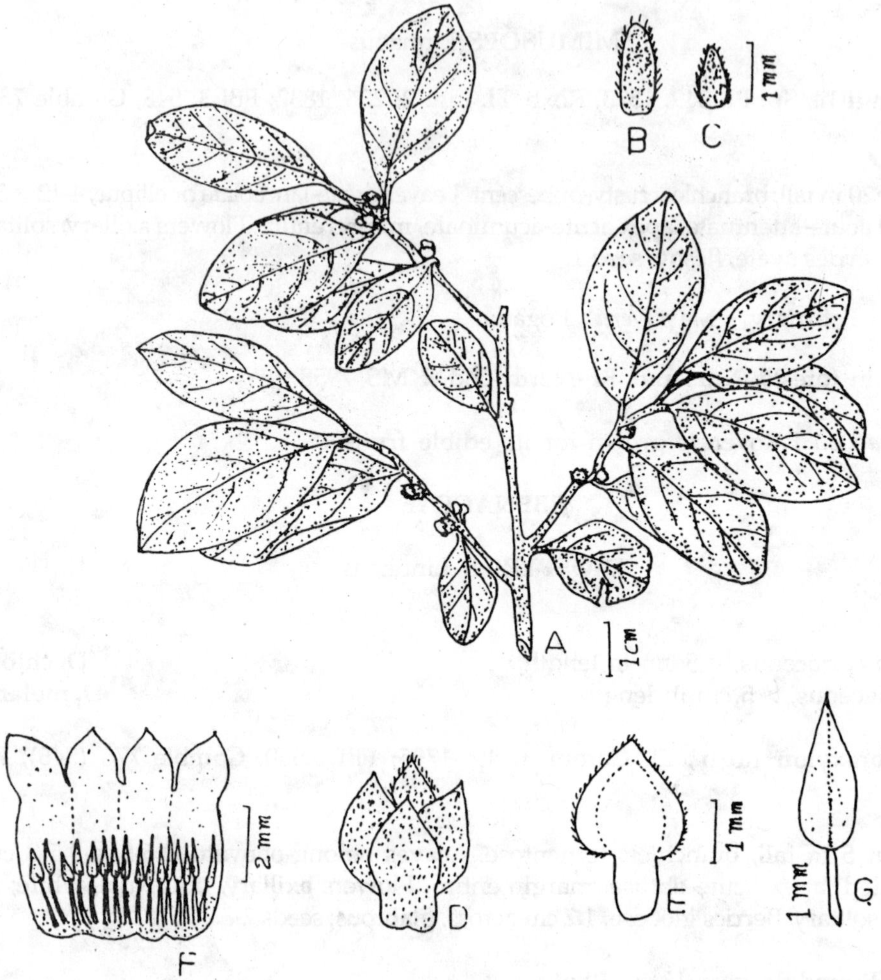

Fig. 6.19: *Diospyros chloroxylon* Roxb. A. Twig, B. Bract, C. Bracteole, D. Calyx, E. Calyx lobe, F. Corolla with stamens, G. Stamens

Fls. & Frs. : March - July. *Vern.* : Beediaku, Tumki.

Common in dry deciduous forests. Narsapur RF, **BR & CP** 1412.

OLEACEAE

1. Climbing shrubs; corolla-tube cream-white **Jasminum**
1. Trees; corolla-tube orange **Nyctanthes**

JASMINUM Linnaeus

1. Leaves simple; calyx-lobes triangular **J. roxburghianum**
1. Leaves 3-foliolate; calyx-lobes ovate **J. auriculatum**

Jasminum auriculatum Vahl, Symb. Bot. 3:1. 1794; FBI 3:600; Gamble 795 (555).

Climbing shrubs; branchlets pubescent. Leaves 3- foliolate, lateral leaflets reduced; central leaflet ovate-orbicular, 1.5-4 × 1-2.5 cm, sparsely pubescent, base truncaate, apex acute-obtuse, margin entire. Flowers in lax corymbose cymes, white. Berries globose, 0.5 cm acros, glabrous, black when ripe; seed 1.

Fls. & Frs.: July - January. *Vern.* : Adavimalle.

Occasional in forests. Nagulabanda, **CP** 11332; Gangapur RF, **TP & CP** 12099.

Jasminum roxburghianum Wall. ex Clarke in FBI 3:595. 1882; Gamble 780 (555).

Climbing shrubs. Leaves ovate-subcordate, 3-5 × 1-2.5 cm, softly pubescent, base rounded, apex acute-shortly acuminate. Flowers axillary or terminal, in lax trichotomous cymes, white. Berries globose, 0.7 cm across; seed 1.

Fls. & Frs. : March - July.

Occasional in deciduous forests. Kondapur, **GNR** 4353.

NYCTANTHES Linnaeus

Nyctanthes arbor-tristis L., Sp. Pl. 6. 1753; FBI 3:603; Gamble 792 (556); Khan in FFH 222.

Trees, about 5 m tall; branchlets quadrangular, hairy. Leaves ovate-obovate, 3.5-9.5 × 1.5-6 cm, scabrous above, pubescent below, base truncate-cuneate, apex acuminate, margin entire. Flowers axillary or terminal, in fascicles, corolla tubes orange, lobes white. Capsules obovate, 1 × 0.2 cm, compressed, glabrous; seeds 2.

Fls. & Frs. : August - December. *Vern.*: Parijatham.

Occasional in villages and forests. Narsapur RF, **TP & CP** 11973.

Jasminum azoricum L., *J. grandiflorum* L. and *J. sambac* (L.) W.T. Aiton are grown for their fragrant flowers.

APOCYNACEAE

1. Climbing shrubs **Ichnocarpus**
1. Erect herbs, shrubs or trees:
   2. Trees:
      3. Corolla-throat with coronary scales; stamens exserted **Wrightia**
      3. Corolla-throat without coronary scales; stamens inserted **Holarrhena**
   2. Herbs or shrubs:
      4. Herbs; leaves decussate; corolla salver-form **Catharanthus**
      4. Shrubs; leaves alternate; corolla funnel-shaped **Cascabela**

CASCABELA Rafinesque

Cascabela thevitia (L.) Lippold in Feddes Repert. 91:52. 1980. *Cerbera thevetia* L., Sp.Pl. 209. 1753. *Thevetia peruviana* (Pers.) Merr. in Phillip. Journ. Sci. 9:130. 1914. *Cerbera peruviana* Pers., Syn. Pl. 1:267. 1809. *Thevetia neriifolia* Juss. ex Steud., Nom. Bot. (ed.2.) 2:680. 1841 (*nereifolia*); Gamble 821 (577); Khan in FFH 229.

Shrub, about 5 m tall; branchlets glabrous. Leaves linear-lanceolate, 4-11 × 0.2-0.8 cm, glabrous, base cuneate, apex acute, margin entire. Flowers axillary, in cymes, yellow. Drupes ellipsoid-oblong, 3 × 2.5 cm, slightly compressed laterally, glabrous.

Fls. & Frs. : Round the year. *Vern.* : Pachhaganneru.

Occasional in villages, often as a hedge plant. Medak **CP** 11743.

CATHARANTHUS G.Don

1. Flowers white; follicles glabrous **C. pusillus**
1. Flowers deep pink or white; follicles pubescent **C. roseus**

Catharanthus pusillus (Murr.) G. Don, Gen Hist. 4:95. 1837. *Vinca pusilla* Murr. in Novi Comment. Soc. Regiae. Sci. Gott. 3:66. t. 2.f. 1. 1773; FBI 3: 640. *Lochnera pusilla* (Murr.) K. Schum. in Engl. Nat. Pflanzenfam. 4(2): 145. 1895; Gamble 809 (568).

Erect herbs, about 25 cm tall; branchlets glabrous. Leaves lanceolate, 2-6 × 1.5 - 2 cm, glabrous, base cuneate, apex acuminate, margin entire. Flowers axillary, solitary or in pairs, white. Follicles linear 2 x 0.2 cm, glabrous; seeds many.

Fls. & Frs. : July - November. *Vern.* : Errimirapa.

Common in cultivated fields. Sangareddy tank, **BR & CP** 11294; Siddipet, **CP** 11374, Venkatapuram, **TP & CP** 14010.

Catharanthus roseus (L.) G. Don, Gen. Hist. 4:95. 1837. *Vinca rosea* L., Syst. Nat. ed. 10. 944. 1759; FBI 3: 640. *Lochnera rosea* (L.) Reichb., Consp. Regni Veg. 134. 1828; Gamble 809 (569).

Erect herbs, about 90 cm tall; branchlets glabrous. Leaves elliptic-obovate, 4-8 × 2-4 cm, base acute, apex mucronate, margin entire. Flowers axillary, solitary or in pairs; deep pink or white. Follicles linear, 4 × 0.2 cm, pubescent; seeds many.

Fls. & Frs. : Round the year. *Vern.* : Billaganneru.

Common in villages and waste lands. Mombojipally, **BR & CP** 11640.

HOLARRHENA Robert Brown

Holarrhena pubescens (Buch.-Ham) Wall ex G. Don, Gen. Hist. 4: 78. 1837. *Echites pubescens* Buch. - Ham. in Trans. Linn. Soc. London 13: 521. 1821. *Holarrhena antidysentarica* (Roxb. ex Flem.) Wall. ex DC., Prodr. 8: 413. 1844; FBI 3: 644; Gamble 811 (570); Khan in FFH 230.

Trees, about 5 m tall; branchlets pubescent. Leaves elliptic-oblong, 5-12 × 2-6 cm, base rounded, apex acuminate, margin entire. Flowers axillary and terminal, in corymbose cymes, white. Follicles in pairs, 12 × 0.5 cm, glabrous; seeds many.

Fls. & Frs. : March - October. *Vern.* : Istari-pala.

Occasional in forests. Narsapur RF, **CP** 11576.

ICHNOCARPUS Robert Brown *nom. cons.*

Ichnocarpus frutescens (L.) R.Br. in W.T. Aiton, Hortus Kew. ed. 2. 2: 69. 1811; FBI 3: 669; Gamble 820 (577); Khan in FFH 235. *Apocynum frutescens* L., Sp. Pl. 213. 1753.

Climbing shrubs; branchlets tomentose. Leaves oblong, 2.5-5 × 1-2.5 cm, glabrous, base cuneate, apex acute-acuminate, margin entire. Flowers axillary and terminal, in dichasial cymes, greenish-white. Follicles in pairs, linear, 1.2 × 0.1 cm, pubescent; seeds many.

Fls. & Frs. : September - April. *Vern.* : Nallateega.

Common in forests. Gangapur RF, **BR & CP** 11436.

WRIGHTIA Robert Brown

1. Flowers white; follicles connate at apex **W. tinctoria**
1. Flowers pale yellow; follicles connate throughout **W. arborea**

Wrightia arborea (Dennst.) Mabb. in Taxon 26: 533. 1977. *Periploca arborea* Dennst. Schlussel Hortus Malab. 13. 23 & 25. 1818. *Wrightia tomentosa* Roemer & Schultes in L., Syst. Veg. 4: 414. 1819; FBI 3:653; Gamble 816 (573); Khan in FFH 233.

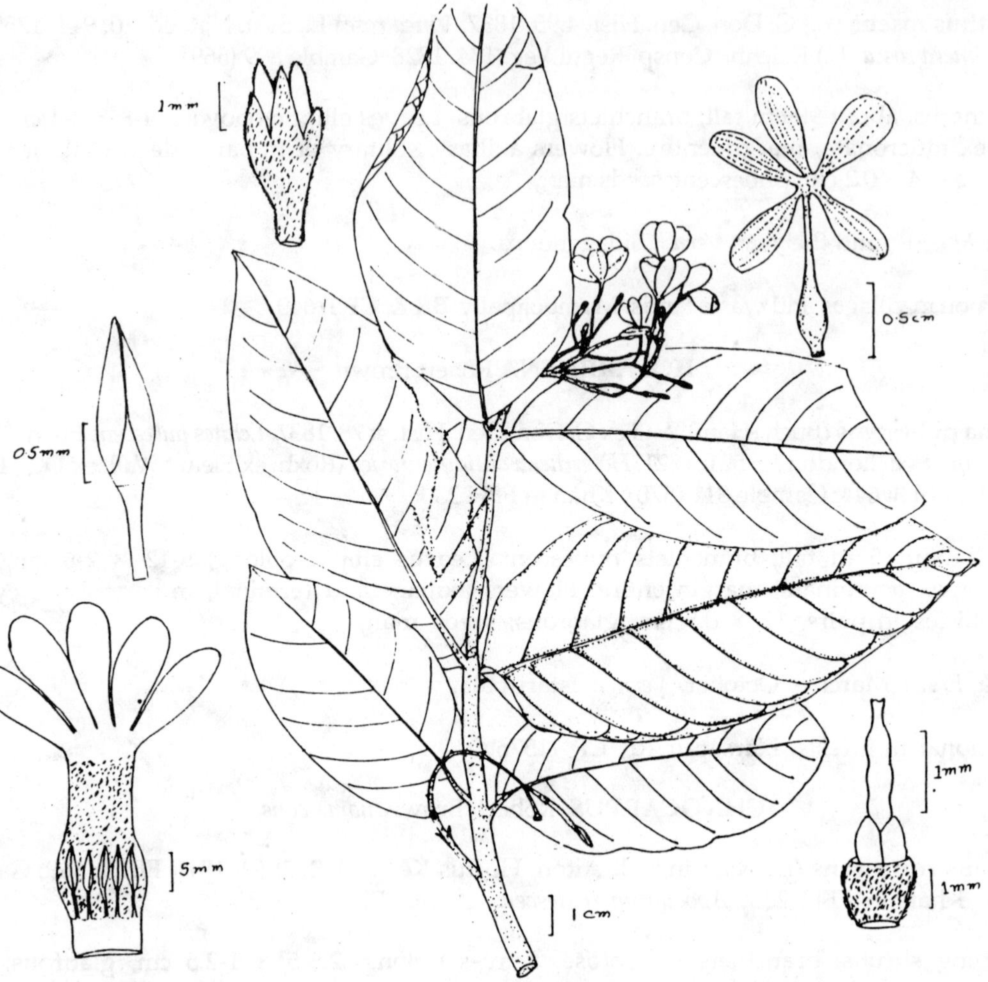

Fig. 6.20: *Holarrhena pubescens* (Ham.) G. Don. A. Twig, B. Corolla, C. Calyx, D. Corolla split open, E. Stamens, F. Pistil

Trees, about 12 m tall; branchlets tomentose. Leaves elliptic-oblong, 5-13.5 × 3-5 cm, thinly pubescent above, tomentose below, base acute, apex acuminate, margin entire. Flowers axillary and terminal, in corymbose cymes, pale yellow. Follicles in pairs, connate throughout, cylindrical, 15 × 10 cm; seeds many.

Fls. & Frs. : April - December.

Occasional in forests. Narsapur RF-North end, **KMS** 7953.

Wrightia tinctoria (Roxb.) R.Br. in Mem. Wern. Nat. Hist. Soc. 1: 47. 1811, var. **tinctoria;** FBI 3: 653; Gamble 815 (573); Khan in FFH 232. *Nerium tinctorium* Roxb. in Orient. Repert. 1:39. 1791.

Trees, about 12 m tall; branchlets pubescent. Leaves elliptic-lanceolate, 5-15 × 2.5 - 8 cm, glabrous above, pubescent below, base obtuse, apex acute-acuminate, margin entire. Flowers terminal, in cymes, white. Follicles in pairs, 20 × 0.5 cm, glabrous, curved, apically jointed; seeds many.

Fls. & Frs. : March - November. *Vern.* : Palakodisha.

Common in forests. Marrikunta, **CP** 11381; Narsapur RF, **TP & CP** 11988.

Plumeria rubra L., *Tabernaemontana divaricata* (L.) R.Br. are planted as ornamentals.

ASCLEPIADACEAE

Note : A.A. Bullock (1959) has separated the genera *Hemidesmus* R. Br., *Cryptolepis* R. Br. and *Cryptostegia* R. Br. from Asclepiadaceae and treated them under Periplocaceae based on stamen morphology. However, in the present study the above three genera are retained under Asclepiadaceae.

1. Climbing shrubs or herbs:
   2. Leaves cordate at base:
      3. Stem hispid; follicles softly echinate — **Pergularia**
      3. Stem glabrous; follicles smooth — **Holostemma**
   2. Leaves not cordate at base:
      4. Inflorescence other than umbel:
         5. Branchlets tomentose or pubescent — **Hemidesmus**
         5. Branchlets glabrous:
            6. Flowers in axillary panicles, greenish-yellow — **Cryptolepis**
            6. Flowers in terminal cymes, purple — **Cryptostegia**
      4. Inflorescence an umbel:
         7. Calyx glandular; corona single, corolline — **Gymnema**
         7. Calyx eglandular; corona double:
            8. Corolla rotate — **Tylophora**
            8. Corolla companulate — **Leptadenia**
1. Erect herbs or shrubs:
   9. Leafy shrubs — **Calotropis**
   9. Leafless herbs or shrubs:
      10. Branchlets pendulous, twining; flowers pale green — **Sarcostemma**
      10. Branchlets angular; flowers purple — **Caralluma**

CALOTROPIS Robert Brown

1. Corolla lobes spreading, dull-white, coronal process hairy — **C. gigantea**
1. Corolla lobes erect, pink-purple; coronal process not hairy — **C. procera**

Calotropis gigantea (L.) R. Br. in W.T. Aiton, Hortus Kew. ed. 2.2:78. 1871; FBI 4:17; Gamble 832(585); Khan in FFH 237. *Asclepias gigantea* L., Sp. Pl. 214. 1753.

Shrubs, about 1.5 m tall; branchlets white. Leaves ovate-obovate, 4-10 × 2-5 cm, pubescent, base amplexicaul, apex obtuse-acuminate, margin entire. Flowers axillary and terminal, in umbellate cymes, dullwhite. Follicles oblong, 6 × 3 cm, pubescent; seeds many.

Fls. & Frs. : Round the year. *Vern.* : Jilledu.

Common along road sides and waste lands. Narsingi, **BR & CP** 11522.

Calotropis procera (W. Aiton) R. Br. in W. Aiton, Hortus Kew ed. 2.2:78 1811; FBI 4:18; Gamble 832 (585); Khan in FFH 238. *Asclepias procera* W. Aiton, Hortus Kew. ed. 1.1 : 305. 1789.

Shrubs, about 2 m tall; branchlets white, woody. Leaves ovate-obovate, 2.5-7 × 1.5-4.5 cm, pubescent, base amplexicaul, apex acute, margin entire. Flowers axillary and terminal, in umbellate cymes, pink. Follicles oblong, 6 × 2.5 cm, sparsely pubescent; seeds many.

Fls. & Frs. : Round the year.

Common in waste lands. Medak, **TP & CP** 14080.

CARALLUMA Robert Brown

Caralluma adscendens (Roxb.) Haw., Syn. Pl. Succ. 47. 1812. var. **attenuata** (Wight) Grav. & Mayuranathan in Bull. Madras Govt. Museum 4:13. 1931. *C. attenuata* Wight, Ic. t. 1268. 1848; FBI 4:76; Gamble 861 (605).

Erect herbs, about 50 cm tall; branchlets 4-angled, fleshy. Leaves o; leaf scars spine-like. Flowers solitary or in pairs, dark purple. Follicles cylindric, 1 × 0.5 cm, apically beaked, glabrous; seeds many.

Fls. & Frs. : July - November.

Rare in waste lands. Medak, **CP** 11795; Pocharam tank, **TP & CP** 12048.

CRYPTOLEPIS Robert Brown

Cryptolepis buchanani Roemer & Schultes, in L. Syst. Veg. 4:409. 1819; FBI 4:5; Gamble 826 (580); Khan in FFH 236.

Climbing shrubs; branchlets glabrous. Leaves elliptic- oblong, 8-9 × 3.5-4.5 cm, glabrous, base acute, apex obtusely acuminate-retuse, margin entire. Flowers axillary, in paniculate cymes, greenish-yellow. Follicles in pairs, lanceolate, 7.5 × 1.5 cm, terete, glabrous; seeds many.

Fls. & Frs. : Round the year. *Vern.* : Palateega.

Common in forests. Pathur RF, **TP & CP** 12114.

CRYPTOSTEGIA Robert Brown

Cryptostegia grandiflora R.Br., Bot. Reg. t. 435. 1820; FBI 4:6 Gamble 862 (606); Khan in FFH 237.

Climbing shrubs; branchlets glabrous. Leaves elliptic- obovate, 5-10 × 3-5 cm, coriaceous, base cuneate, apex subacute, margin entire. Flowers terminal, in cymes, purple. Follicies in pairs, ovate-oblong, 12 × 2 cm, angled; seeds many.

Fls. & *Frs.* : Round the year.

Occasional in waste lands. Doulatabad, **GNR** 4788.

GYMNEMA Robert Brown

Gymnema sylvestre (Retz.) R.Br. ex Roemer & Schultes in L. Syst. Veg. 6:57. 1820; FBI 4: 29; Gamble 839 (590). *Periploca sylvestris* Retz., Obs. Bot. 2:15. 1781.

Climbing shrubs; branchlets tomentose. Leaves ovate-oblong 1-5 × 0.5-2.5 cm, pubescent, base rounded, apex acute-acuminate, margin entire. Flowers axillary, in umbellate cymes, pale yellow. Follicles lanceolate, 7 × 0.5 cm, glabrous, one follicle often supressed; seeds many.

Fls. & *Frs.* June - October.

Common in forests. Marrikunta, CP 11390; Sardhana, **BR & CP** 11619; Narsapur RF, **TP & CP** 11989.

HEMIDESMUS Robert Brown

1. Branchlets and leaves glabrous **H. indicus** var. **indicus**
1. Branchlets and leaves pubescent **H. indicus** var. **pubescens**

Hemidesmus indicus (L.) R.Br. in W.T. Aiton. Hortus Kew. ed. 2. 2: 71. 1811; FBI 4: 5; Gamble 825(580); Khan in FFH 239. var. **indicus.** *Periploca indica* L., Sp. Pl. 211. 1753.

Prostrate or twining shrubs; branchlets striate. Leaves elliptic, 4-5 × 1.5-2 cm, glabrous, base rounded-subcordate, apex mucronate, margin entire. Flowers axillary, in cymose clusters, purple. Follicles in pairs. cylindric, 8 × 0.5 cm, striate, glabrous; seeds many.

Fls. & *Frs.* : July - December. *Vern.* Sugandipala.

Common in forests, often on hedges. Sivanoor, **BR & CP** 11547; Pochammaralu, **CP** 11764.

Hemidesmus indicus (L.) R.Br. var. **pubescens** (Wight & Arn.) Hook.f. in FBI 4:5. 1883; Gamble 825(580). *H. pubescens* Wight & Arn. in Wight Contrib. Bot. 63. 1834.

Branchlets pubescent. Leaves elliptic-ovate, 2-4.5 × 1.8-4 cm, softly pubescent above, velvety tomentose below, base broadly acute, apex apiculate. All other characters similar to above variety.

Fls. & Frs. : July - December.

Common in forests and on hedges. Gangapur RF, **TP & CP** 12101; Pathur, **TP & CP** 12105.

HOLOSTEMMA Robert Brown

Holostemma ada-kodien Schultes in L., Syst. Veg. 6:95. 1820. *Sarcostemma annulare* Roth, Nov. Pl. Spec. 178. 1821. *Holostemma annulare* (Roxb.) K. Schum. in Engl. Nat. Pflanzenfam 4.2:250. 1895; Gamble 834 (586). *H. rheedii* Wall, Pl. As. Rar. 2:51. 1831; FBI 4:21; Khan in FFH 239.

Twining shrubs; branchlets glabrous. Leaves ovate, 4-6 × 3-3.5 cm, glabrous, base cordate, apex acuminate, margin entire. Flowers sub-lateral, in lax cymes, purple. Follicles linear-oblong, 10 × 0.5 cm, glabrous; seeds many.

Fls. & Frs. July - November.

Occasional in hedges. Muttayp[illegible]ly, **CP** 11781.

LEPTADENIA Robert Brown

Leptadenia reticulata (Retz.) Wight & Arn. in Wight Contrib. Bot. 47. 1834; FBI 4: 63; Gamble 850 (598). *Cynanchum reticulatum* Retz, Obst. Bot. 2: 15. 1781.

Climbing shrubs. Leaves ovate, 4-8 x 2-5.5 cm, pubescent, base obtuse, apex acute, margin entire. Flowers axillary and terminal, in umbellate cymes, yellowish-green. Follicles in pairs, cylindric; seeds oblong.

Fls. & Frs. : May - December.

Occasional in waste lands and forests. Narsapur, **GNR** 4360.

PERGULARIA Linnaeus

Pergularia daemia (Forssk.) Chiov., Result. Sci. Miss. Stefanpaoli. Somal. Hal. 1:15. 1916. *Asclepias daemia* Forssk., Fl. Aegypt.-Arab. 51 1775. *Cynanchum extensum* Jacq., Misc. Austr. 2:353. 1782. *Daemia extensa* (Jacq.) R.Br. in W.T. Aiton, Hortus Kew. ed. 2.2:76. 1811; FBI 4: 20. *Pergularia extensa* (Jacq.) N.E.Br. in Harvey & Sonder, Fl. Cap. 7: 758. 1908; Gamble 836 (588).

Twining shrubs; branchlets hispid. Leaves ovate, 1-4 × 1.5-3 cm, pubescent, base cordate, apex acuminate, margin entire. Flowers axillary, in umbellate cymes, green. Follicles in pairs, lanceolate, 6 × 2 cm, softly echinate; seeds many.

Fls. & Frs. : Round the year.

Common on hedges and waste lands. Marrikunta, **CP** 11383; Mombojipally, **BR & CP** 11643; Siddipet-Medak Road, **CP** 11709.

SARCOSTEMMA Robert Brown *emend*. R.W. Holm

Sarcostemma acidum (Roxb.) Voigt. Hort. Sub. Calc. 542. 1845. *Asclepias acida* Roxb. Fl. Ind. 2: 31. 1832. *Sarcostemma brevistigma* Wight & Arn., in Wight Contrib. Bot. 59. 1834; FBI 4: 26; Gamble 838 (589).

Shrubs; branchlets pendulous, twining, fleshy. Leaves 0. Flowers terminal, in umbellate cymes, pale green. Follicles lanceolate, in pairs; seeds many.

Fls. & Frs.: June-December. *Vern.*: Kondamalle, Pandiri-jamudu.

Rare in forests. Pochammaralu, **CP** 11754.

TYLOPHORA Robert Brown

1. Stems erect, more or less climbing in the branchlets; coronal lobes ovate **T. macrantha**
1. Stems always climbing; coronal lobes gibbous **T. indica**

Tylophora indica (Burm.f.) Merr. in Phillipp. Journ. Sci. 19:373. 1921. *Cynanchum indicum* Burm.f. Fl. Ind. 70. 1768. *Tylophora asthmatica* (L.f.) Wight & Arn. in Wight Contrib. Bot. 51. 1834; FBI 4: 44; Gamble 843 (593).

Twining herbs; branchlets densely pubescent. Leaves elliptic-ovate, 2-8 × 1-3.5 cm, pilose, base rounded- cordate, apex acute-apiculate, margin entire. Flowers lateral, in umbellate cymes, greenish-yellow. Follicles in pairs, lanceolate, 7 × 1.2 cm, glabrous, seeds many.

Fls. & Frs. : Round the year. *Vern.* : Verripala.

Common on hedges. Nagulabanda, **CP** 11348; Narsingi, **BR & CP** 11537; Pocharam tank, **TP & CP** 12021.

Tylophora macrantha Hook.f. in Fl. Brit. India. 4:40. 1883; Gamble 842 (592).

Erect herbs; branchlets twining. Leaves oblong- lanceolate, 6-8 × 2-3 cm, glabrous, base cordate, apex acuminate, margin entire. Flowers axillary, in umbellate cymes, greenish-red. Follicles in pairs, oblong, acuminate; seeds many.

Fls. & Frs. : July - December.

Rare in hedges. Narsapur, **GNR** 4022.

LOGANIACEAE

STRYCHNOS Linnaeus

Strychnos nux-vomica L., Sp. Pl. 189. 1753; FBI 4:90; Gamble 868 (610); Khan in FFH 241.

Trees, about 15 m tall; branchlets glabrous. Leaves orbicular-ovate, 5-11.5 × 3-6.5 cm, coriaceous, base attenuate, apex acute-apiculate, margin entire. Flowers axillary and terminal, in cymes, green. Berries globose, 1.5 cm across, bright orange when ripe; seeds 1-4.

Fls. & Frs. : March - August. *Vern.* : Mushti.

Occasional in forests. Gangapur RF, **BR & CP** 11425.

GENTIANACEAE

1. Flowers tetramerous:
   2. Stamens equal; flowers blue — **Exacum**
   2. Stamens unequal; flowers other than blue:
      3. Flowers greenish-yellow, sessile — **Hoppea**
      3. Flowers pink-rose, pedicellate — **Canscora**
1. Flowers pentamerous:
   4. Flowers in dichasial cymes — **Centaurium**
   4. Flowers in axillary clusters — **Enicostema**

CANSCORA Lamarck

1. Stem 4-winged — **C. decurrens**
1. Stem not winged — **C. diffusa**

Canscora decurrens Dalz. in Hook., Journ. Bot. Kew. Gard. Misc. 2:136. 1850; FBI 4:103; Gamble 878 (617).

Erect herbs, about 30 cm tall; Stems winged, glabrous. Leaves elliptic, 2-3.5 × 0.5-1.9 cm, membranous, base attenuate, apex acute, margin entire. Flowers terminal, in paniculate cymes, pink. Capsules oblong, 1 × 0.2 cm, 2-valved; seeds many.

Fls. & Frs. : Round the year.

Occasional in moist places. Pochammaralu, **BR & CP** 11488; Gangapur RF, **TP & CP** 12084.

Canscora diffusa (Vahl) R.Br., Prodr. 451. 1810; FBI 4:103; Gamble 878 (617). *Gentiana diffusa* Vahl, Symb. Bot. 3:47. 1794. *Exacum diffusum* Willd., Sp. Pl. 1:637; 1798.

Erect herbs, about 30 cm tall; stem narrowly 4-angled, Leaves obovate, 1-2.5 × 0.5-1 cm, base rounded, apex acute, margin entire. Flowers axillary and terminal, in lax panicles, rose. Capsules oblong, 0.6 × 0.2 cm; seeds many.

Fls. & Frs. : January - April.

Occasional along stream banks. Laxmapur, **BR & CP** 11448.

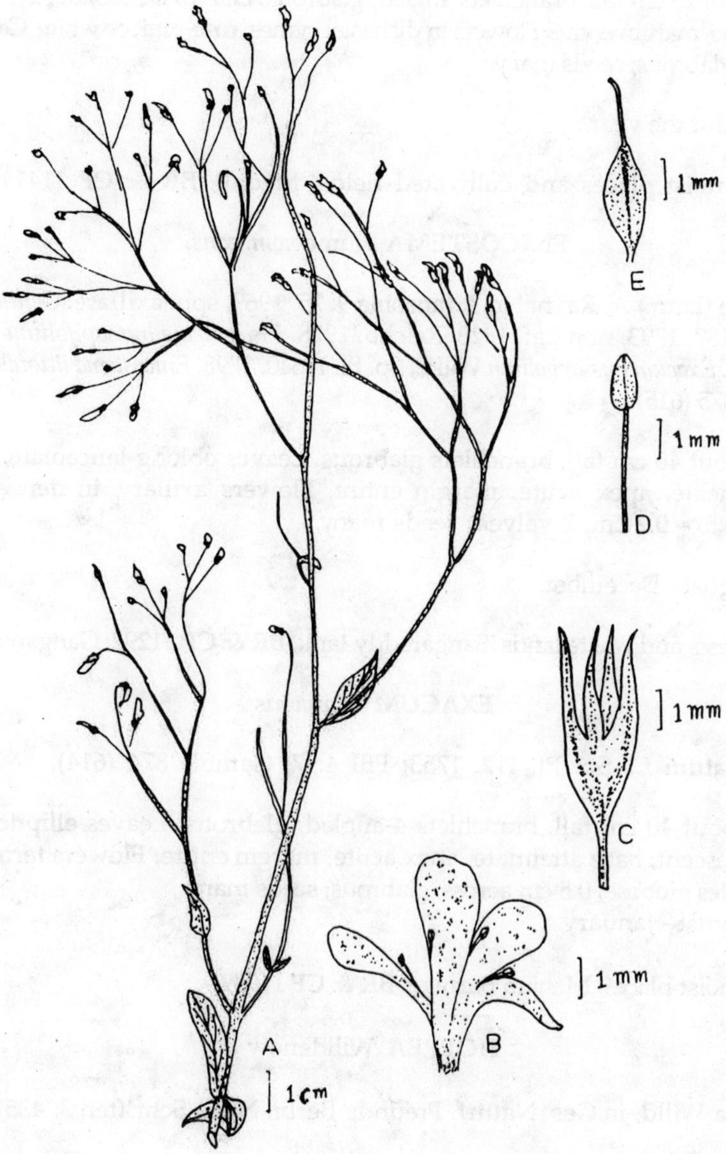

Fig. 6.21: *Canscora decurrens Dalz.* A. Twig, B. Flower, C. Calyx, D. Stamens, E. Pistil

CENTAURIUM J. Hill

Centaurium roxburghii (G.Don) Druce in Bot. Exch. Club Soc. Brit. Isles 4:614. 1916. *Erythraea roxburghii* G. Don, Gen. Hist. 4:206. 1837 - 1838; FBI 4:102; Gamble 875 (615).

Erect herbs, about 25 cm tall; branchlets angled, glabrous. Leaves lanceolate, 2-4 × 0.2-0.5 cm, base subcordate apex acute, margin entire. Flowers in dichasial cymes; rose-pink or white. Capsules oblong, 1.5 × 0.3 cm, 2-valved, glabrous; seeds many.

Fls. & Frs. : Round the year.

Occasional in moist places and cultivated fields. Medak, **BR & CP** 11419.

ENICOSTEMA Blume *nom. cons.*

Enicostema axillare (Lam.) A. Raynal in Adansonia 9: 75. 1969. spp. **axillare.** *Gentiana axillaris* Lam., Tabl. Encycl. 1 (2): 487. 1793, non Raf. 1828. Reichb. 1928. *Enicostema hyssopifolium* (Willd.) Verdc. in Bothalia 7: 462. 1961. *Exacum hyssopifolium* Willd., Sp. Pl. 1: 640. 1798. *Enicostema littorale* auct. non Blume; FBI 4: 101; Gamble 875 (615).

Erect herbs, about 40 cm tall; branchlets glabrous. Leaves oblong-lanceolate, 2-6 × 0.3-0.8 cm, glabrous, base cuneate, apex acute, margin entire. Flowers axillary, in dense clusters, white. Capsules oblong, 0.5 × 0.2 cm, 2-valved; seeds many.

Fls. & Frs. : August - December.

Common in forests and waste lands. Sangareddy tank, **BR & CP** 11292; Gangapur CP 11792.

EXACUM Linnaeus

Exacum pedunculatum L., Sp. Pl. 112. 1753; FBI 4:97; Gamble 874 (614).

Erect herbs, about 40 cm tall; branchlets 4-angled, glabrous. Leaves elliptic-lanceolate, 1-3.5 × 0.4-0.8 cm, glabrescent, base attenuate, apex acute, margin entire. Flowers terminal, in dichasial cymes, blue. Capsules globose, 0.5 cm across, glabrous; seeds many.
Fls. & Frs. : August - January.

Occasional in moist places. Manjira Barrage, **BR & CP** 14089.

HOPPEA Willdenow

Hoppea dichotoma Willd. in Ges. Naturf. Freunde Berlin Neue Schriften 3: 435. 1801; FBI 4: 100; Gamble 877 (616).

Erect herbs, about 10 cm tall; branches 4-angled. Leaves ovate, 0.2-0.6 × 0.1-0.4 cm, glabrous, base cuneate, apex apiculate, margin entire. Flowers in dichotomous cymes, greenish- yellow. Capsules globose, 0.4 cm across, glabrous; seeds many.

Fls. & Frs. : Round the year.

Common in waste lands. Pochammaralu, **BR & CP** 11480.

MENYANTHACEAE

NYMPHOIDES Seguier

Note : Although most of the Indian floristic workers treated this taxon under the name *Limnanthemum* Gmel. (1770) *Nymphoides* Seguier (Pl. Vernon. 3:121. 1754) or *Nymphoides* Hill. (Brit. Herb. 77. 1756) have priority over it. Since Seguiers publication is earlier than Hill (Ann. Miss. Bot. Gard. 56 : 29. 1969) the name *Nymphoides* Seguier is retained. *Nymphoides* Seguier is separated from the family Gentianaceae and included under the family Menyanthaceae based on aquatic habit, alternate phyllotaxy, valvate aestivation (Hutchinson, 1959; Cronquist, 1968).

Nymphoides hydrophylla (Lour.) O. Kuntze, Rev. Gen. Pl. 2: 429. 1891. *Menyanthes hydrophlla* Lour., Fl. Cochinch. 1: 129. 1790. *Nympyoides cristatum* (Roxb.) O. Kuntze, Rev. Gen. Pl. 2: 429, 1891. *Menyanthes cristata* Roxb., Pl. Corom. t. 105. 1799. *Limnanthemum cristatum* (Roxb,) Griseb. Gen. Sp. Gent. 342. 1839; FBI 4: 131; Gamble 883 (620).

Aquatic herbs; rhizome short, branches petiole-like. Leaves orbicular, 4-8.5 × 4-11 cm, scaberulous, base deeply cordate, apex obtuse, margin entire. Flowers at nodes, in dense clusters, white. Capsules ovate, 1 × 0.5 cm, glabrous; seeds many.

Fls. & Frs. : July - February. *Vern.* : Kukkatamara.

Common in ponds and tanks. Patancheru tank, **BR & CP** 11321; Narsapur tank, **TP & CP** 12007.

BORAGINACEAE

1. Flowers solitary or in cymes:
   2. Prostrate herbs; flowers cream **Coldenia**
   2. Erect herbs: flowers blue-pale violet **Trichodesma**
1. Flowers in spikes or racemes **Heliotropium**

COLDENIA Linnaeus

Coldenia procumbens L., Sp. Pl. 125. 1753; FBI 4:144; Gamble 892 (627).

Prostrate herbs; branchlets densely pilose. Leaves elliptic-ovate, 1-2.5 × 0.4-1 cm, pilose base attenuate-decurrent, apex rounded, margin coarsely serrate-crenate. Flowers axillary, solitary, cream. Drupe 4-lobed, 0.3 cm across, pubescent; seed 1 in each lobe.

Fls. & Frs. : August - March.

Common weed of waste lands and cultivated fields. Ramayampet, **BR & CP** 11518.

HELIOTROPIUM Linnaeus

1. Anthers connivent by their tips:
   2. Bracts inconspicuous **H. scabrum**
   2. Bracts conspicuous:
      3. Branchlets softly sericeous;bracts lanceolate **H. bracteatum**
      3. Branchlets glabrescent; bracts linear **H. zeylanicum**
1. Anthers free by their tips:
   4. Leaves > 3 cm in length **H. indicum**
   4. Leaves < 3 cm in length:
      5. Leaves petiolate, spathulate-obovate **H. ovalifolium**
      5. Leaves sessile, oblong-lanceolate **H. strigosum**

Heliotropium bracteatum R.Br. Prodr. 493. 1810; FBI 4:151; Gamble 897 (630).

Herbs, about 50 tall; branchlets softly sericeous. Leaves linear-lanceolate, 1-2.5 × 0.2-0.4 cm, chartaceous, base cuneate, apex acute, margin entire. Flowers terminal, in racemes, white. Drupes of 4 nutlets; nutlets smooth.

Fls. & Frs. : March - August.

Common in dry deciduous forests. Narsapur RF, **TP & CP** 14019; Kulcharam RF, **TP & CP** 14078.

Heliotropium indicum L., Sp. Pl. 130 1753; FBI 4:152; Gamble 896 (630).

Erect herbs, about 75 cm tall; branchlets hirsute. Leaves ovate, 3-12.5 × 1.5-5.5 cm, sparsely hairy along nerves, base cordate-rounded, apex acute, margin undulate. Flowers extra-axillary or terminal, in hispid spikes, white or pale-violet. Drupes conical, 0.3 cm across, deeply 2-lobed, glabrous; nutlets 4.

Fls. & Frs. : Round the year. *Vern.* : Nagadanthi.

Common in waste lands. Narsapur, **TP & CP** 12000.

Heliotropium ovalifolium Forssk., Fl. Aegypt. - Arab. 38. 1775; FBI 4:150; Gamble 896 (629).

Erect or ascending herbs, about 20 cm tall; branchlets with silkyhairs. Leaves obovate-spathulate, 1-2.8 × 0.5-1.5 cm, silky-pubescent, base attenuate, apex obtuse, margin entire. Flowers axillary and terminal, in simple or paired spikes, white. Drupes conical, 0.1 cm across, silky; nutlets 4.

Fls. & Frs. : Round the year.

Common in waste lands and cultivated fields. Rampur, **BR & CP** 11651.

Heliotropium scabrum Retz. Obs. Bot. 2:8. 1781; FBI 4: 152. Gamble 897 (630). *H. marifolium* Retz. Obs. Bot. 2:8. 1781; FBI 4: 152. p.p.

Prostrate herbs, branchlets strigose. Leaves lanceolate-oblong, 1-1.5 × 0.3-0.4 cm, scabrous, base cuneate, apex acute, margin entire. Flowers terminal in racemes, white. Nutlets shortly bristly.

Fls. & Frs. : July - December.

Occasional in sandy rocky plains. Annaram, GNR 4149.

Heliotropium strigosum Wild., Sp. Pl. 1:743. 1798; FBI 4:151. ssp. **brevifolium** (Wall.) Kazmi in Journ. Arnold Arbor. 51:153. 1970. *H. brevifolium* Wall. in Roxb., Fl. Ind. 2:2 1824; Gamble 897 (630). *H. strigosum* Willd. var. *brevifolium* (Wall.) C.B.Clarke in FBI 4: 151. 1883.

Ascending herbs; about 40 cm tall; branchlets strigose- hispid. Leaves oblong-lanceolate, 0.5-1.2 × 0.2-0.4 cm, hirsute base acute, margin reflexed. Flowers axillary and terminal, in hispid spikes, white. Drupes conical, 0.3 × 0.2 cm, hispid; nutlets 4.

Fls. & Frs. : Round the year.

Occasional in waste lands. Gnanapur, **CP** 11786; Pocharam RF, **TP & CP** 12058.

Heliotropium zeylanicum (Burm.f.) Lam. Encycl. 3:94. 1789; Gamble 896 (630). ssp. **zeylanicum.** *H. paniculatum* R.Br. Prodr. 494. 1810; FBI 4: 151. *H. curassavicum* L. var *zeylanicum* Burm. f. Fl. Ind. 41 t. 16. f. 2. 1768.

Erect herbs, about 50 cm tall; branchlets glabrescent. Leaves linear-lanceolate, 1-2.5 × 0.3-0.5 cm, chartaceous, scabrous above, appressed strigose below, base cuneate, apex acute, margin revolute. Flowers axillary and terminal, in profusely branched racemes, white. Nutlets hairy.

Fls. & Frs. : September - February.

Occasional weed along road sides. Annaram, **GNR** 4155.

TRICHODESMA Robert Brown *nom. cons.*

1. Leaves sessile; corolla throat with glandular spots:
   2. Flowers pinkish-white **T. sedgwickianum**
   2. Flowers blue **T. indicum**
1. Leaves petiolate; corolla throat without glandular spots **T. zeylanicum**

Trichodesma indicum (L.) R.Br. Prodr. 149. 1810; FBI 4:153; Gamble 899 (631). *Borago indica* L., Sp. Pl. 137. 1753.

Erect herbs, about 60 cm tall; branchlets hispid. Leaves lanceolate-oblong, 1.5-4 × 0.5-1.5 cm, hirsute-pubescent below, base cordate, apex acute, margin entire. Flowers terminal, solitary or in simple cymes, pinkish-white. Drupes pyramidal, 0.6 × 0.4 cm, smooth; nutlets 4.

Fls. & Frs. : August - March.

Common in waste lands, cultivated fields and hedges. Ponnala, **CP** 11366; Venkatapuram, **TP** & **CP** 14008.

Trichodesma sedgwickianum S.P. Banerjee in Bull. Bot. Soc. Bengal 16: 10. 1962. *T. amplexicaule* auctt. non Roth: DC., Prodr. 10: 172. 1846; FBI 4: 154; Gamble 899 (631).

Erect herbs, about 75 cm tall; branchlets hispid. Leaves elliptic, 2-9 × 0.5-4 cm, scabrous above, sparsely hispid below, base cordate, apex acute, margin entire. Flowers terminal, solitary or in simple cymes, pinkish-white. Drupes pyramidal, 0.6 × 0.4 cm, smooth; nutlets 4.

Fls. & *Frs.* : July - December.

Occasional in waste lands. Medak, *CP* 11730.

Trichodesma zeylanicum (Burm.f.) R.Br., Prodr. 496. 1810; FBI 4: 154; Gamble 899 (632). *Borago zeylanica* Burm.f. Fl. Ind. 41. 1768.

Erect herbs, about 70 cm tall; branchlets strigosely hispid. Leaves elliptic-lanceolate, 5-12.5 × 0.5-4 cm, pubescent, base truncate-subcordate, apex acute, margin entire. Flowers terminal, in drooping paniculate cymes, blue. Drupes pyramidal, 0.6 × 0.4 cm, nutlets 4.

Fls. & *Frs.* : December - February.

Occasional in waste lands. Siripuram, **GNR** 4721.

CORDIACEAE

1.	Shurbs; flowers solitary or in pairs	**Carmona**
1.	Trees; flowers in cymes	**Cordia**

CARMONA Cavanilles

Carmona retusa (Vahl) Masam. in Trans. Nat. Hist. Soc. Taiwan 30: 61. 1940. *Cordia retusa* Vahl. Symb. Bot. 2: 42. 1791. *Ehretia buxifolia* Roxb., Pl. Corom. t.. 57. 1796; FBI 4: 144. *E. microphylla* Lam., Tabl. Encycl. 1: 425. 1792; Gamble 891 (626).

Shrubs, about 1 m tall; branchlets pubescent. Leaves ovate, 1-2.5 × 0.5-1 cm, glabrous, base cuneate, apex truncate, margin entire. Flowers axillary, solitary or in pairs, white. Drupes globose. 0.4 cm across, glabrous; seeds 4.

Fls. & *Frs.* : Round the year.

Common in forests and waste lands. Pocharam, **BR** & **CP** 11612; Pochammaralu, **CP** 11756; Rampur, **TP** & **CP** 14074.

CORDIA Linnaeus

1. Leaves tomentose below; fruiting calyx ribbed **C. domestica**
1. Leaves glabrescent below; fruiting calyx pubescent **C. obliqua**

Cordia domestica Roth, Nov. Pl. Sp. 123. 1821; Gamble 888 (624). *Cordia myxa* L. var. *domestica* (Roth) C.B. Clarke in FBI 4: 137. 1883.

Trees, about 6 m tall; Leaves broadly ovate, 2-6.5 × 1 -2.5 cm, white-pubescent, base rounded, apex obtuse-acute, margin entire. Flowers terminal, in corymbose cymes. cream. Drupes ovate, fruiting calyx ribbed.

Fls. & Frs. : January - July.

Occasional in scrub jungles. Siripuram, **GNR** 4720.

Cordia obliqua Willd., Phytographia 4.t. f. 1 1794; FBI 4:137. Gamble 887 (623). *C. myxa* auct. non L: Khan in FFH 243.

Trees, about 10 m tall; branchlets glabrous. Leaves ovate-elliptic, suborbicular, 3-6.5 × 3-5 cm, glabrous, base subcordate-rounded, apex obtuse, margin entire. Flowers axillary or terminal, in paniculate cymes, white. Drupes ovate, 2 × 1.8 cm, glabrous.

Fls. & Frs. : March - July. *Vern.* : Irki.

Common in forests. Narsapur RF, **CP** 11579.

CONVOLVULACEAE

1. Pollen grains smooth:
   2. Flowers blue **Evolvulus**
   2. Flowers other than blue:
      3. Stigma lobes elongate; capsule 8-valved **Jacquemontia**
      3. Stigma biglobose; capsule 4-valved **Merremia**
1. Pollen grains spinulose:
   4. Fruit indehiscent:
      5. Flowers white; corolla salver-shaped **Rivea**
      5. Flowers red or purple; corolla funnel-shaped **Argyreia**
   4. Fruit dehiscent **Ipomoea**

ARGYREIA Loureiro

1. Leaves hairy above; bracts orbicular or broadly ovate **A. cymosa**
1. Leaves glabrous above; bracts ovate-lanceolate **A. nervosa**

Argyreia cymosa (Roxb.) Sweet, Hort. Brit. (ed. 2.) 373. 1830; FBI 4:190; Gamble 908 (637) *Lettsomia cymosa* Roxb., Fl. Ind. 2: 82. 1824.

Climbing shrubs; branchlets thinly pubescent. Leaves cordiform, 6-10 × 4-8 cm, thinly pubescent, base cordate, apex acute-acuminate, margin entire. Flowers axillary, in corymbose cymes, red. Berries globose, 1.5 cm across; seeds 4.

Fls. & *Frs*. July - November.

Common in forests. Dundigal, **GNR** 4652.

Argyreia nervosa (Burm.f.) Bojer, Hort Maurit. 224. 1837. *Convolvulus nervosus* Burm.f. Fl. Ind. 48. t. 20. f. 1. 1768. *C. speciosus* L.f., Suppl. 137. 1781. *Argyreia speciosa* (L.f.) Sweet, Hort. Brit. 1: 289. 1827; FBI 4: 185; Gamble 907 (637).

Climbing shrubs, branchlets densely pubescent. Leaves cordate-broadly ovate, 5-18 × 3-12 cm, glabrous above, silky white-pubescent below, base cordate, apex acute, margin entire. Flowers axillary, in cymes, purple. Berries sub-globose, 1 × 0.5 cm; seeds 4.

Fls. & *Frs.* : July - December.

Occasional along forests edges. Sathaihpalle, **GNR** 4518.

EVOLVULUS Linnaeus

1. Leaves elliptic-lanceolate, apex apiculate; flowers light blue **E. alsinoides**
1. Leaves suborbicular, apex obtuse; flowers cream **E. nummularius**

Evolvulus alsinoides (L.) L., Sp. Pl. ed. 2. 393. 1762; FBI 4:220; Gamble 923 (648); Khan in FFH 249. *Convolvulus alsinoides* L., Sp. Pl. 157. 1753.

Prostrate or ascending herbs; branchlets sericeous. Leaves elliptic-lanceolate or oblong, 0.4-1.5 × 0.2-0.5 cm, sericeous, base acute, apex apiculate, margin entire. Flowers solitary or in pairs, light-blue. Capsules globose, 0.5 cm across, glabrous; seeds 2-4.

Fls. & *Frs.* : Round the year. *Vern* : Vishnukrantha.

Common in all localities. Pillutla RF, **BR & CP** 11259.

Evolvulus nummularius (L.) L., Sp. Pl. ed. 2. 391. 1762. *Convolvulus nummularius* L., Sp. Pl. 157. 1753.

Prostrate herbs, branchlets hairy. Leaves suborbicular, 0.5 -2 x 0.5-1.5 cm, glabrous, base subcordate - truncate, apex obtuse, margin entire. Flowers solitary or 2-3 in clusters cream. Capsules subglobose, 0.2 x 0.3 cm; seeds 2-4.

Fls. & *Frs.* : Round the year.

Occasional near moist places. Narsapur tank, **TP & CP** 12005.

IPOMOEA Linnaeus

1. Erect shrubs **I. carnea**
1. Twining or prostrate herbs:
   2. Leaves entire:
      3. Flowers in sessile capitate clusters **I. eriocarpa**
      3. Flowers in pedunculate cymes or solitary:
         4. Aquatic plants, rooting at nodes **I. aquatica**
         4. Terrestrial plants, not rooting at nodes:
            5. Flowers scarlet-orange **I. hederifolia**
            5. Flowers not as above:
               6. Corolla white with purple throat **I sepiaria**
               6. Corolla yellowish-white **I. obscura**
   2. Leaves lobed:
      7. Leaves distinctly 3-lobed **I. nil**
      7. Leaves palmately 5-7-lobed:
         8. Stem hirsute **I. pes-tigridis**
         8. Stem glabrous **I. cairica**

Ipomoea aquatica Forssk., Fl. Aegypt-Arab. 44.1775; FBI 4:210. *I. reptans* Poir. in Lam., Encycl. (Suppl.3) : 460. 1814, non L; Gamble 910 (643).

Prostrate herbs; branchlets glabrous. Leaves lanceolate, 3-6 × 1-2.5 cm, glabrous, base sagittate, apex acuminate, margin entire. Flowers solitary or in simple cymes, pink. Capsules globose, 0.8 cm across, glabrous; seeds 3-4.

Fls. & *Frs.* : Round the year.

Common in ponds, stagnant waters, often floating on surface. Medak, **BR & CP 11464.**

Ipomoea cairica (L.) sweet, Hort. Brit. 287. 1827; Gamble 918 (645). *Convolvulus cairicus* L., Syst. Nat. ed. 10. 922. 1759. *Ipomoea palmata* Forssk., Fl. Aegypt-Arab. 43. 1775; FBI 4:214.

Twining herbs; branchlets slender, glabrous. Leaves 5-7-foliolate, lobes oblanceolate. 3-5 × 1-2 cm, glabrous, base attenuate, apex acute-mucronate, margin entire. Flowers solitary or 2-3 in cymes, dull violet. Capsules subglobose, 0.8 cm, across; seeds 4.

Fls. & *Frs.* : Round the year.

Common on the hedges. Akkannapet, **BR & CP** 11554; Patancheru, **CP** 11684.

Ipomoea carnea Jacq., Enum. Syst. Pl. 13. 1760; Gamble 919 (645). ssp. **fistulosa** (Mart. ex Choisy) D. Austin in Taxon 26:237. 1977. *I. fistulosa* Mart. ex Choisy in DC., Prodr. 9:349. 1845.

Erect shrubs, about 2.5 m tall; branchlets hollow, glabrous. Leaves ovate-hastate, 8-12 × 3-5 cm, glabrous, base cordate, apex gradually acuminate, margin entire. Flowers axillary and terminal, in umbellate cymes, pink-rose. Capsules ovate, 2 × 1.5 cm, glabrous, mucronate; seeds 2-4.

Fls. & Frs. : September - April.

Common in hedges, along the streams. Narsapur tank, **BR & CP** 11247; Gnanapur **CP** 11790.

Ipomoea eriocarpa R. Br., Prodr. 484. 1810; FBI 4:204. *Ipomoea hispida* (Vahl) Roemer & Schultes in L., Syst. Veg. 4:238. 1819. non Zucc. 1809; Gamble 915 (643). *Convolvulus hispidus* Vahl, Symb. Bot. 3:29. 1794.

Twining herbs; branchlets hirsute. Leaves oblong-ovate, 2.5-8 × 2-3 cm, glabrous, base cordate, apex gradually acute, margin entire. Flowers axillary , in clusters, pink. Capsules globose, 0.6 cm across, hairy; seeds 3-4.

Frs. & Fls. : September - March.

Occasional in waste lands. Bonthapalle, **GNR** 4476.

Ipomoea hederifolia L., Syst. Nat. ed. 10. 925. 1759. *Quamoclit phoenicea* (Roxb.) Choisy in Mem. Soc. Phys. Geneve 6: 433. 1834; Gamble 919 (645). *Ipomoea coccinea* auct. non. L: FBI 4:199; Khan in FFH 249; *I. angulata* Lam., Tabl. Encycl. 1: 464. 1793.

Twining herbs; branchlets glabrous. Leaves ovate- cordiform, 2-9.5 × 2.5-11 cm, base cordate, apex acuminate , margin entire or obscurely lobed. Flowers axillary, in cymes, scarlet orange. Capsules globose, 0.8 cm across, apiculate; seeds 4.

Fls. & Frs. : May - December. *Vern.* : Kasiratnam.

Occasional on field hedges. Narsapur forest nursery, **BR & CP** 11224; Narsingi, **BR & CP** 11525.

Ipomoea nil (L.) Roth, Catal. Bot. 1:36. 1797. *Convolvulus nil* L., Sp. Pl. ed. 2. 1:29. 1762. *Ipomoea hederacea* auct. non (L.) Jacq.: FBI 4:199; Gamble 917 (644); Khan in FFH 249.

Climbers; stem villous. Leaves 3-lobed, ovate, 4.5-8 × 5-10 cm, strigose-hispid, base cordate, apex acuminate, margin entire. Flowers in cymose clusters, pink-purple Capsules globose, 1 cm across, glabrous; seeds 5-7.

Fls. & Frs. : August - December.

Occasional in forests. Sardhana, **BR & CP** 11629; Gangapur RF, **TP & CP** 12102.

Ipomoea obscura (L.) Ker. - Gawl., Bot. Reg. 3 : t. 239. 1817; FBI 4:207; Gamble 916 (643). *Convolvulus obscurus* L., Sp. Pl. ed. 2. 220. 1762.

Twining herbs; branchlets sparsely-hairy. Leaves ovate, 1-6 × 1.5-6.5 cm, pubescent, base cordate, apex acuminate, margin entire. Flowers solitary or in simple lax cymes, pale yellow. Capsules subglobose, 0.7 cm across, glabrous, apiculate; seeds 4-6.

Fls. & Frs. : Round the year. *Vern.* : Nallakokkita.

Common on the field hedges. Narsapur, **BR & CP** 11239; Akkanapet, **BR & CP** 11552; Pocharam sancturay, **CP** 11753; Choutkoor, **TP & CP** 11911.

Ipomoea pestigridis L., Sp. Pl. 162. 1753; FBI 4: 204; Gamble 918 (644).

Twining herbs; branchlets hirsute. Leaves palmately 5- 7-lobed; lobes obovate, 3-5 × 1-2.5 cm, base attenuate, apex acumiate, margin entire. Flowers axillary, in capitate clusters, whitish- pink. Capsules globose, 1 cm across, glabrous; seeds 4.

Fls. & Frs. : August - September. *Vern.* : Mokamaduga.

Occasional in waste lands and on hedges. Narsapur forest nursery, **BR & CP** 11223; Pathur, **TP & CP** 12106.

Ipomoea sepiaria J. Koening ex Roxb. Fl. Ind. 2:90. 1824; FBI 4:209; Gamble 916 (643).

Climbing herbs; branchlets hairy. Leaves triangular-cordiform, 2-4.5 × 2-3 cm, chartaceous, glabrous, base cordate, apex acute apiculate, margin entire. Flowers axillary, in subumbellate cymes, white. Capsules globose, 0.8 cm across; seeds many.

Fls. & Frs. December - May.

Occasional on bushes, along roadsides and waste lands. Patancheru, **CP** 11689.

JACQUEMONTIA J.D. Choisy

Jacquemontia paniculata (Burm.f.) Hallier. f. Bot. Jahrb. Syst. 16:541. 1893; Gamble 926 (650). *Ipomoea paniculata* Burm. f. Fl. Ind. 50. t. 21. f. 3. 1768. *Convolvulus parviflorus* Vahl, Roxb. Fl. Ind. 2:51. 1824; FBI 4:220.

Climbing herbs. Leaves cordate, 3-5 × 2-3 cm, chartaceous, sparsely pubescent, base subcordate, apex acuminate, margin entire. Flowers axillary, in corymbose cymes, white. Capsules globose, 0.5 cm across; seeds 4.

Fl.s & Frs. : December - April.

Occasional in forests. Narsapur, **TP & CP** 11999.

MERREMIA Dennstedt ex Endlicher *nom. cons.*

1. Leaves petiolate, reniform-ovate, margin crenate **M. emarginata**
1. Leaves subsessile, oblong-obovate, margin entire **M. tridentata**

Merremia emarginata (Burm.f.) Hallier f. in Bot. Jahrb. Syst. 16:552. 1893; Gamble 928 (652). *Evolvulus emarginatus* Burm.f., Fl. Ind. 77. t. 30. f. 1. 1768. *Ipomoea reniformis* (Roxb.) Choisy in Mem. Soc. Phys. Geneve 6:446. 1834; FBI 4:206. *Convolvulus reniformis* Roxb., Fl . Ind. 2:67. 1824. *Merremia gangetica* (L.) Cuf. in Bull. Jard. Bot. Etat. 31. (Suppl.) : 743. 1961. *Convolvulus gangeticus* L., Cent. Pl. 2:9. 1756.

Prostrate herbs; branchlets sparsely hispid. Leaves reniform-ovate, 0.6-2 × 0.5-2 cm, chartaceous, base cordate, apex obtuse, margin crenate. Flowers axillary, solitary, yellow. Capsules subglobose, 0.8 × 0.6 cm, glabrous; seeds 4.

Fls. & Frs. : September - December.

Common in dried cultivated fields and waste lands. Pochammaralu, **BR & CP** 11493.

Merremia tridentata (L.) Hallier f. in Bot. Jahrb. Syst. 16:552. 1893; Gamble 928 (652). *Convolvulus tridentatus* L., Sp. Pl. 157. 1753. *Ipomoea tridentata* (L.) Roth in Roemer in Arch. Bot. (Leipzig) 1.2: 38. 1798; FBI 4: 205. *Convolvulus hastatus* Desr. in Lam, Encycl. 3:542. 1792. non Forsskal 1715. *Merremia tridentata* ssp. *hastata* (Desr.) Van Ooststroom in Blumea 3:317. 1939. *Merremia hastata* (Desr.) Hallier f. in Bot. Jahrb. Syst. 16:552. 1893; Gamble 929 (652).

Prostrate herbs; branches rusty brown. Leaves oblong- obovate, 0.5-1.5 × 0.2-0.4 cm, glabrescent, base auriculate, apex acute-3-toothed, margin entire. Flowers axillary, solitary, yellow. Capsules globose, 0.4 cm across, glabrous; seeds 4.

Fls. & Frs. : Round the year.

Common in waste lands and cultivated fields. Sangareddy tank, **BR & CP** 11300; Thornal, **CP** 11398; Patancheru, **CP** 11689.

Note : Austin (Fl. Ceylon. 1980. Vol. 1 :353) treated this species in broad sense. Accordingly the species exhibit ecophenotypic variations which may not be useful in distinguishing the taxon into subspecies and varieties. The same concept is followed here.

RIVEA Choisy

Rivea hypocrateriformis (Desr.) Choisy in Mem. Soc. Phys. Geneve 6: 408. 1834; FBI 4: 184; Gamble 903 (635); Khan in FFH 248. *Convolvulus hypocrateriformis* Desr. in Lam., Encycl. 3: 561. 1792.

Woody climbers; branchlets silky-pubescent. Leaves cordate, 2.5-4 × 3-4 cm, glabrous above, pubescent below, base cordate, apex obtuse-mucronate, margin entire. Flowers axillary, solitary or in simple cymes, white. Capsules globose, 1.3 cm across, glabrous; seeds 4.

Fls. & Frs. : March - December. *Vern.* : Bodditeega.

Common in hedges and on bushes in open forests. Pochammaralu, **CP** 11761.

Ipomoea batatas (L.) Lam. is cultivated for its tubers.

CUSCUTACEAE

CUSCUTA Linnaeus

Cuscuta reflexa Roxb., Pl. Corom. t. 104. 1799. 1832; FBI 4:225; Gamble 931 (654).

Parasitic twiners; stem glabrous, yellow. Leaves O. Flowers in racemes, creamish-white. Capsules globose, 0.5 cm across, glabrous, fleshy; seeds 4.

Fls. & Frs. : November - June. *Vern.* : Pasiteega.

Occasional on shrubs and trees in villages. Patancheru, **CP** 11694.

SOLANACEAE

1. Fruit a berry; corolla sub-rotate:
   2. Fruiting calyx inflated, covering the berry **Physalis**
   2. Fruiting calyx not as above:
      3. Flowers yellow **Lycopersicon**
      3. Flowers other than yellow **Solanum**
1. Fruit a capsule; corolla funnel-shaped **Datura**

DATURA Linnaeus

Datura innoxia Miller, Gard. Dict. (ed.8) no.5. 1768. *D. metel* auct. non L: Sims in Curtis, Bot. Mag. t. 1440. 1812; FBI 4:243; Gamble 941 (660).

Erect herbs, about 75 cm tall; branchlets pubescent. Leaves ovate, 4.5-10 × 3-4.5 cm, pubescent, base unequally truncate-rounded, apex acuminate, margin bluntly-toothed; Flowers axillary or terminal , solitary, white. Capsules globose, 3.5 cm across, prickly; seeds many.

Fls. & Frs. : Round the year. *Vern.* : Ummetha.

Common weed of waste lands. Ramayampet, **BR & CP** 11524.

LYCOPERSICON P. Miller

Lycopersicon esculentum Miller, Gard. Dict. (ed.8) no. 2. 1768 *nom. cons.*; FBI 4: 237; Gamble 942 (661). *Solanum lycopersicum* L., Sp. Pl. 185. 1753. *Lycopersicon lycopersicum* (L.) Karsten, Deut. Fl. 966. 1882.

Erect or diffuse herbs, about 1 m tall; branchlets glandular pubescent. Leaves pinnatifid, lobes 1-4 × 0.5-2 cm, obliquely traingular-ovate, terminal lobes acute. Flowers extra axillary, solitary or in drooping racemes, yellow. Berries depressed - globose, 3 × 5 cm; seeds many.

Fls. & Frs. : Round the year.

Commonly cultivated, often as an escape, Medak, **BR & CP** 11458.

PHYSALIS Linnaeus

Physalis minima L., Sp. Pl. 183. 1753; FBI 4:238; Gamble 939 (659).

Erect herbs, about 60 cm tall; branchlets glabrous. Leaves elliptic-ovate, 2–5 × 1-2.5 cm, sparsely strigose, base acute, apex acuminate, margin entire Flowers axillary, solitary, yellow. Berries globose, 0.7 cm across, with accrescent calyx; seeds many.

Fls. Frs. : July - February.

Common in cultivated fields and wastelands. NSF, **BR & CP** 11668; Lingareddypally CP 11718.

SOLANUM Linnaeus

1. Plants unarmed:
   2. Herbs; berries black **S. nigrum**
   2. Shrubs; berries yellowish-green **S. erianthum**
1. Plants armed **S. surattense**

Solanum eriantham D. Don Prodr. Fl. Nepal 96. 1825; Roe in Taxon 17:176. 1968. *S. verbascifolium* auct. non L: FBI 4: 230; Gamble 936 (657).

Shrubs, about 2 m tall; branchlets fulvous-tomentose. Leaves elliptic-oblong, 12-16 × 4.5-8 cm, woolly-tomentose, base obliquely acute, apex acute-acuminate, margin entire. Flowers terminal, in corymbose panicles, creamish-white. Berries globose, 1.5 cm across; seeds many.

Fls. Frs. : Round the year.

Occasional in forests. Narsapur RF, **TP & CP** 12013.

Solanum nigrum L., Sp. Pl. 186. 1753; FBI 4:229; Gamble 936 (657).

Erect herbs, about 1 m tall; branchlets glabrous. Leaves broadly ovate-elliptic, 2-6 × 1-3.5 cm, glabrous, base rounded-decurrent, apex acute, margin entire or bluntly toothed. Flowers extra-axillary, in umbellate cymes, white. Berries globose, 0.5 cm across, glabrous; seeds many.

Frs. Fls. : Round the year.

Common in waste lands and fields. Mombojipally, **BR & CP** 11634.

Solanum surattense Burm.f., Fl. Ind. 57. 1768; *S. xanthocarpum* Schrad. & Wendl., Sert. Hannov. 1:8.t.2. 1795; FBI 4:236; Gamble 938 (658).

Diffuse herbs; branchlets and leaves with yellow prickles. Leaves lacerate, 3-6 × 1.5-4 cm, chartaceous, base attenuate, apex acute, margin deeply and closely sinuate. Flowers extra-axillary, in cymes, purple. Berries globose, 1.3 cm across; glabrous; seeds many.

Fls. Frs. : Round the year.

Common weed of waste lands. Mombojipally, **BR & CP** 11450.

Capsicum annum L, *C. frutescens* L., *Solanum melongena* L., are cultivated for their fruits and *Nicotiana tabaccum* L., for leaves.

SCROPHULARIACEAE

1. Leaves pinnatifid or capillaceo-multifid:
   2. Leaves alternate **Verbascum**
   2. Leaves opposite or whorled, atleast the lower:
      3. Anthers 1-celled:
         4. Corolla-tube bent from above the middle **Striga**
         4. Corolla tube not bent **Sopubia**
      3. Anthers 3-celled:
         5. Leaf base auriculate; stigma 2-lobed **Stemodia**
         5. Leaf base not auriculate; stigma 2-lamellate **Limnophila**
1. Leaves not as above:
   6. All leaves opposite:
      7. Leaves serrate-denatate **Lindernia**(*p.p.*)
      7. Leaves entire:
         8. Capsules globose: seeds reticulate **Dopatrium**
         8. Capsules ovate- oblong; seeds rugose **Lindernia** (*p.p.*)
   6. Leaves not as above:
      9. Leaves oblong-spathulate, entire **Bacopa**
      9. Leaves elliptic-lanceolate, serrate **Scoparia**

BACOPA Aublet *nom. cons.*

Bacopa monnieri (L.) Pennell in Proc. Acad. Nat. Sci. Philadelphia 98: 94. 1946. *Lysimachia monnieri* L., Cent. Pl. 11:9. 1756. *Moniera cuneifolia* Michaux, Fl. Bor-Amer. 2:22. 1803, *"Monniera"*. Gamble 953 (669). *Herpestis monnieria* (L.) Kunth, Nov. Gen. Sp. 2:366. 1818, *nom. illegit*; FBI 4: 272. *Gratiola monnieri* (L.) L., Syst. Nat. (ed.10) 851. 1759 & Amoen. Acad. 4:306. 1759 "*monnieria*".

Prostrate herbs, about 20 cm tall; branchlets ascending, fleshy, glabrous. Leaves oblong-spathulate, 0.5-1.2 × 0.1-0.3 cm, glabrous, base attenuate, apex obtuse, margin entire. Flowers axillary, solitary, white. Capsules ovate, 0.5 × 0.2 cm, glabrous; seeds many.

Fls. Frs.: Round the year.

Common in wet places. Ramayampet, **BR & CP** 11503; Patancheru, **CP** 11686.

DOPATRIUM Hamilton ex Bentham

Dopatrium junceum (Roxb.) Buch. - Ham. ex Benth., Scroph. Ind. 31. 1835; FBI 4:274; Gamble 954 (670). *Gratiola juncea* Roxb., Pl. Corom. t.129. 1800 & Fl. Ind. 1: 142. 1832.

Erect herbs, about 20 cm tall; branchlets fleshy, glabrous. Leaves oblong 1-3 × 0.4-1 cm, membranous, base decurrent, apex acute, margin entire. Flowers axillary, solitary, pink. Capsules globose, 0.3 cm across; seeds many.

Fls. & Frs. : December - March.

Occasional in moist places. Dosalagunda lake, **KMS** 6795.

LIMNOPHILA Robert Brown *nom. cons.*

Limnophila indica (L.) Druce in Bot. Exch. Club Soc. Brit. Isles 3: 420. 1914. *Hottonia indica* L., Syst. Nat. (ed.10) 919. 1759. *Limnophila gratioloides* R, Br., Prodr. 442. 1810; FBI 4:271; Gamble 952 (668).

Diffuse herbs, about 15 cm tall; branchlets glabrous. Leaves whorled, linear-lanceolate, submerged leaves capillaceo multifid, 0.3-1 × 0.1-0.3 cm, glabrous. Flowers axillary, solitary, creamish-white. Caplsules ovate-globose, 0.3 cm across; seeds many.

Fls. & Frs. : Round the year.

Common in shallow water ponds and ditches. Narsapur tank, **TP & CP** 11997.

LINDERNIA Allioni

Note : Pennell in his Scrophulariaceae of East Temp. North Amer. (Monographs, Academy of Nat. Sci. of Philadelphia, No. 1. 1935) reduced *Vandellia* L., *Ilysanthes* Raf. and *Bonnaya* Link & Otto into *Lindernia* Allioni. It is characterised by the remarkable uniform corolla, curiously recurving anterior filaments and by similar septicidal dehiscence of capsule. Indian species of this genus have been revised by Sivarajan and Mathew (1983) who recognised 22 species.

1. Leaves penni-nerved, margin serrate-dentate:
   2. Flowers in racemes; fruiting calyx smaller than capsule **L. ciliata**
   2. Flowers solitary; fruiting calyx longer than capsule **L. crustacea**
1. Leaves 3-nerved from base, margin entire **L. parviflora**

Lindernia ciliata (Colsm.) Pennell in Brittonia 2:182. 1936. *Gratiola ciliata* Colsm., Prodr. Desc. Gratiol. 14. 1793. *Bonnaya brachiata* Link & Otto. Ic. Pl. Select 25.t. 11. 1820; FBI 4:284. *Ilysanthes serrata* (Roxb.) Urban in Ber. Deutsch. Bot. Ges. 2:436. 1884; Gamble 962 (675). *Gratiola serrata* Roxb., Fl. Ind. 1: 140. 1820.

Erect herbs, about 15 cm tall; branchlets glabrous. Leaves oblong-elliptic, 1-2 × 0.2-0.8 cm, glabrous, base cuneate, apex acute, margin sharply aristate-dentate. Flowers terminal, in racemes, white. Capsules linear-lanceolate, 1 × 0.1 cm, glabrous; seeds many.

Fls. & Frs. : August - November.

Common in forests and in wet places. Pillutla RF, **BR & CP** 11261; Nagulabanda, **CP** 11341; Narsapur RF, **TP & CP** 14024.

Lindernia crustacea (L.) F. Muell., Syst. Census Austral. Pl. 97. 1887. *Caprarea crustacea* L., Mant. Pl. 87. 1767. *Vandellia crustacea* (L.) Benth. Scroph. Ind. 35. 1835; FBI 4:279; Gamble 959 (673).

Erect herbs, about 20 cm tall; branchlets glabrous. Leaves ovate, 0.5-1 × 0.3-0.8 cm, glabrous, base truncate, apex acute, margin serrate. Flowers axillary, solitary, purple-rose. Capsules oblong, 0.4 x 0.2 cm, glabrous; seeds many.

Fls. & Frs. : July - November.

Common in moist places. Gangapur, **TP** & **CP** 12078.

Lindernia parviflora (Roxb.) Haines Bot. Bihar & Orissa 635. 1922. *Gratiola parviflora* Roxb. Pl. Corom. t. 203. 1811. *Ilysanthes parviflora* (Roxb.) Benth. in DC., Prodr. 10.419. 1846; FBI 4:283; Gamble 961 (675).

Herbs, about 30 cm tall; branchlets glabrous. Leaves oblong-elliptic, 0.5-2 × 0.2-1 cm, base rounded, apex acute, margin entire. Flowers axillary or terminal, solitary, pinkish-white capsules ovate-oblong, 0.4 × 0.2 cm; seeds many.

Fls. & Frs. : August - March.

Occasional in moist places. Narsapur, **BR** & **CP** 14114.

SCOPARIA Linnaeus

Scoparia dulcis L., Sp. Pl. 116. 1753; FBI 4: 289; Gamble 964 (678).

Erect herbs, about 50 cm tall; branchlets glabrous. Leaves decussate or whorled, elliptic-lanceolate, 1.5 - 2.3 × 0.4-0.8 cm, glabrous, base attenuate, apex acute, margin coarsely serrate. Flowers axillary, solitary or in pairs, white-yellow. Capsules globose, 0.2 cm across, glabrous; seeds many.

Fls. & Frs. : Round the year.

Common in waste lands. Pillutla RF, **BR** & **CP** 11260.

SOPUBIA F. Hamilton ex D. Don

Sopubia delphinifolia (L.) G. Don, Gen. Hist. 4:560. 1837; FBI 4 : 302; Gamble 970 (682). *Gerardia delphinifolia* L., Cent. Pl. 2:21. 1756.

Erect herbs, about 60 cm tall; branchlets glabrous. Leaves linear-trifid above, pinnatifid below; lobes filiform, glabrous. Flowers terminal, in racemes, purple. Capsules oblong, 0.6 × 0.4 cm, hairy; seeds many.

Fls. & Frs. : August - November.

Common in wet places and cultivated fields. Narayanapur RF, **BR** & **CP** 11275; Poosanpalle, **TP** & **CP** 11957.

STEMODIA Linnaeus *nom. cons.*

Stemodia viscosa Roxb., Pl. Corom. t. 163. 1802; FBI 4:265; Gamble 949 (666).

Erect herbs, about 60 m tall; branchlets glandular- hairy. Leaves oblong, 0.5-3 x 0.2 -0.4 cm, pubescent, base auriculate, apex acute, margin serrate. Flowers axillary, solitary, blue-violet. Capsules globose, 0.3 cm across; seeds many.

Fls. Frs.: Round the year.

Common weed of cultivated fields and moist places. Medak, **BR** & **CP** 11422; Narsapur tank, **TP** & **CP** 12009.

STRIGA Loureiro

1. Flowers white, calyx 5-ribbed — **S. densiflora**
1. Flowers white-yellow, calyx 10-15-ribbed:
   2. Calyx 10-ribbed, lateral ribs ending just above sinuses — **S. asiatica**
   2. Calyx 15-ribbed, lateral ribs converging at apex — **S. angustifolia**

Striga angustifolia (D.Don) C.J. Saldanha in Bull. Bot. Surv. Ind. 5: 70. 1963. *Buchnera angustifolia* D.Don, Prodr. Fl. Nepal 91. 1825. *Striga euphrasioides* Benth., in Comp. Bot. Mag. 1:364. 1835; FBI 4: 299; Gamble 968 (680).

Erect herbs, about 40 cm tall; stem and leaves scarbrous. Leaves linear, 1-1.5 × 0.2 cm, base narrowed, apex acute, margin entire. Flowers terminal, in spikes, white. Capsules oblong, 0.6 × 0.3 cm, enclosed in persistent calyx; seeds many.

Fls. Frs. : Round the year.

Common parasitic herb in all places. Pandivagu, **BR** & **CP** 11205.

Striga asiatica (L.) O. Kuntze, Rev. Gen. Pl. 2:466. 1891. *Buchnera asiatica* L., Sp. Pl. 630. 1753. *Striga lutea* Lour., Fl. Cochinch. 22. 1790; FBI 4 : 299; Gamble 968 (680).

Erect herbs, about 35 cm tall; stem scabrid. Leaves linear, 1-4 × 0.3-1 cm, base narrowed, apex acute, margin entire. Flowers terminal, in spikes, whitish- yellow. Capsules oblong, 0.4 x 0.2 cm; seeds many.

Fls. Frs.: Round the year.

Common parasite on grasses. Nagulabanda. **CP** 11353; Narsapur RF, **TP** & **CP** 11983; Pocharam tank, **TP** & **CP** 12032.

Striga densiflora (Benth.) Benth., in Comp. Bot. Mag. 1:363. 1835; FBI 4: 299; Gamble 967 (680). *Buchnera densiflora* Benth., Scroph. Ind. 41. 1835.

Erect herbs, about 10 cm tall. Leaves narrowly oblong, 1-2 × 0.1-0.3 cm, base decurrent, apex acute, margin entire. Flowers axillary, solitary, white. Capsules oblong, 0.5 x 0.2 cm; seeds many.

Fls. & Frs. : Round the year.

Occasional in waste lands. Narsapur, **GNR** 3980.

VERBASCUM Linnaeus

Verbascum chinense (L.) Santapau, Fl. Purandhar 90. 1958. *Scrophularia chinensis* L., Mant. Pl. 250. 1771. *Celsia coromandeliana* Vahl. Symb. Bot. 3: 79. 1794; FBI 4: 251; Gamble 945 (663).

Erect herbs, about 1 m tall; branchlets densely pubescent. Leaves oblanceolate-pinnatifid, 2-5 × 1-2 cm, pubescent, base cuneate, apex acute, margin crenate-serrate. Flowers terminal, in racemes, yellow. Capsules ovate, 0.6 × 0.4 cm, pubescent; seeds many.

Fls. & Frs. : January - August.

Common weed of waste lands and moist places. Pochammaralu, **BR & CP** 11489; Akkannapet, **BR & CP** 11567.

LENTIBULARIACEAE

UTRICULARIA Linnaeus

Utricularia stellaris L.f., Suppl. Pl. 86. 1781; FBI 4:328; Gamble 980 (689).

Aquatic stoloniferous floating herb; stolons filiform. Leaves in whorls, submerged, cut into pectinate segments. Flowers in racemes, yellow. Capsules globose, 0.4 cm across; seeds many.

Fls. & Frs. : August - October.

Rare in stagnant waters. Pocharam tank, **TP & CP** 12027; Sardhana, **CP** 11770.

BIGNONIACEAE

1. Leaflets orbicular-obovate; calyx spathaceous — **Dolichandrone**
1. Leaflets and calyx not as above — **Stereospermum**

Dolichandrone (Fenzl) B. Seemann *nom. cons.*

1. Leaves even-pinnate; leaflets > 4 cm in length — **D. falcata**
1. Leaves odd-pinnate; leaflets < 4 cm in length — **D. atrovirens**

Dolichandrone atrovirens (Roth) Sprague in Bull. Misc. Inform. 1919: 307. 1919; Gamble 996 (700). *Bignonia atrovirens* Heyne ex Roth, Nov. Pl. Sp. 284. 1821. *Dolichandrone crispa* Seemann in J. Bot. 8: 381. 1870; FBI 4:379; Khan in FFH 256.

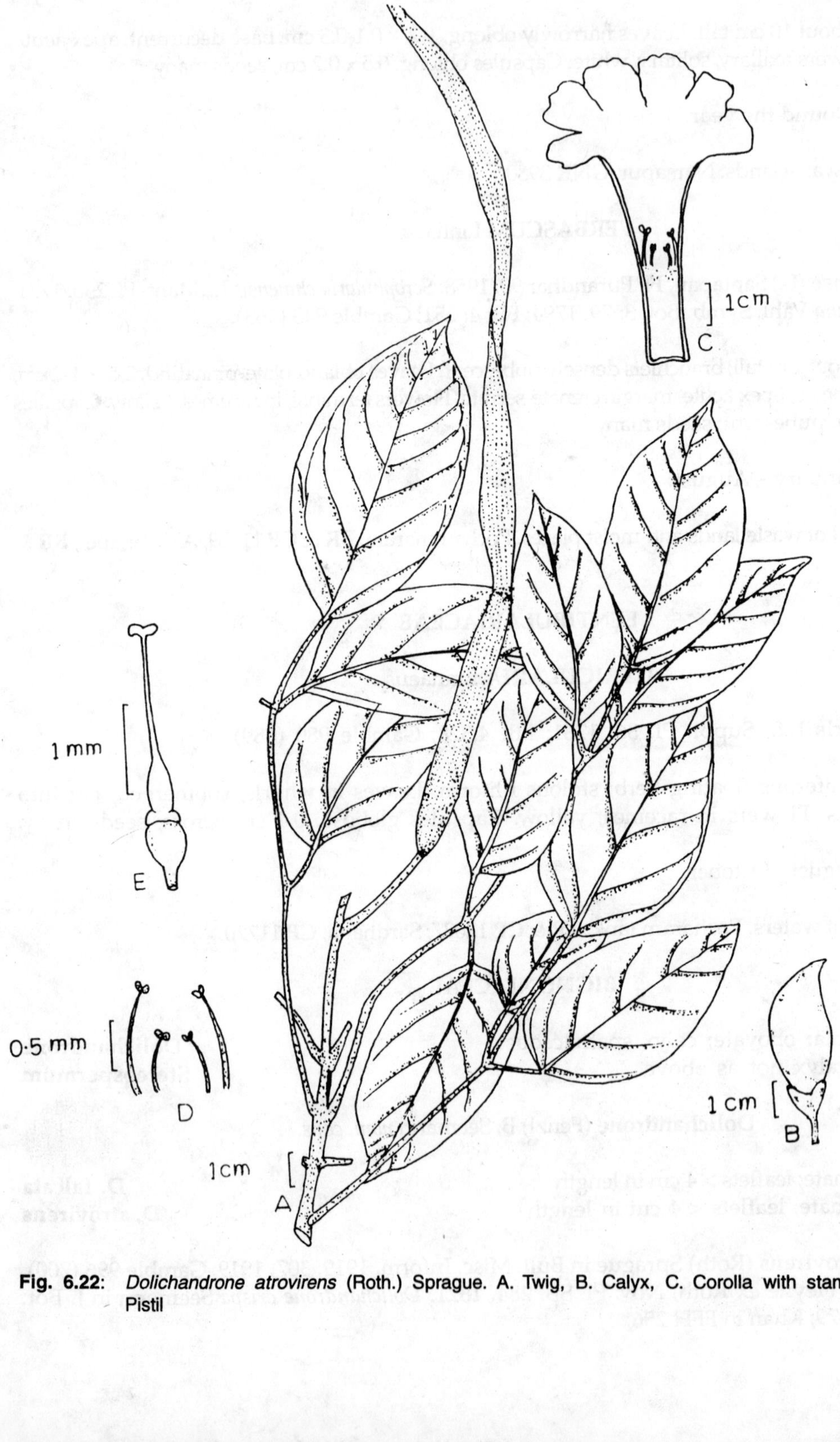

Fig. 6.22: *Dolichandrone atrovirens* (Roth.) Sprague. A. Twig, B. Calyx, C. Corolla with stamens, D. Stamens, E. Pistil

Trees, about 6 m tall; branchlets granulate. Leaves even-pinnate; leaflets 3-5, broadly ovate, 7-10.5 × 3-4.5 cm, thin-coriaceous, glabrescent, base obliquely truncate, apex acuminate, margin entire. Flowers in corymbose racemes, white. Capsules subterete with speckles, 25 × 1 cm; seeds winged.

Fls. & Frs. : July - December. *Vern.* : Neeuddi.

Common in deciduous forests. Narsapur RF, **TP & CP** 11985.

Dolichandrone falcata (DC.) Seem. Fl. Vill. - Nov. 151. 1880; FBI 4: 380; Gamble 996 (700); Khan in FFH 257. *Spathodea falcata* DC., Prodr. 9: 206. 1845.

Trees, about 10 m tall; branchlets glabrous. Leaves odd-pinnate, leaflets 3-5, orbicular-obovate, 1.5-4 × 1-3 cm, pubescent, base cordate, apex rounded-shortly acute, margin entire. Flowers in corymbs, white. Capsules compressed, 30 × 2.5 cm, falcate; seeds many.

Fls. & Frs. : April - December. *Vern.* : Wodi.

Occasional in dry deciduous forests. Gangapur, **BR & CP** 11428; Akkannapet, **BR & CP** 11565.

STEREOSPERMUM Chamisso

Stereospermum colais (Dillw.) Mabberley in Taxon 27 : 553. 1978. var. **angustifolium** (Haines) Bennet & Raizada in Ind. Journ. For. 4 : 67. 1981. *Bignonia colais* Bunch. - Ham. ex Dillw., Rev. Ref. Hort. Mal. 28. 1839. *Stereospermum chelonoides* DC. var. *angustifolium* Haines, Deser. List. Trees etc. South circle of C.P. 169. 1916. *S. anguistifolium* (Haines) Haines in Kew Bull. 1922. 121. 1922; Gamble 998 (702).

Trees, about 10 m tall; branchlets glabrous. Leaves 1- pinnate, 5-7-foliolate; leaflets oblong-elliptic, 6-15 × 4-6.5 cm, pubescent, base cuneate, apex acuminate, margin entire. Flowers terminal, in panicles, pale yellow. Capsules linear, 50 × 1.4 cm, woody; seeds many.

Fls. & Frs. : March - May. *Vern.* : Kappagargu.

Occasional in forests. Narsapur, **KMS** 8016; Kondapur, **GNR** 4236.

PEDALIACEAE

1. Leaves palminerved; flowers in racemes **Martynia**
1. Leaves penninerved; flowes solitary **Sesamum**

Martynia annua L., Sp. Pl. 618. 1753; Gamble 1003 (705). *M. diandra* Gloxin, Observ. Bot. 14.t.1.1785; FBI 4: 386; Khan in FFH 262.

Erect herbs, about 1 m tall; branchlelts glandular- hairy. Leaves broadly ovate, 4.5-15 × 3-14.5 cm, pubescent, base cordate, apex acute, margin sinuate. Flowers axillary, in racemes, purple-rose. Drupes ovate, 2.5 × 1.2 cm, with 2 apical hooks, hairy; seeds many.

Fls. & Frs. : August - December. *Vern.* : Telukondikaya.

Common in waste lands and road sides. Nagulabanda, **CP** 11342.

SESAMUM Linnaeus

Sesamum orientale L., Sp. Pl. 634. 1753. *S. indicum* L., Sp. Pl. 634. 1753; FBI 4: 387; Gamble 1002 (204); Khan in FFH 261.

Erect herbs, about 70 cm tall; branchlets sparsely hairy. Leaves elliptic-lanceolate, 2-6 × 1-1.5 cm, pubescent, base rounded-obtuse, apex acute, margin serrate-lobed. Flowers axillary, solitary, white-pink. Capsules oblong 2.5 × 1 cm, densely pubescent; seeds many.

Fls. & Frs. : July - December. *Vern.* : Nuvvulu.

Cultivated, rarely as an escape. Medak, **CP** 11719; Kaulcharam RF, **TP** & **CP** 14077.

ACANTHACEAE

1. Climbing herbs **Thunbergia**
1. Erect or prostrate plants:
   2. Armed:
      3. Corolla lobes 5, subequal **Barleria**
      3. Coralla 2-lipped **Hygrophila**
   2. Unarmed:
      4. Leaves 4, in a whorl **Blepharis**
      4. Leaves 2, opposite :
         5. Flowers in racemes or panicles :
            6. Capsules more than 4-seeded **Andrographis**
            6. Capsules 4-seeded :
               7. Branchlets hispid **Indoneesiella**
               7. Branchlets glabrous **Peristrophe**
         5. Flowers solitary, in clusters or in spikes:
            8. Stamens 4:
               9. Flowers in spikes:
                  10. Bracts spine-tipped **Lepidagathis**
                  10. Bracts not spine-tipped **Hemiadelphis**
               9. Flowers solitary or in clusters :
                  11. Leaves entire **Dipteracanthus**
                  11. Leaves crenate-serrate **Hemigraphis**
            8. Stamens 2:
               12. Bracts 4-farious **Rungia**
               12. Bracts not 4-farious **Justicia**

ANDROGRAPHIS Wallich ex C.G.D. Nees

Andrographis paniculata (Burm.f.) Wall. ex Nees in Wall. Pl. As. Rar. 3: 116. 1832; FBI 4: 501; Gamble 1048 (734). *Justicia paniculata* Burm.f. Fl. Ind. 9. 1768.

Undershrubs, about 1 m tall; branchlets tetragonous, terete. Leaves obovate-linear, 5-8 × 1-2 cm, base acuminate, apex acuminate, margin entire. Flowers axillary and terminal, in racemose panicles, white with pink tinge. Capsules oblong, minutely hairy; seeds 8-10.

Fls. Frs .: December - March.

Often planted as an ornamental. Donthi, **GNR** 4056.

BARLERIA Linnaeus

1. Unarmed: flowers pink **B. cristata**
1. Armed: flowers yellow **B. prionotis**

Barleria cristata L., Sp. Pl. 636. 1753; FBI 4:488; Gamble 1060 (743).

Undershrubs, about 1 m tall; branchlets pubescent. Leaves lanceolate-oblong, 3-10 × 0.8-3 cm, sericeous, base cuneate, apex mucronate, margin entire. Flowers axillary and terminal in spikes, violet-pink. Capsules oblong, 1.2 × 0.2 cm, glabrous; seeds 4.

Fls. Frs. : August - July.

Common in forests. Narsapur RF, **BR** & **CP** 14085.

Barleria prionotis L., Sp. Pl. 636. 1753; FBI 4: 482; Gamble 1058 (741).

Undershrubs, about 1 m tall; branchlets glabrous; Leaves ovate-obovate, 4-12 × 2-4.5 cm, sparsely hirsute, base attenuate, apex acuminate, margin entire. Flowers axillary and terminal, solitary or in clusters, yellow. Capsules ovate, 1.5 × 0.8 cm, beaked; seeds 2.

Fls. Frs .: September - January. *Vern.* : Mullagorinta.

Occasional in hedges and waste lands, often grown as an ornamental. Narsapur-Hyderabad Road, **BR** & **CP** 11210.

BLEPHARIS A.L. Jussieu

1. Leaves elliptic-lanceolate, membranous, > 2.5 cm in length **B. maderaspatensis**
1. Leaves oblong-obovate, thick, < 2.5 cm in length **B. repens**

Blepharis maderaspatensis (L.) Heyne ex Roth, Nov. Pl. Spec. 320. 1821. *Acanthus maderaspatensis* L., Sp. Pl. 639. 1753. *Blepharis boerhaaviaefolia* Pers., Syn. Pl. 2: 180. 1806; FBI 4:478; Gamble 1013 (712).

Prostrate herbs, about 1 m tall; branchlets hispid. Leaves elliptic-lanceolate, 1.5-7.5 × 1 - 2.5 cm, sparsely scabrous, base cuneate, apex acute-apiculate, margin entire. Flowers axillary and terminal, in clusters, blue-pink. Capsules oblong, 0.6 × 0.4 cm; seeds 2.

Fls. Frs .: September - March.

Occasional in shady localities. Medak, **BR & CP 11417.**

Blepharis repens (Vahl) Roth Nov. Pl. Spec. 321. 1821. *Acanthus repens* Vahl, Symb. Bot. 2: 76 1791. *Blepharis molluginifolia* Pers., Syn. Pl. 2: 180. 1806; FBI 4: 479; Gamble 1013 (712).

Prostrate herbs, about 1 m tall; branchlets strongly hispid. Leaves oblong-obovate, 0.5-2 × 0.4-0.6 cm, hispid, base cuneate, apex obtuse, margin entire. Flowers axillary and terminal, in clusters, blue white. Capsules oblong, 0.5 × 0.3 cm; seeds 2.

Fls. Frs .: September - March.

Occasional in waste lands. Toopran, **BR & CP** 11277; NSF, **BR & CP** 11664.

DIPTERACANTHUS C.G.D.Nees

1. Erect herbs; corolla tube < 1.5 cm long **D. patulus**
1. Prostrate herbs; corolla tube > 1.5 cm long **D. prostratus**

Dipteracanthus patulus (Jacq.) Nees in Wall., Pl. As. Rar. 3:82. 1832. *Ruellia patula* Jacq. Misc. Austriac. 2:358. 1781; FBI 4: 412; Gamble 1017 (714).

Erect herbs, about 40 cm tall; branchlets pubescent. Leaves elliptic-ovate. 1-2 × 0.5-1.5 cm, pubescent above, tomentose below, base acute, apex subacute, margin entire. Flowers axillary, solitary, purple. Capsules elliptic, 1.5 × 0.5 cm; seeds many.

Fls. Frs. : July - March.

Common weed along roadsides and in cultivated fields. Zaheerabad, **CP 11590.**

Dipteracanthus prostratus (Poir.) Nees in Wall., Pl. As. Rar. 3:81. 1832. *Ruellia prostrata* Poir. in Lam., Encycl. 6: 349. 1804; FBI 4: 411; Gamble 1017 (714).

Prostrate herbs, about 50 cm tall; branchlets sparsely hairy. Leaves elliptic-ovate, 1-3 × 0.5-1 cm, strigose, base acute, apex obtuse, margin entire. Flowers axillary, solitary, blue-violet. Capsules oblong, 1.5 × 0.6 cm, thinly pubescent; seeds 10-15.

Fls. Frs. : February - October. *Vern.* : Neelambaram.

Common in forests and waste lands. Narsingi, **BR & CP** 11534.

HEMIADELPHIS C.G.D. Nees

Hemiadelphis polysperma (Heyne ex Roth) Nees in Wall., Pl. As. Rar. 3: 75. 80. 1832. *Ruellia polysperma* Heyne ex Roth, Nov. Pl. Spec. 305. 1821. *Justicia polysperma* Roxb. Fl. Ind. 1: 119. 1820. *Hygrophila polysperma* (Roxb.) T. Anderson in Journ. Linn. Soc. London 8: 456. 1867; FBI 4: 406; Gamble 1015 (713).

Procumbent herbs, about 20 cm tall; branchlets rooted at base. Leaves oblong-ovate, 0.5-2 × 0.2-0.6 cm, pubescent, base tapering, apex obtuse, margin entire. Flowers terminal, in spikes, light blue. Capsules oblong, 0.7 × 0.2 cm, glabrous; seeds many.

Fls. & Frs. : June - January.

Occasional in wet places. Gangapur RF, **TP & CP** 12068.

Note : Cramer (1989) stated that this taxon is misidentified by Indian authors as *Hygrophila polysperma* (Roxb.) T. Anderson, based on its amphibious habit, terminal spikes and many-seeded capsules. By a critical study of the representative material Cramer is of the opinion that this species must be reinstated in *Hemiadelphis*, a valid monotypic genus. The genus differs from *Hygrophila* in having 2 stamens and the filament with a staminodal appendage towards base.

HEMIGRAPHIS C.G.D. Nees

Hemigraphis latebrosa (Heyne ex Roth) Roxb., Fl. Ind. 3:46. 1832; FBI 4:423; Gamble 1018 (715). *Ruellia latebrosa* Heyne ex Roth, Nov. Pl. Spec. 307. 1821.

Erect herbs, about 30 cm tall; branchlets quadrangular, pubescent. Leaves lanceolate-ovate 2-5 × 0.5 - 2 cm, sparsely hairy, base cuneate, apex acuminate, margin serrate. Flowers axillary and terminal, in capitate heads, blue. Capsules oblong, 0.5 × 0.2 cm; seeds 6.

Fls. & Frs. : November - February.

Occasional in forests. Pochammaralu, **BR & CP** 11471.

HYGROPHILA Robert Brown

Hygrophila auriculata (Schum) Heine in Kew Bull. 16:172. 1962. *Barleria auriculata* Schum. in Schum. & Thonn. Beskr. Guin. Pl. 285. 1827. *Asteracantha longifolia* (L.) Nees in Wall., Pl. As. Rar. 3:90. 1832; Gamble 1015 (712). *Barleria longifolia* L., Cent. Pl. 2:22. 1756. *Hygrophila spinosa* T. Anderson in Thw., Enum. Pl. Zeyl. 225. 1860; FBI 4: 408.

Erect armed undershrubs, about 75 cm tall; branchlets hispid. Leaves in whorls, oblong-lanceolate, 3-10 × 0.5-2.5 cm, hispid, base cuneate, apex acute, margin minutely dentate. Flowers axillary, in whorls, purple- blue. Capsules linear-oblong, 0.6 × 0.3 cm, seeds 4-6.

Fls. & Frs. : June - January. *Vern.* : Neetimalle.

Common weed of wet places and near cultivated field hedges. Medak, **BR & CP** 11409 & 11410.

INDONEESIELLA C.P. Sreemadhavan

Indoneesiella echioides (L.) Sreemadh. in Phytologia 15:271. 1967. *Justicia echioides* L., Sp. Pl.16. 1753. *Andrographis echioides* (L.) Nees in Wall., Pl. As. Rar. 3: 117. 1832. *p.p.*; FBI 4: 505; Gamble 1051 (756).

Erect herbs, about 40 cm tall; branchlets quadrangular, hispid. Leaves oblong, 1-5 × 0.5-1.5 cm, sparsely hairy, base cuneate, apex acute, margin entire. Flowers axillary, in glandular-hairy racemes, pinkish-white. Capsules elliptic, compressed, 1 × 0.4 cm, hairy; seeds 4.

Fls. & Frs. : July - March.

Common in waste lands and moist places. Sangareddy, **BR & CP** 11297; Siddipet, **CP** 11373.

JUSTICIA Linnaeus

1. Herbs; calyx lobes 4 :
    2. Spikes cylindric; > 1.5 cm in length:
        3. Branchlets pubescent **J. procumbens**
        3. Branchlets sparsely hairy **J. micrantha**
    2. Spikes linear; < 1.5 cm in length **J. prostrata**
1. Shrubs or undershrubs; calyx lobes 5:
    4. Bracts smaller than calyx, lanceolate **J. gendarussa**
    4. Bracts larger than calyx, ovate :
        5. Leaves lanceolate-ovate; bracts scarious **J. betonica**
        5. Leaves elliptic-ovate: bracts herbaceous **J. glauca**

Justicia betonica L., Sp. Pl. 15. 1753; FBI 4:525; Gamble 1078 (755).

Shrubs, about 1.5 m tall; branchlets pubescent. Leaves lanceolate-ovate, 2.5-7 × 1.5-3 cm, pubescent, base cuneate, apex acuminate, margin entire-crenate. Flowers terminal, in spikes, white with purple lines. Capsules oblong, 1.2 × 0.5 cm, pubescent; seeds 4.

Fls. & Frs. : November - April.

Occasional in waste lands and forests. Narsapur RF, **BR & CP** 14110.

Justicia gendarussa Burm.f., Fl. Ind. 10. 1768; FBI 4:532; Gamble 1079 (755); Khan in FFH 265. *Gendarussa vulgaris* Nees in Wall. Pl. As. Rar. 3:104. 1832.

Shrubs, about 1.5 m tall; branchlets glabrous. Leaves lanceolate, 3-7 × 0.5-2.8 cm, pubescent, base acute, apex acute, margin entire. Flowers terminal, in spikes, white with purple spots. capsules oblong, 1.2 × 0.2 cm, glabrous; seeds 4.

Fls. & Frs. : June - February.

Common hedge plant, often as an escape. Ramyampet, **BR & CP** 11570.

Justicia glauca Rottler in Ges. Naturf. Freunde Berlin Neue Schriften 4:219. 1803; FBI 4:529; Gamble 1079 (755).

Undershrubs, about 50 cm tall; branchlets angular, pube-scent. Leaves ovate-elliptic, 2-4.5 × 1-2 cm, sparsely pubescent, base rounded-acute, apex acute, margin entire. Flowers axillary and terminal, in spikes. pink. Capsules oblong, 1 × 0.5 cm, glabrous; seeds 4.

Fls. & Frs. : July - December.

Common in waste lands and forests outskirts. Medak, **CP** 11721.

Justicia micrantha Wall. ex C.B. Clarke in FBI 4:536. 1885. Gamble 1080 (756). *Rostellaria crinita* Nees in Wall, Pl. As. Rar. 3: 101. 1832.

Erect herbs, about 30 cm tall; branchlets terete. Leaves obovate, 1.5-3.5 × 0.7-1.5 cm, pubescent, base acute-attenuate, apex acute, margin entire. Flowers terminal, in spikes at the forks of dichotomy, white with voilet stripes. Capsules oblong, 0.4 × 0.2 cm; seeds 4.

Fls. & Frs. : June - December.

Occasional in forests and waste lands. Pandivagu, **BR** & **CP** 11208.

Justicia procumbens L., Sp. Pl. 15. 1753; FBI 4:539; Gamble 1080 (756).

Diffuse herbs, about 20 cm tall; branchlets pubescent. Leaves elliptic, 1-5 × 0.2-1 cm, densely pubescent, base obtuse, apex acute, margin entire. Flowers terminal, in dense spikes, pink. Capsules oblong, 0.4 × 0.2 cm, sparsely hairy; seeds 4.

Fls. & Frs. : June - December.

Common in moist places. Narsapur, **BR** & **CP** 11234.

Justicia prostrata (C.B.Clarke) Gamble, Fl. Pres. Madras 1081 (757). 1924 *J. diffusa* Willd. var. *prostrata* C.B. Clarke in FBI 4:538. 1885.

Prostrate herbs; branchlets striate, pubescent. Leaves ovate-suborbicular, 0.5-1 × 0.3-0.5 cm, hispid, base rounded-acute, apex acute, margin entire. Flowers in linear spikes, pink. Capsules oblong, 0.4 × 0.2 cm, glabrous; seeds 4.

Fls. & Frs. : Round the year.

Common in waste lands and fields. Sardhana, **BR** & **CP** 11626.

LEPIDAGATHIS Willdenow

Lepidagathis cristata Willd., Sp. Pl. 3:400 1800; FBI 4:516; Gamble 1066 (747).

Erect herbs, about 25 cm tall; branchlets quadrangular, glabrous. Leaves linear-lanceolate, 2-4 × 0.5-1 cm, pubescent, base acute, apex acute, margin entire. Flowers in condensed spikes, pink; Capsules oblong, 0.4 × 0.2 cm, compressed; seeds 2.

Fls. & Frs. : July - March.

Common in waste lands and forests. Medak, **BR** & **CP** 11414.

PERISTROPHE C.G.D. Nees

Peristrophe paniculata (Forssk.) Brummitt in Kew Bull. 39 : 451. 1983. *Dianthera paniculata* Forssk., Fl. Aegypt. - Arab. 7. 1775. *Peristrophe bicalyculata* (Retz.) Nees in Wall., Pl. As. Rar. 3:113. 1832; FBI 4:554; Gamble 1084 (759). *Dianthera bicalyculata* Retz., in Acta Holm. 1775:297. t.9. 1776.

Erect herbs, about 75 cm tall; branchlets quadrangular, hispid. Leaves elliptic-lanceolate, 2-8 × 0.6-4 cm, pubescent, base acute, apex acute-acuminate, margin entire. Flowers axillary and terminal, in lax panicles, white with pink streaks. Capsules oblong, 2 × 0.4 cm, stipitate; seeds 4.

Fls. & Frs. : Round the year.

Common weed of waste lands and forests. Choutkoor, **TP** & **CP** 11902; Narsapur RF, **TP** & **CP** 11972.

RUNGIA C.G.D. Nees

1. Leaves crenulate; spikes 1-3 cm long **R. pectinata**
1. Leaves entire-undulate; spikes 3-10 cm long **R. repens**

Rungia pectinata (L.) Nees in DC., Prodr. 11.469. 1847. *p.p. Justicia pectinata* L., Amoen.-Acad. 4: 299. 1759. *Rungia parviflora* (Retz.) Nees var. *pectinata* (L.) C.B. Clarke in FBI 4:550. 1885; Gamble 1070 (250).

Straggling herbs; branchlets terete, glabrous. Leaves elliptic- oblong, 1.5-7.5 × 1-2.5 cm, glabrous, base acute, apex obtuse, margin crenulate. Flowers axillary and terminal, in spikes, blue. Capsules ovate, 0.2 × 0.1 cm, pubescent at tip: seeds 2.

Fls. & Frs. : November - February.

Common in moist places. Choutkoor, **TP** & **CP** 11913.

Rungia repens (L.) Nees in Wall., Pl. As. Rar. 3: 110. 1832; FBI 4: 549; Gamble 1070 (750). *Justicia repens* L., Sp. Pl. 15. 1753.

Procumbent herbs, about 30 cm tall; branchlets densely pubescent. Leaves elliptic-lanceolate, 2-8 × 0.5-2 cm, puberulous, base rounded-acute, apex acute, margin entire. Flowers terminal, in spikes, pink. Capsules elliptic-oblong, 0.5 × 0.3 cm, pubescent; seeds 2-4.

Fls. & Frs. : September - February.

Occasional in waste lands. Narsapur, **BR & CP** 14100.

THUNBERGIA Retzius *nom. cons.*

Thunbergia fragrans Roxb. Pl. Corom. t. 67. 1796 & Fl. Ind. 3:33. 1832; FBI 4:390; Gamble 1007 (708).

Climbing herbs; branchlets sparsely hispid. Leaves broadly elliptic-ovate, 3.5-7 × 3-4.5 cm, hispid, base subcordate, apex apiculate-acute, margin entire. Flowers axillary, solitary, white. Capsules depressed-globose, 0.9 cm across, beaked, scabrid; seeds 4.

Fls. & Frs. : August - February.

Common in forest borders and on hedges. Pochammaralu, **BR & CP** 11487.
Crossandra infundibuliformis (L.) Nees. and *Ruellia tuberosa* L. are planted as ornamentals.

VERBENACEAE

1. Prostrate herbs **Phyla**
1. Trees or erect shrubs:
    2. Inflorescence centripetal:
        3. Calyx obscurely lobed; stamens 4 **Lantana**
        3. Calyx long, tubular; stamens 2 **Stachytarpheta**
    2. Inflorescence centrifugal:
        4. Leaves simple:
            5. Corolla regular **Tectona**
            5. Corolla 2-lipped:
                6. Trees **Gmelina**
                6. Erect shrubs **Clerodendrum**
        4. Leaves 3-5-foliolate **Vitex**

CLERODENDRUM Linnaeus

Clerodendrum serratum (L.) Moon, Cat. Pl. Ceylon 46. 1824; FBI 4: 592; Gamble 110 (770). *Volkameria serrata* L., Mant. Pl. 90. 1767.

Shrubs, about 1 m tall; branchlets angular, pubescent. Leaves obovate-lanceolate, 2-20 × 1-8 cm, glabrous, base cuneate, apex acute-acuminate, margin coarsely serrate. Flowers terminal, in pyramidal panicles, blue. Drupes globose, 0.6 cm across, seeds oblong.

Fls. & Frs. : July - November.

Occasional in dry deciduous forests. Narsapur RF, **TP & CP** 14031.

GMELINA Linnaeus

Gmelina arborea Roxb., Pl. Corom. t. 246. 1815; FBI 4:581; Gamble 1097 (768).

Trees, about 6 m tall; branchlets quadrangular, pubescent. Leaves broadly ovate-cordate, 14-20 × 5-9 cm, glabrous above, fulvous-tomentose below, base truncate-cordate, apex obtuse-acute, margin entire. Flowers axillary and terminal, in paniculate cymes, brownish-yellow. Drupes pyriform, 2.5 × 1 cm, smooth, orange-yellow when ripe.

Fls. & Frs.: February - April. *Vern.*: Gummudu-teku, Pedda gummudu.

Occasional in dry deciduous forests. Thornal, **CP** 11707.

LANTANA Linnaeus

Lantana camara L., Sp. Pl. 627. 1753. var. **aculeata** (L.) Moldenke in Torreya 34:9. 1934. *L. aculeata* L., Sp. Pl. 627. 1753; FBI 4:562; Gamble 1087 (761).

Armed shrubs, about 2.5 m tall; branchlets quadrangular. Leaves ovate, 3.5-6.5 × 2-4 cm, pubescent, base rounded-truncate, apex acute, margin coarsely crenate-serrate. Flowers axillary, in pedunculate heads, pink. Drupes globose, 0.4 cm across. fleshy, purple when ripe.

Fls. & Frs.: Round the year. *Vern.*: Akshintalu.

Abundant weed of waste lands and scrub jungles. Narsapur-Hyderabad Road, **BR & CP** 11211.

PHYLA Loureiro

Phyla nodiflora (L.) E. Greene in Pittonia 4: 46. 1899. *Verbena nodiflora* L., Sp. Pl. 20. 1753. *Lippia nodiflora* (L.) A. Rich in Michaux, Fl. Bor. - Amer. 2:15, 1803; FBI 4:563; Gamble 1088 (762).

Prostrate herbs, stem pubescent. Leaves obovate-spathulate, 1-2 × 0.2-0.5 cm, pubescent, base cuneate, apex obtuse, margin sharply serrate. Flowers axillary, in globose heads, whitish-pink. Drupes subglobose, 0.2 cm across, pyrenes 2; seed 1 per pyrene.

Fls. & Frs.: Round the year.

Common in moist places. Sangareddy tank, **BR & CP** 11456.

STACHYTARPHETA Vahl *nom. cons.*

1. Leaves glabrous; flowers blue **S. jamaicensis**
1. Leaves softly tomentose; flowers pink **S. mutabilis**

Stachytarpheta jamaicensis (L.) Vahl, Enum. Pl. 1:206. 1804. *Verbena jamaicensis* L., Sp. Pl. 19. 1753. *Stachytarpheta indica* auct. non Vahl : FBI 4:564; Gamble 1090 (763).

Undershrubs, about 75 cm tall; branchlets glabrous. Leaves elliptic, 2-8 × 1-3.5 cm, glabrous, base tapering and decurrent into the petiole, apex acute, margin serrate. Flowers terminal, in spikes, subtended by bracts, blue. Drupes oblong, 0.4 × 0.2 cm; pyrenes 2; seed 1 per pyrene.

Fls. & Frs. : July - December.

Occasional in waste lands. Manjira Barrage, **BR** & **CP** 14094.

Stachytarpheta mutabilis (Jacq.) Vahl in Enum, Pl. 1:209. 1804; FBI 4: 564; Gamble 1090 (763). *Terpena mutabilis* Jacq., Coll. 2:334. 1789.

Undershrubs, about 1.5 m tall; branchlets tetragonal, pubescent. Leaves ovate, 4-12.5 × 2-6 cm, softly tomentose, base decurrent, apex acute, margin crenate. Flowers terminal, in spikes, pink. Drupes cylindric, pyrenes 2, plano-convex; seed 1 per pyrene.

Fls. & Frs. : Round the year.

Occasional in waste lands. Dundigal, **GNR** 4645.

TECTONA Linnaeus f. *nom. cons.*

Tectona grandis L.f., Suppl. Pl. 151. 1781; FBI 4:570; Gamble 1092 (765); Khan in FFH 266.

Trees, about 30 m tall; branchlets quadrangular, tomentose. Leaves broadly elliptic-obovate, 10-22 × 5- 11.5 cm, thin-coriaceous, glabrous above, tawny tomentose below, base cuneate, apex acute-acuminate, margin entire. Flowers terminal, in trichotomous panicles, white-yellow. Drupes globose. 1.5 cm across, epicarp inflated, stellate-pubescent.

Fls. & Frs, : July - December. *Vern.* : Teku.

Abundant in dry deciduous forests. Mutaypally, **CP** 11783

VITEX Linnaeus

Vitex negundo L., Sp. Pl. 638. 1753; FBI 4:583; Gamble 1102 (771); Khan in FFH 273.

Shrubs, about 6 m tall; branchlets grey-pubescent. Leaves 3-5-foliolate; leaflets lanceolate, laterals 3- 7.5 × 0.5-0.2 cm, median 5-11 × 1-2.2 cm, glabrous above, glaucous-pubescent below, base acute, apex acuminate, margin entire-crenate. Flowers terminal, in panicles, purple. Drupes globose, 0.5 cm across; pyrenes 1-3.

Fls. & Frs. : Round the year. *Vern.* : Vavili.

Common in waste lands, road sides, hedges and villages. Siddipet, **CP** 11377; Medak, **BR** & **CP** 11412.

Duranta repens L. is grown for ornamental purpose.

LAMIACEAE (nom. alter. LABIATAE)

1. Calyx 8-10-toothed:
    2. Undershrubs; flowers orange **Leonotis**
    2. Herbs; flowers white **Leucas**
1. Calyx 5-toothed:
    3. Calyx campanulate:
        4. Calyx-lobes lanceolate; flowers pink **Anisomeles**
        4. Calyx-lobes linear; flowers blue **Hyptis**
    3. Calyx 2- lipped:
        5. Upper lip of calyx 1-lobed, broader than lower lobes:
            6. Flowers crowded in capitate heads **Anisochilus**
            6. Flowers in whorls on long racemes or panicles:
                7. Lower lip of calyx 2-mucronate **Becium**
                7. Lower-lip of calyx 4-toothed:
                    8. Calyx as long as corolla tube **Ocimum**
                    8. Calyx shorter than corolla tube **Plectranthus**
        5. Upper lip of calyx 3-lobed, as broad as lower lobes:
            9. Calyx without appendages, persistent **Salvia**
            9. Calyx with appendages, deciduous **Scutellaria**

ANISOCHILUS Wallich ex Bentham

Anisochilus carnosus (L.f.) Wall., Pl. As. Rar. 2:18. 1831; FBI 4:627; Gamble 1126 (788). *Lavandula carnosa* L.f., Suppl. Pl. 273. 1781.

Erect herbs, about 40 cm tall; branchlets angular, sparsely hairy. Leaves ovate, 1-3.5 × 0.5-1.2 cm, glandular, base cordate, apex acute, margin serrate. Flowers terminal, in cylindrical spikes, purple. Nutlets orbicular, brown.

Fls. & Frs.: August - December.

Rare in forests. Way to Peddhakutta, **KMS 6761**.

ANISOMELES Robert Brown

1. Branchlets acutely tetragonous; fruiting calyx hispid **A. indica**
1. Branchlets obtusely tetragonous; fruiting calyx softly villous **A. malabarica**

Anisomeles indica (L.) O. Kuntze, Rev. Gen. Pl. 2:512, 1891; Gamble 1140 (797). *Nepeta indica* L., Sp. Pl. 571. 1753. *Anisomeles ovata* R.Br. in N.J. Alton, Hortus Kew. 3:364. 1811; FBI 4:672.

Shrubs, about 1.5 m tall; branchlets quadrangular, tomentose. Leaves ovate-obovate, 2-7 × 1-4.5 cm, densely pubescent, base truncate, apex acute, margin crenate-serrate. Flowers axillary, in dense whorls on spikes, purplish-pink. Nutlets ovate, black.

Fls. & Frs.: July - December.

Common in forests. Pillutla RF, **BR & CP** 11267.

Anisomeles malabarica (L.) R.Br. ex Sims in Curtis, Bot. Mag. t. 2071. 1819; FBI 4:673; Gamble 1140 (797). *Nepeta malabarica* L., Mant. Alt. 566. 1771.

Undershrubs, about 1.5 m tall; branchlets tetragonous, silvery tomentose. Leaves lanceolate, 5-9 × 1-2 cm, tomentose, base rounded, apex acute, margin crenate- serrate. Flowers axillary, in dense whorls, pale purple. Nutlets lenticular.

Fls. & Frs. : Round the year.

Occasional in waste lands along fences. Narsapur, **GNR** 3945.

BECIUM J. Lindley

Note : R.M. Harley (1983) transfered *Ocimum adscendens* Willd. to the African genus *Becium* Lindl. based on calyx character. The genus *Ocimum* is distinguished from *Becium* by having shorter, but distinct teeth of lateral calyx, where these are obsolete and replaced by a row of minute teeth or by a row of dense cilia in the latter.

Becium filamentosum (Forssk.) Chiov., Nuovo Giorn. Bot. Ital. 26: 162. 1919. *Ocimum filamentosum* Forssk., Fl. Aegypt. - Arab. 108. 1775. *O. adscendens* Willd., Sp. Pl. 3:166. 1800; FBI 4:609; Gamble 1112 (778); Khan in FFH 276.

Erect herbs, about 50 cm tall; branchlets angular, pubescent. Leaves elliptic-oblanceolate, 1-3 × 0.5-2.5 cm, chartaceous, base attenuate, apex acute, margin serrulate. Flowers terminal, in distant whorls on racemes, white. Nutlets erect.

Frs. & Frs. : July - December.

Common near shady wastelands. Narsingi, **BR & CP** 11523; Ranjhole, **CP** 11679; Pocharam sanctuary, **CP** 11752.

HYPTIS N.J. Jacquin *nom. cons.*

Hyptis suaveolens (L.) Poit, in Ann. Mus. Natl. Hist. Nat. 7: 472.t.29. f.2.1806; FBI 4: 630; Gamble 1129 (789). *Ballota suaveolens* L., Syst. Nat. ed. 10. 1110. 1759.

Erect undershrubs, about 2 m tall; branchlets quandrangular, hispid. Leaves ovate, 1.2-5 × 1-4 cm, strigose, base truncate, apex acute, margin irregularly serrate. Flowers axillary and terminal, in whorls, blue. Nutlets ovate-oblong.

Fls. & Frs. : December - April.

Common weed of waste lands. Toopran, **BR & CP** 11284.

LEONOTIS Robert Brown

Leonotis nepetifolia (L.) R.Br., Prodr. 504. 1810; FBI 4:691; Gamble 1155 (807) ("*nepetaefolia*"). *Phlomis nepetifolia* L., Sp. Pl. 586. 1753 ("*nepetaefolia*").

Erect undershrubs, about 1.5 m tall; branchlets quadrangular, pubescent. Leaves ovate, 2.5-6 × 1.5-5 cm, softly pubescent, base truncate, apex acute, margin deeply crenate-serrate. Flowers axillary, in dense whorls, orange. Nutlets oblong, trigonous.

Fls. & Frs. : October - March. *Vern.* : Ranaberi.

Common in hedges, waste lands and riverbanks. Pocharam tank, **TP** & **CP** 12054.

LEUCAS Robert Brown

1. Flowers 1-4, not in whorls **L. biflora**
1. Flowers more than 6, in whorls:
    2. Calyx mouth straight **L. stricta**
    2. Calyx mouth oblique :
        3. Whorls very large **L. cephalotes**
        3. Whorls small:
            4. Calyx tube erect, not bent in the middle:
                5. Calyx tube glabrous in the lower half, hispid in the upper **L. aspera**
                5. Calyx tube uniform **L. zeylanica**
            4. Calyx tube bent in the middle **L. nutans**

Leucas aspera (Willd.) Link, Enum. Hort. Berol. Alt. 2:113. 1822; FBI 4:690; Gamble 1150 (803). *Phlomis aspera* Willd., Enum. Hort. Berol. 621. 1809.

Diffuse herbs, about 50 cm tall; branchlets quadrangular, hispid. Leaves linear-lanceolate, 1-3 × 0.2-0.5 cm, pubescent, base cuneate, apex acute, margin entire - serrate. Flowers terminal, in verticillasters, white. Nutlets trigonous.

Fls. & Frs. : October - May. *Vern.* : Thummi.

Common weed of cultivated fields. Nagulabanda, **CP** 11324; Pathur, **TP** & **CP** 12120.

Leucas biflora (Vahl) R.Br., Prodr. 504. 1810; FBI 4:683; Gamble 1150 (804). *Phlomis biflora* Vahl, Symb. Bot. 3:42. 1794.

Straggling or diffuse herbs, about 50 cm tall; branchlets quadrangular, pubescent. Leaves ovate- elliptic, 1-4 × 0.6-2 cm, pubescent, base cuneate, apex acute, margin crenate-serrate. Flowers axillary, 2-4 per node, white. Nutlets trigonous.

Fls. & Frs. : December - April.

Common near water courses, in waste lands, and forest borders. Pochammaralu, **BR** & **CP** 11494; NSF, **BR** & **CP** 11656; Gangapur RF, **TP** & **CP** 12091.

Leucas cephalotes (Roth) Spreng. in L., Syst. Veg. 2: 743. 1825; FBI 4: 689; Gamble 1150 (804). *Phlomis cephalotes* Roth, Nov. Pl. Spec. 262. 1821.

Erect herbs, about 75 cm tall; branchlets quadrangular, pubescent. Leaves ovate-lanceolate, 2-8 x 1-3 cm, pubescent, base cuneate, apex obtuse-acute, margin crenate-serrate. Flowers terminal, in dense globose heads, white. Nutlets ovate, brown.

Fls. & Frs. : July - February.

Occasional in forests, cultivated fields and waste lands. Narsapur RF, **TP** & **CP** 14027.

Leucas nutans Spreng. in L, Syst. Veg. ed. 16. 2:743. 1825; FBI 4: 688; Gamble 1149 (803).

Erect herbs, about 40 cm tall; branchlets quadrangular. hispid. Leaves lanceolate, 1.5-4 × 0.5-1 cm, hispid, base cuneate, apex acute, margin serrate. Flowers axillary and terminal, in globose heads, white. Nutlets ovate.

Fls. & Frs. : June - December.

Occasional in forests. Narsapur RF, **TP** & **CP** 14053.

Leucas stricta Benth. in Wall., Pl. As. Rar. 1:61. 1824; FBI 4:688; Gamble 1150 (804).

Erect herbs, about 30 cm tall; branchlets quadrangular, grooved. Leaves oblong-lanceolate, 2.5-5 × 0.5-1 cm, hispid, base narrow, apex obtuse, margin entire. Flowers terminal, in globose heads, white. Nutlets oblong.

Fls. & Frs. : July - December.

Rare in rocky hills. Gummadidala, **GNR** 4509.

Leucas zeylanica R. Br., Prodr. 504. 1810; FBI 4:689; Gamble 1150 (804).

Erect herbs, about 40 cm tall; branchlets quadrangluar. hispid. Leaves linear-lanceolate, 3-7.5 × 0.2 -1 cm, hispid, base tapering, apex obtuse, margin entire- serrulate. Flowers terminal, in globose heads, white. Nutlets obovate-oblong.

Fls. & Frs. : June - October.

Rare along road sides. Narsapur, **GNR** 3997.

OCIMUM Linnaeus

1. Calyx with glands **O. gratissimum**
1. Calyx without glands:
    2. Peduncle 1 or 2 times branched; calyx glabrous within **O. tenuiflorum**
    2. Peduncle 3 times branched; calyx hairy within **O. americanum**

Ocimum americanum L., Cent. Pl. 1: 15. 1755. *O. canum* Sims in Curtis, Bot. Mag. t. 2452. 1823; FBI 4: 607; Gamble 1111 (777); Khan in FFH 276.

Erect herbs, about 60 cm tall; branchlets quadrangular, pubescent. Leaves elliptic-lanceolate, 1-3 × 0.5-1 cm, sparsely pubescent, base rounded, apex acute, margin entire. Flowers terminal spiciform racemes, white. Nutlets trigonous, mucilaginous when wet.

Fls. & Frs. : Round the year. *Vern.* : Tulasi.

Common in fields and waste lands. Medak, **BR & CP** 11420.

Ocimum gratissimum L., Sp. Pl. 1197. 1753; FBI 4:608; Gamble 1111 (777).

Erect herbs, about 50 cm tall; branchlets quadrangular, glabrous. Leaves obovate, 2-8 × 1.2-6 cm, pubescent, base attenuate, apex acuminate, margin distantly serrate. Flowers terminal in trichotomous racemes, white. Nutlets trigonous, mucilaginous when wet.

Fls. & Frs. : Round the year.

Common in waste lands. Narsapur-Hyderabad road, **BR & CP** 11213.

Ocimum tenuiflorum L., Sp. Pl. 597. 1753. *O. sanctum* L., Mant. Pl. 85. 1767; FBI 4: 609; Gamble 1111 (778); Khan in FFH 277.

Erect herbs, about 60 cm tall; branchlets terete, hispid. Leaves elliptic-oblong, 2-3.5 × 0.6-1 cm, pubescent, base cuneate, apex acute, margin serrate. Flowers terminal, in trichotomous racemes, white. Nutlets ellipsoid.

Fls. & Frs. : Round the year. *Vern.* : Tulasi.

Common in villages. Cultivated as a Sacred plant. Narsapur, **GNR** 3921.

PLECTRANTHUS L'Heritier *nom. cons.*

Plectranthus mollis (Aiton) Spreng in L. Syst. Veg. ed. 16.2: 690. 1825. *Ocimum molle* Aiton, Hort. Kew. 2: 322. 1789. *Plectranthus incanus* Link, Enum. Hort. Berol. Alt. 2: 1822; FBI 4:621; Gamble 1121 (784).

Erect herbs, about 80 cm tall, branchlets quadrangular, puberulous. Leaves ovate, 2-8 × 0.5-4.5 cm, glabrous above, puberulous below, base cordate, apex acute, margin crenate. Flowers terminal, in trichotomous racemes, pale-blue. Nutlets trigonous.

Fls. & Frs. : June - December.

Occasional in waste lands. Pillutla RF, **BR** & **CP** 11270.

SALVIA Linnaeus

Salvia plebeia R.Br., Prodr. 501. 1810; FBI 4: 665; Gamble 1156 (808).

Erect herbs, about 50 cm tall; branchlets quadrangular, pubescent. Leaves oblong-lanceolate, 1-6 × 0.8-2 cm, sparsely hairy, base decurrent, apex acute, margin irregularly crenate. Flowers terminal, in simple or trichotomous racemes, yellow. Nutlets ovate.

Fls. & Frs. : September - February.

Common along river banks. Medak, **BR** & **CP** 11411.

SCUTELLARIA Linnaeus

Scutellaria violacea Heyne ex Benth. in Wallich, Pl. As. Rar. 1:66. 1830; FBI 4: 668; Gamble 1142 (798).

Herbs, about 30 cm tall; branchlets hirsute. Leaves ovate, 3.5-7 × 2-6 cm, chartaceous, base cordate, apex acute-acuminate, margin crenate. Flowers terminal, in racemes, violet. Nutlets transverse.

Fls. & Frs. : Round the year.

Occasional in fields. Venkatapuram, **TP** & **CP** 14007.

Mentha piperata L. is cultivated for its leaves.

PLANTAGINACEAE

PLANTAGO Linnaeus

Plantago asiatica L., Sp. Pl. 113. 1753; FBI 4:705; Gamble 1160 (811).

Herbs. Leaves radical, in rosettes, elliptic-ovate, 5-11 × 5-6.5 cm, chartaceous, base rounded acute, apex obtuse, margin entire. Flowers terminal, in spikes, cream. Capsules ellipsoid, 0.4 × 0.2 cm; seeds 25.

Fls. & Frs. : Round the year.

Occasional in moist places. Ranjhole, **CP** 11600.

NYCTAGINACEAE

1. Flowers distinctly pedicelled; anthocarps 10-ribbed **B. chinensis**
1. Flowers sessile; anthocarps 5-ribbed **B. diffusa**

Boerhavia chinensis (L.) Aschers. & Schweinf. Beitr. Fl. Aeth. 1:167. 1867 ("*Boerhaavia*"). *Valerina chinensis* L., Sp. Pl. 33. 1753. *Boerhavia repanda* Willd., Sp. Pl. 1:22. 1797; FBI 4:709; Gamble 1163 (814).

Diffuse herbs, about 40 cm tall; branchlets glabrous. Leaves ovate, 1-5 - 5.5 × 1-5 cm, glabrous, base cordate, apex acute, margin sinuate. Flowers axillary and terminal, in umbels, pink. Anthocarps club-shaped, viscid-glandular; seed 1.

Fls. & Frs. : September - February.

Occasional in forests. Narsapur RF, **TP & CP** 12016.

Boerhavia diffusa L., Sp. Pl. 3. 1753; Gamble 1162 (814); Khan in FFH 277. *B. repens* L., Sp. Pl. 3. 1753; FBI 4:709.

Diffuse herbs, about 1 m tall; branchlets sparsely pubescent. Leaves ovate-orbicular, 0.5-2 × 0.5-1 cm, glabrous, base truncate-rounded, apex acute, margin entire-undulate. Flowers axillary, in umbels, pink-violet. Anthocarps ellipsoid, glandular hairy; seed 1.

Fls. & Frs. : Round the year. *Vern.* : Atikamamidi.

Common weed in waste lands and hedges. Narsingi, **BR & CP** 11526.

Bougainvillea glabra Choisy, *B. spectabills* Willd. and *Mirabills jalapa* L. are grown as ornamentals.

AMARANTHACEAE

1. Leaves linear, not exceeding 0.1 cm **Trichurus**
1. Leaves not as above :
    2. Leaves alternate :
        3. Tepals persistent on fruiting spikes **Allmania**
        3. Tepals falling off with fruits:
            4. Branchlets woolly - tomentose **Aerva**
            4. Branchlets glabrous :
                5. Tepals variously modified in fruit **Digera**
                5. Tepals not so modified :
                    6. Flowers unisexual, green **Amarnathus**
                    6. Flowers bisexual, white **Celosia**
    2. Leaves opposite:
        7. Prostrate herbs **Alternanthera**
        7. Erect herbs or undershrubs:
            8. Flowers in globose of oblong heads **Gomphrena**

8. Flowers in spikes :
   9. Fruits with bristles **Pupalia**
   9. Fruits without bristles **Achyranthes**

ACHYRANTHES Linnaeus

Achyranthes aspera L., Sp. Pl. 204. 1753. var. **aspera**: FBI 4:730; Gamble 1176 (823).

Erect subshrubs, about 1 m tall. Leaves obovate-orbicular, 2-6.5 × 1-4 cm, pubescent, base attenuate, apex broadly acute, margin entire. Flowers terminal, in spikes, green. Utricles ovate, 0.4 × 0.2 cm, tip spiny; seed 1.

Fls. & Frs. : Round the year. *Vern.* : Uthareni.

Common weed of waste lands. Narsapur, **BR** & **CP** 11241.

AERVA Forsskal *nom. cons.*

Aerva lanata (L.) A.L. Juss. in Ann. Mus. Natl. His. Nat. 2:131. 1803; FBI 4:728; Gamble 1178 (825). *Achyranthes lanata* L., Sp. Pl. 204. 1753.

Erect herbs, about 70 cm tall. Leaves ovate-orbicular, 1.5-3 × 0.8-1.5 cm, pubescent above, white woolly below, base cuneate, apex acute, margin entire. Flowers axillary and terminal, in spikes, white. Utricles ovate, 0.1 cm across, tip acute; seed 1.

Fls. & Frs. : September - March.

Common weed of waste lands. Ponnala, **CP** 11368.

ALLMANIA Robert Brown

Allmania nodiflora (L.) R. Br. ex Wight in Hook., Journ. Bot. 1:227. t. 128. 1834; FBI 4:716; Gamble 1167 (817). *Celosia nodiflora* L., Sp. Pl. 205. 1753.

Erect herbs, about 25 cm tall; branchlets angled, hispid. Leaves oblong-spathulate or elliptic, 2-6 × 1-1.8 cm, sparsely pubescent, base cuneate-attenuate, apex apiculate, margin entire. Flowers axillary, in globose heads, yellow-brown. Utricles subglobose, 0.3 cm across, membranous; seed 1.

Fls. & Frs. : June - December.

Common in all places. Nagulbanda, **CP** 11330; Narsapur RF, **TP** & **CP** 14059

ALTERNANTHERA Forsskal

1. Bracts and bracteoles glabrous :
   2. Leaves obliquely elliptic-orbicular; flower heads prickly **A. pungens**

2. Leaves lanceolate-spathulate; flower heads not prickly **A. sessilis**
1. Bracts and bracteoles barbellate hairy **A. tenella**

Alternanthera pungens Kunth, Nov. Gen. Sp. 2:206. 1818. *A. echinata* auct. non Smith : Tadulingam & Venkatanarayana, Handb. S. Ind. Weeds 274. t. 114. 1932.

Prostrate herbs; branchlets pubescent. Leaves obliquely elliptic-orbicular, 0.5-2.5 × 0.5-1.8 cm, pubescent, base attenuate, apex acute-apiculate, margin entire. Flowers axillary in spikes yellow. Utricles ovate, 0.4 × 0.2 cm, awned; seed 1.

Fls. & Frs. : Round the year.

Common in waste lands. Narsapur, **BR** & **CP** 11235.

Alternanthera sessilis (L.) R.Br. ex DC., Cat. Pl. Hort. Bot. Monsp. 4:77. 1813; FBI 4:731. *Gomphrena sessilis* L., Sp. Pl. 225. 1753. *Alternanthera triandra* Lam., Encycl. 1:95. 1783; Gamble 1179 (825).

Prostrate herbs, about 50 cm tall; branchlets ascending, glabrous. Leaves laneolate, 1-4 × 0.5-1 cm, glabrous, base cuneate, apex obtuse-acute, margin entire. Flowers axillary, in spikes, white. Utricles cordate, 0.1 × 0.2 cm, glabrous; seed 1.

Fls. & Frs. : Round the year. *Vern.* : Ponnagantaku

Common in moist waste places. Narsapur tank, **BR** & **CP** 11246; Medak, **BR** & **CP** 11463; Akkannapet, **BR** & **CP** 11557.

Alternanthera tenella Colla, Mem. Acad. Sci. Terine Furin 33:131. t. 9. 1820; Mears. Proc. Acad. Nat. Sci. Philadelphia 129: 19. 1977. *A. polygonoides* var. *erecta* Mart. Mears, *loc. cit*. in Syn. non sensu Shirvastava & Santapau. 1955.

Prostrate herbs; branchlets white-villous. Leaves elliptic- oblong, 1.5-3.5 × 0.6-1.5 cm, hairy above, base narrowed, apex subacute, margin entire. Flowers axillary, in ovate heads. Utricles suborbicular, indehiscent.

Fls. & Frs. : Round the year.

Occasional weed of cultivated fields and waste lands. Patancheru, **BR** & **CP** 11319; Zaheerabad, CP 11595.

AMARANTHUS Linnaeus

1. Plants armed **A. spinosus**
1. Plants unarmed:
   2. Prostrate herbs; flowers in axillary clusters **A. roxburghianus**
   2. Erect or ascending herbs; flowers in axillary and terminal panicles:
      3. Leaves < 2 cm in length; bracteoles longer than bracts **A. tricolor**
      3. Leaves > 2 cm in length; bracteoles shorter than bracts **A. viridis**

Amaranthus roxburghianus Nevski in Act. Inst. Bot. Acad. Sci. USSR Ser. 1.4:311. 1937. in Obs ("*roxburghiana*"); Nair in J. Bombay Nat. Hist. Soc. 73:60. 1976. *A. graecizans* L., Sp. Pl. 990. 1753. *A. polygamus* auct. non L: FBI 4: 721 *p.p.*; Gamble 1171 (820).

Prostrate herbs, Leaves obovate, 0.5-1.5 × 0.5-1 cm, base obtuse, apex notched, margin entire. Flowers axillary, in clusters, green. Utricles ovate, 0.2 × 0.1 cm, rugose; seed 1.

Fls. & Frs. : December - April.

Common weed of waste lands and cultivated fields. Dundigal, **GNR** 4639.

Amaranthus spinosus L., Sp.Pl. 991. 1753; FBI 4: 718; Gamble 1170 (819); Khan in FFH 279.

Erect herbs, about 50 cm tall. Leaves elliptic-lanceolate, 1-3.5 × 0.5-1.5 cm, glabrous, base acute, apex acute-emarginate, margin entire. Flowers axillary and terminal, in green panicles. Utricles ovate, 0.2 × 0.1 cm, circumscissile; seed 1.

Fls. & Frs. : Round the year. *Vern.* : Nalladoggata.

Common in waste lands and fields. Mombojipally, **BR** & **CP** 11637.

Amaranthus tricolor L., Sp. Pl. 989. 1753. *A. mangostanus* L., Cent Pl. 1:32 1755; FBI 4:720; Gamble 1171 (820). *A. gangeticus* L., Syst. Nat. ed. 10.2 : 1268. 1759; FBI 4:719.

Diffuse herbs, about 40 cm tall; branchlets glabrous. Leaves ovate-obovate, 0.5-1.5 × 0.3-0.7 cm, glabrous, base attenuate, apex retuse, margin entire. Flowers axillary and terminal, in fascicles, yellowish-green. Utricles ovate, 0.1 cm across, circumscissile; seed 1.

Fls. & Frs. : Round the year. *Vern.* : Chirraku.

Occasional in waste lands and fields. Rampur, **BR** & **CP** 11654.

Amaranthus viridis L., Sp. Pl. ed. 2:1405. 1763; FBI 4:720 Gamble 1171 (820).

Erect herbs, about 1 m tall; branchlets glabrous. Leaves ovate-rhomboid, 2-4 × 0.5-2.5 cm, glabrous, base cuneate, apex retuse, margin entire. Flowers axillary or terminal, in spikes, greenish-yellow. Utricles ovate, 0.2 cm across, circumscissile; seed 1.

Fls. & Frs. : Round the year. *Vern.* : Thotakura.

Common weed in waste lands, fields and moist places. NSF, **BR** & **CP** 1161.

CELOSIA Linnaeus

Celosia argentea L, Sp. Pl. 205. 1753; FBI 4: 714; Gamble 1166 (816); Khan in FFH 278.

Erect herbs, about 1.5 m tall; branchlets grooved, glabrous. Leaves ovate below linear-lanceolate above, 3-6.5 × 0.5-1.8 cm, glabrous, base cuneate, apex acute, margin entire. Flowers terminal, in dense spikes, dullwhite. Utricles globose, 0.4 cm across; seeds 4-10.

Fls. & Frs. : July - November. *Vern.* : Garugaku.

Common weed in cultivated fields. Narsapur, **BR & CP** 11216; Medak, **CP** 11796.

DIGERA Forsskal

Digera muricata (L.) C. Martius, Beitr. Amar. 77. 1825. *Achyranthes muricata* L., Sp. Pl. ed. 2:295. 1762. *Digera arvensis* Forssk., Aegypt. - Arab. 65. 1775; FBI 4:717; Gamble 1169 (818).

Erect herbs, about 75 cm tall; branchlets terete, glabrous. Leaves elliptic-ovate, 2-4 × 1-3 cm, glabrous, base truncate, apex acute, margin entire. Flowers axillary and terminal, in spikes, pink. Utricles ovate 0.3 × 0.2 cm; seed 1.

Fls. & Frs. : June - November.

Common weed in cultivated fields and waste places. Lingareddypally, **CP** 11717.

GOMPHRENA Linnaeus

Gomphrena serrata L., Sp. Pl. 224. 1753; Mears in Taxon 29:86. 1980. *G. decumbens* Jacq., Hort Schoenbr. t. 484. 1805; Gamble 1179 (825). *G. celosioides* auct. non C. Martius : Bennet, Fl. Howrah 142. 1979.

Erect herbs, about 30 cm tall; branchlets pubescent. Leaves lanceolate-spathulate, 1-2.5 × 0.2-0.7 cm, glabrous, base cuneate, apex acute, margin entire. Flowers axillary, in globose heads, white. Utricles globose, 0.2 cm across; seed 1.

Fls. & Frs. : June - December.

Common in waste lands and cultivated fields. Sangareddy tank, **BR & CP** 11289.

Note : According to Mears (*loc. cit.*) the species to which the name *G. serrata* L. applies was re-named by Jacquin (1805) as *G. decumbens* without reffering Linnaeus taxon. Hence the former name is retained.

PUPALIA A.L. Jussieu *nom. cons.*

Pupalia lappacea (L.) A.L. Juss. in Ann. Mus. Natl. Hist. Nat. (Paris) 2: 132. 1803; FBI 4: 724; Gamble 1173 (821). var. **lappacea**. *Achyranthes lappacea* L., Sp. Pl. 204. 1753. *Pupalia atropurpurea* (Lam.) Moq. in DC., Prodr. 13(2): 331. 1849; FBI 4: 723; Gamble 1173 (821). *Achyranthes atropurpurea* Lam., Encycl. 1:546. 1785.

Undershrubs, about 1 m tall; branchlets pubescent. Leaves obovate, 2-10 x 1-7 cm, glabrous, base cuneate, apex apiculate acuminate, margin entire. Flowers axillary and terminal, in spikes, green. Utricles globose, 0.3 cm across; seed 1.

Fls. & Frs. : July - December.

Common in field hedges and road sides. Nathyanipally, **BR** & **CP** 11250; Medak, **CP** 11797.

TRICHURUS C. Townsend

Trichurus monsoniae (L.f.) Towns. in Kew Bull. 29:466. 1974. *Illecebrum monsoniae* L.f., Suppl. Pl. 161. 1781. *Aerva monsoniae* (L.f.) C. Martius in Nov. Act. Phys. - Med. Acad. Caes. - Leop. Carol. Nat. Curis 13(1): 291. 1826; FBI 4:729; Gamble 1178 (825).

Prostrate herbs, about 50 cm tall; branchlets woolly-tomentose. Leaves linear, 1-2.5 × 0.1 cm, fascicled at nodes, subglabrous, spine-tipped. Flowers in solitary spikes, palepink. Utricles ovate, 0.3 × 0.2 cm, circumscissile; seed 1.

Fls. & Frs. : July - November.

Common in all places. Sangareddy, **BR** & **CP** 11287; Pochammaralu, **BR** & **CP** 11479.

Gomphrena globosa L., is grown for ornamental purpose.

CHENOPODIACEAE

CHENOPODIUM Linnaeus

Chenopodium album L., Sp. Pl. 219. 1753; FBI 5: 3; Gamble 1181 (827).

Erect herbs, about 1 m tall; branchlets grooved, glabrous. Leaves ovate-lanceolate, 1-5 × 0.5-2 cm, glabrous, base cuneate, apex obtuse, margin irregularly lobed. Flowers axillary or terminal, in panicles, green. Utricles ovate, 0.2 cm across; seed 1.

Fls. & Frs.. : August - March. *Vern.* : Pappukura.

Occasional weed of cultivated fields. Rampur, **BR** & **CP** 11652.

Spinacia oleracea L. is cultivated for its leaves.

BASELLACEAE

BASELLA Linnaeus

Basella alba L., Sp. Pl. 272. 1753. *B. rubra* L., Sp. Pl. 272. 1753; FBI 5: 20; Gamble 1185 (830).

Herbaceous vine, stem fleshy, glabrous. Leaves ovate- oblanceolate, 1-4 × 1.25 cm, glabrous, fleshy, base attenuate, apex acute-acuminate, margin entire. Flowers axillary, in spikes, dull white-red. Utricles globose, 0.3 cm across, glabrous; seed 1.

Fls. & Frs. : July - November. *Vern.* : Bachhali.

Common in villages. Mambojipally, **BR** & **CP** 11641; Choutkoor, **TP** & **CP** 11925.

POLYGONACEAE

1. Tendril climbers **Antigonon**
1. Erect or prostrate herbs **Polygonum**

ANTIGONON Endlicher

Antigonon leptopus Hook. & Arn., Bot. Beechey Voy. 308. t. 69. 1841; Gamble 1192 (835).

Climbing shrubs; branchlets angled, pubescent. Leaves hastate, 3-10 × 2-4.5 cm, glabrous, base cordate, apex acute-subacuminate, margin entire. Flowers in panicles, rose. Nutlets trigonous, 0.8 × 0.6 cm, glabrous.

Fls. & Frs. : Round the year.

Occasional in waste lands, villages and towns. Medak, **CP** 11744.

POLYGONUM Linnaeus *nom. cons.*

1. Prostrate herbs; leaves; < 0.3 cm broad **P. plebeium**
1. Erect undershrubs; leaves > 1 cm broad **P. glabrum**

Polygonum glabrum Willd., Sp. Pl. 2: 447. 1799; FBI 5: 34; Gamble 1189 (832). var. **glabrum**.

Undershrubs, about 1.5 m tall; branchlets glabrous. Leaves lanceolate, 8-18 × 0.5-2 cm, glabrous, base attenuate, apex gradually acuminate, margin entire. Flowers axillary and terminal, in racemes, pink. Nutlets globose, 0.3 cm across, biconvex, glabrous.

Fls. & Frs. : Round the year.

Common near drainage canals and margins of tanks. Medak, **BR** & **CP** 11465, **CP** 11728.

Polygonum plebeium R. Br. Prodr. 420. 1810: FBI 5:27; Gamble 1188 (832).

Prostrate herbs, branchlets striate, glabrous. Leaves oblong, 0.6-1.3 × 0.1-0.3 cm, glabrous. base attenuate, apex acute, margin entire. Flowers axillary, solitary or 2-3 in clusters, rose. Nutlets trigonous, 0.2 × 0.1 cm, pubescent.

Fls. & Frs. : Round the year.

Common in moist places. Ramayampet, **BR** & **CP** 11517.

Rumex vesicarius L. is cultivated for its leaves.

ARISTOLOCHIACEAE

ARISTOLOCHIA Linnaeus

Aristolochia indica L., Sp. Pl. 960. 1753; FBI 5: 75; Gamble 1202 (841).

Twining shrubs, branchlets grooved, glabrous. Leaves oblong- obovate, 8-11 x 4-5.5 cm glabrous, base truncate-rounded, apex acuminate, margin entire. Flowers axillary, in racemes, greenish-white. Capsules oblong, 4 x 2 cm, 6-valved; seeds many.

Fls. & Frs. : August - February.

Occasional in forests. Nagulabanda, **CP** 11334.

LAURACEAE

CASSYTHA Linnaeus

Cassytha filiformis L., Sp. Pl. 35. 1753; FBI 5:188; Gamble 1241 (868); Khan in FFH 282.

Twining parasites, stem dark green, glabrous. Leaves O. Flowers axillary or terminal, in spikes, yellow. Drupes globose, 0.4 cm across, enclosed in a fleshy perianth tube, crowned by lobes, glabrous.

Fls. & Frs. : August - December.

Common on shrubs and small trees. Pochammaralu, **BR & CP** 11481; Akkannapet, **BR & CP** 11555.

LORANTHACEAE

1. Flowers actinomorphic; bracts enclosing calyx-base **Dendrophthoe**
1. Flowers zygomorphic; bracts free from calyx:
    2. Corolla lobes 4; calyx margin truncate **Scurrula**
    2. Corolla lobes 5; calyx margin toothed **Taxillus**

DENDROPHTHOE C.F.P. Martius

Dendrophthoe falcata (L.f.) Ettingsh. in Akad. Wiss. Math. Naturwiss. Kl. Denkschr. 32: 68.t. 13.f. 14 & 15. 1872. *Loranthus falcatus* L.f., Suppl. Pl. 211. 781. *L. longiflorus* Desr. in Lam., Encycl. 3:598. 1792; FBI 5:214; Gamble 1253(877): Khan in FFH 284.

Parasitic shrubs; branchlets glaucous. Leaves elliptic-lanceolate, 6-12 × 2-3.5 cm, glabrous, base cuneate, apex acute- rounded, margin entire. Flowers axillary, in racemes, yellow- orange. Berries oblong, 1 × 0.6 cm, glabrous; seed 1.

Fls. & Frs. : Round the year. : Wajinika.

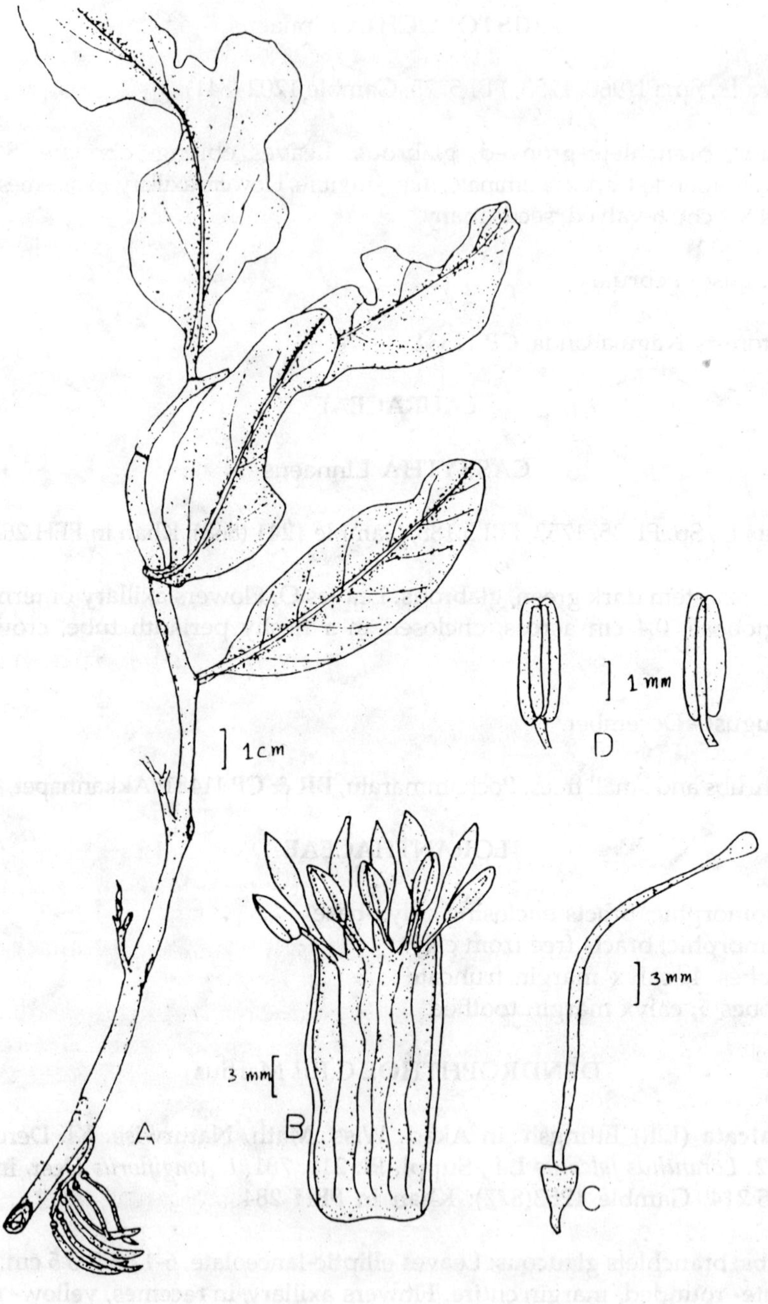

Fig. 6.23: *Dendrophthoe falcata* (L.f.) Ett. A. Twig, B. Corolla split open, C. Pistil, D. Stamens

Common in all forests on trees. Gangapur RF, **BR & CP** 11442 and 11443; Narsapur RF, **TP & CP** 11971.

SCURRULA Linnaeus

Scurrula parasitica L., Sp. Pl. 110. 1753. *Loranthus buddleioides* Desr. in Lam., Encycl. 3:600. 1792; Gamble 1251 (876). *L. scurrula* (L.) L., Sp. Pl. (ed.2) 472. 1762; Khan in FFH 285.

Stem parasites; branchlets rusty-tomentose. Leaves elliptic- ovate, 3-7 × 2-4.6 cm, tomentose, base rounded, apex rounded, margin entire. Flowers axillary, in fascicles, scarlet. Berries pyriform, 0.8 × 0.4 cm, tomentose; seed 1.

Fls. & Frs. : August - December.

Common in forests on trees, Ponnala, **CP** 11322; Gangapur RF, **BR & CP** 11441 and **TP** & **CP** 12074.

TAXILLUS van Tieghem

Taxillus bracteatus (Wall.) Tieghem in Bull. Soc. Bot. France 42. 256. 1895. *Loranthus bracteatus* Heyne ex Wall. in Roxb. Fl. Ind. 2:220. 1824; FBI 5:213; Gamble 1252 (876).

Parasitic shrubs; branchlets tawny-tomentose. Leaves orbicular-obovate, 1.5-4.5 × 1-4 cm, softly white grey, base rounded-cuneate. Flowers axillary, in cymes or umbels, greenish-yellow. Berries oblong, 1 × 0.5 cm; seed 1.

Fls. & Frs. : Round the year.

Occasional in forests on deciduous trees. Narsapur, **GNR** 4732.

VISCACEAE

VISCUM Linnaeus

1. Leaves absent or reduced, internodes flattened **V. nepalense**
1. Leaves present, internodes terete **V. heyneanum**

Viscum heyneanum DC., Prodr. 4: 278. 1830. *V. verruculosum* Wight & Arn., Prodr. 379. 1834; FBI 5: 224; Gamble 1259(881). *V. orbiculatum* Wight, Ic. t. 1016. 1845; FBI 5: 224; Gamble 1259 (881).

Semiparasitic herbs; branchlets terete. Leaves obovate-oblong, coriaceous, base attenuate, apex obtuse-acute. Flowers axillary, in fascicles. Berries oblong, warty.

Fls. & Frs. : November - May. *Vern.* : Badanika.

Rare in forests. Sathaiapalle, **GNR** 4529.

Viscum nepalense Spreng., Syst. Curi-post. 47. 1827. *V. articulatum* auct. non Burm.f: FBI 5: 226; Gamble 1259 (881); Khan in FFH 285.

Parasitic shrubs, dichotomously branched, branches jointed; internodes 3-6 × 0.2-0.6 cm, flattened, green, terete. Leaves O. Flowers at nodes, in fascicles purple. Berries globose, 0.4 cm across, glabrous; seed 1.

Fls. & Frs. : Round the year.

Rare in forests. Akkannapet, **BR & CP** 11558.

SANTALACEAE

SANTALUM Linnaeus

Santalum album L., Sp. Pl. 349. 1753; FBI 5: 231; Gamble 1261 (883); Khan in FFH 286.

Trees, about 5 m tall. Leaves elliptic-ovate, 2-5 × 1.5-2.5 cm, glabrous, base acute, apex acute, margin entire. Flowers axillary and terminal, in cymes, purple. Drupes globose, glabrous.

Fls. & Frs. : June - March. *Vern.* : Chandan.

Occasional in field hedges. Zaheerabad, **CP** 11593.

EUPHORBIACEAE

1. Inflorescene a cyathium **Euphorbia**
1. Inflorescence not a cyathium:
    2. Stem with stinging hairs **Tragia**
    2. Stem without stinging hairs:
        3. Petals present in both sexes:
            4. Trees or shrubs:
                5. Branchlets fulvous - tomentose **Givotia**
                5. Branchlets glabrous or pubescent:
                    6. Leaves cordate at base **Jatropha**
                    6. Leaves not cordate at base:
                        7. Fruit a drupe, ovary 2-celled **Bridelia**
                        7. Fruit a capsules, ovary 3-celled **Cleistanthus**
            4. Herbs:
                8. Leaves lobed **Chrozophora**
                8. Leaves serrate **Croton**
        3. Petals absent in both sexes:
            9. Locules with 1 ovule:
                10. Female flowers with accrescent leafy bracts **Acalypha**
                10. Female flowers not as above **Sebastiania**
            9. Locules with 2 ovules:

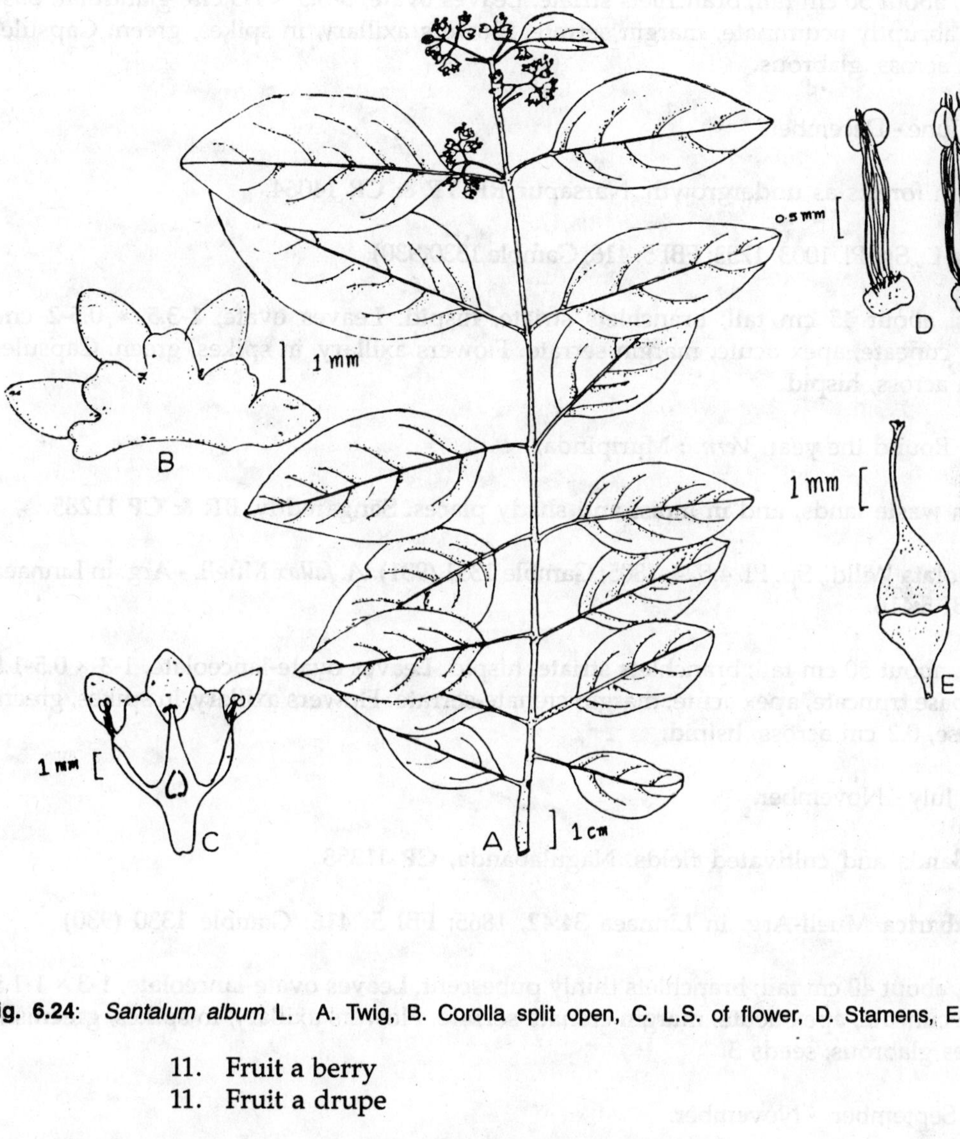

Fig. 6.24: *Santalum album* L. A. Twig, B. Corolla split open, C. L.S. of flower, D. Stamens, E. Pistil

11. Fruit a berry **Phyllanthus**
11. Fruit a drupe **Breynia**

ACALYPHA Linnaeus

1. Capsules longer than the bracts:
   2. Capsules hispid, bracts with glandular hairs **A. malabarica**
   2. Capsules glabrous; bracts without glandular hairs **A. lanceolata**
1. Capsules shorter than the bracts
   3. Spikes < 3 cm in length; capsules glabrous **A. ciliata**
   3. Spikes > 4 cm in length; capsules hispid **A. indica**

Acalypha ciliata Forssk., Fl. Aegypt.-Arab. 162. 1775; FBI 5:417; Gamble 1330 (930).

Erect herbs, about 50 cm tall; branchlets striate. Leaves ovate, 3-5.5 × 1-5 cm, glandular, base rounded, apex abruptly acuminate, margin serrate. Flowers axillary, in spikes, green. Capsules globose, 0.3 cm across, glabrous.

Fls. & Frs. : June - December.

Occasional in forests as undergrowth. Narsapur RF, **TP & CP** 14064.

Acalypha indica L., Sp. Pl. 1003. 1753; FBI 5: 416; Gamble 1330(930).

Erect herbs, about 45 cm tall; branchlets striate, hispid. Leaves ovate, 1-3.5 × 0.5-2 cm, glandular, base cuneate, apex acute, margin serrate. Flowers axillary, in spikes, green. Capsules globose, 0.2 cm across, hispid.

Fls. & Frs. : Round the year. *Vern.* : Muripinda.

Common in waste lands, and in moist and shady places. Sangareddy, **BR & CP** 11285.

Acalypha lanceolata Willd., Sp. Pl. 4:524, 1805; Gamble 1331 (931). *A. fallax* Muell. - Arg. in Linnaea 34: 43. 1865; FBI 5:416.

Erect herbs, about 50 cm tall; branchlets striate, hispid. Leaves ovate-lanceolate, 1-3 × 0.5-1.5 cm, glandular, base truncate, apex acute, margin crenate-serrate. Flowers axillary, in spikes, green. Capsules globose, 0.2 cm across, hsipid.

Fls. & Frs. : July - November.

Rare in waste lands and cultivated fields. Nagulabanda, **CP** 11358.

Acalypha malabarica Muell-Arg. in Linnaea 34:42, 1865: FBI 5: 416; Gamble 1330 (930).

Erect herbs, about 40 cm tall; branchlets thinly pubescent. Leaves ovate-lanceolate, 1-3 × 1-1.5 cm hispid, base cuneate, apex acute, margin crenate-serrate. Flowers axillary, in spikes, greenish-yellow. Capsules glabrous; seeds 3.

Fls. & Frs. : September - November.

Occasional in moist shady places. Dundigal, **GNR** 4640.

BREYNIA J.R. Forster et J.G.A. Forster *nom. cons.*

Breynia vitisidaea (Burm, f.) C. Fischer in Bull. Misc. inform. 1932: 65 1932. *Rhamnus vitis-idiaea* Burm.f., Fl. Ind. 61. 1768. *Breynia rhamnoides* (Retz.) Muell - Arg. in DC., Prodr. 15(2) : 440. 1866; FBI 5: 330; Gamble 1304 (912). *Phyllanthus rhamnoides* Retz., Obs. Bot. 5:30. 1788.

Shrubs, about 3 m tall; branchlets glabrous. Leaves elliptic- ovate, 2-4 × 1.5-3 cm, membranous, base rounded, apex obtuse-acute, margin entire. Flowers axillary, solitary or in clusters, male yellow, female red. Capsules globose, 0.6 cm across, glabrous; seeds 6.

Fls. & Frs. : Round the year. *Vern.* : Yellari.

Occasional in forests and waste lands. Way to Rusthempet, **KMS** 6790.

BRIDELIA Willdenow

1. Shrubs; Leaves with less than 15 pairs of main nerves **B. montana**
1. Trees; leaves with more than 15 pairs of main nerves **B. retusa**

Bridelia montana Willd., Sp. Pl. 4:978. 1806; FBI 5: 269 *p.p.*, Gamble 1290 (896). *B. hamiltoniana* Wall. ex Muell-Arg. in Linnaea 34: 77 1865; FBI 5: 271; Khan in FFH 292.

Shrubs, about 3 m tall; branchlets glabrous. Leaves ovate- obovate, 3-8.5 × 1.5-5 cm, glabrous, base cuneate, apex obtuse- rounded or retuse, margin entire. Flowers axillary, in clusters, purple. Drupes globose, 0.2 cm across, glabrous; seeds 1-2.

Fls. & Frs. : May - September. *Vern.* : Panchothkam.

Common near streams in forests. Nagulabanda, **CP** 11349.

Bridelia retusa (L.) Spreng. in L., Syst. Veg. (ed.16) 3:48. 1826; FBI 5:268 p.p.; Gamble 1280 (896); Khan in FFH 292. *Cluytia retusa* L., Sp. Pl. 1042. 1753.

Trees, about 10 m tall; branchlets tomentose. Leaves elliptic-oblong, 8-13 × 4-6 cm, glabrous above, pubescent below, base rounded, apex obtuse-retuse, margin obscurely serrate. Flowers axillary and terminal, in tomentose spikes, purple-yellow. Drupes globose, 0.5 cm across, glabrous; seeds 2.

Fls. & Frs. : May - November. *Vern.* : Dura-maddi, Mulumaddi.

Common in forests. Narsapur **RF, TP & CP** 11992, 14058.

CHROZOPHORA A.H.L. Jussieu *nom. cons.*

1. Prostrate herbs; capsules black, stellate-tomentose **C. prostrata**
1. Erect herbs; capsules red, stellate-hairy **C. rottleri**

Chrozophora prostrata Dalz. in Daiz & Gibs, Bombay Fl.233. 1861. var. **parvifolia** (Klotzch ex Schweinf) Bala. in Bull. Bot. Surv. Ind. 15:7. 1973. *C. parvifolia* Klotzsch ex Schweinf Pl. Quaed. Nilot. 11. 1862; Gamble 1317 (921). *C. plicata* forma 2. FBI 5: 410.

Prostrate herbs; branchlets woolly-tomentose. Leaves broadly ovate-orbicular, 1.5 - 3 × 1-2.5 cm,

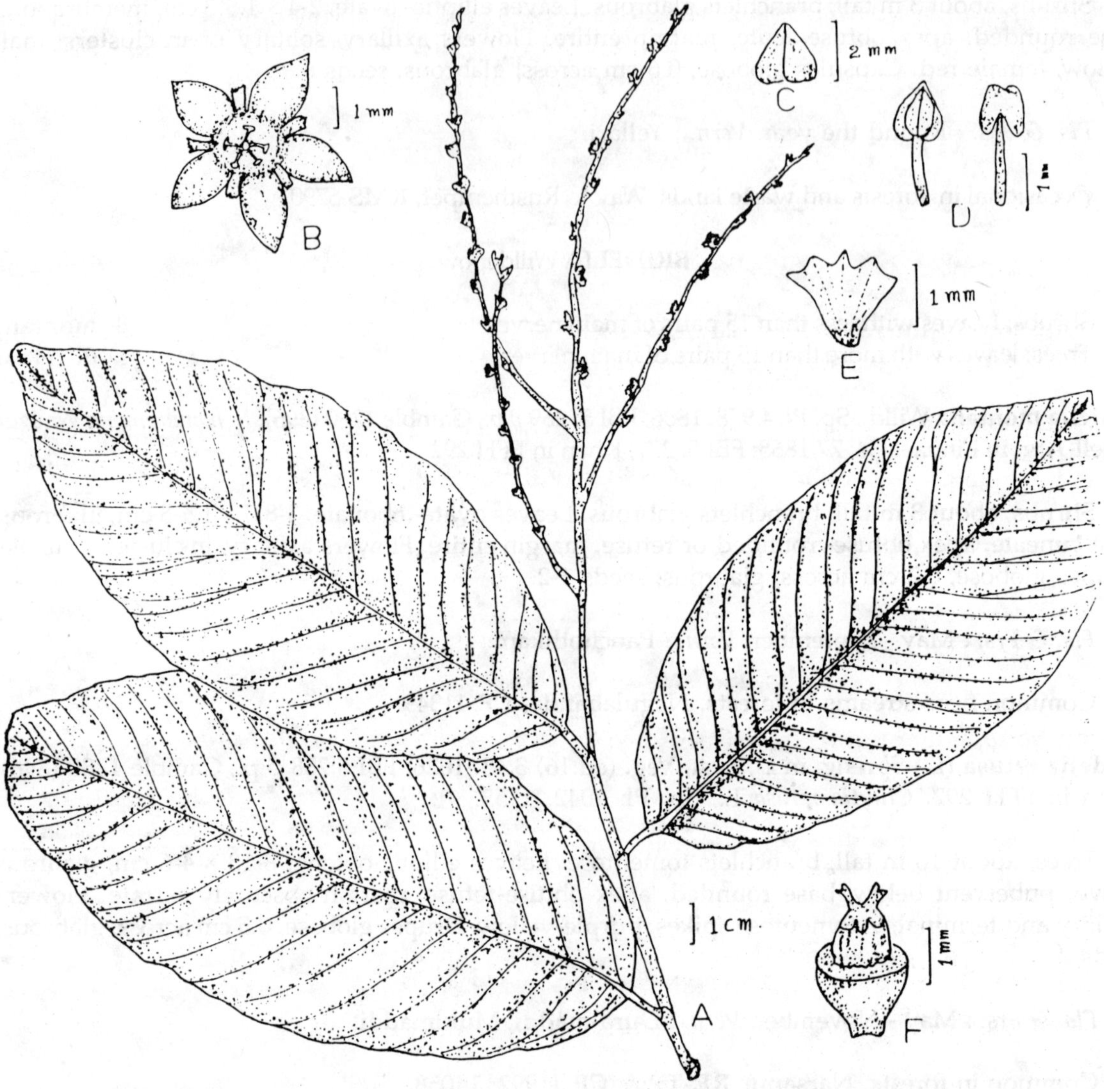

Fig. 6.25: *Briedelia retusa* (L.) A. Twig, B. Flower, C. Calyx lobe, D. Stamens, E. Petal, F. Pistil

hairy, base cordate-truncate, apex obtuse, margin sinuate. Flowers axillary, in racemes, yellow. Capsules 3-lobed, 0.8 cm across, stellate-tomentose; seeds 3.

Fls. & Frs. : June - August

Occasional in dried irrigated fields. Annaram, **GNR 4133.**

Chrozophora rottleri (Geis.) Adr. Juss., Euphorb. Gen. 28. 1824; Gamble 1317 (921). *Croton rottleri* Geis., Crot. Monogr. 54. 1807. *Chrozophora plicata* Hook.f. in FBI 5: 409. 1887

Erect herbs, about 60 cm, tall; branchlets tomentose. Leaves ovate-orbicular, 2-6 × 1.5-5 cm, pubescent, base truncate, apex obtuse, margin sinuate-undulate. Flowers axillary and terminal, in racemes, yellowish-green. Capsules 3-lobed, 0.6 cm across, stellate-tomentose; seeds 3.

Fls. & Frs. : Round the year.

Occasional weed in cultivated fields. Near Ooracheru lake, **KMS** 7992.

CLEISTANTHUS J.D. Hooker ex Planchon

Cleistanthus collinus (Roxb.) Benth. ex Hook.f. in FBI 5:274, 1887; Gamble 1282 (897); Khan in FFH 293. *Cluytia collina* Roxb., Pl. Corom.t. 169. 1802.

Trees, about 12 m tall; branchlets glabrous. Leaves elliptic- ovate, 2.5-7 × 2-5 cm, glabrous, base rounded, apex obtuse, margin entire. Flowers axillary, in clusters, pale green. Capsules globose, 2 cm across, 3-valved; seeds 3.

Fls. & Frs. : March - October, *Vern.* : Nallakodsha.

Common in forests. Gangapur RF, **BR** & **CP** 11430.

CROTON Linnaeus

Croton bonplandianum Baill. in Adansonia 4: 339. 1864. *C. sparsiflorous* Morong. in Ann. New York Acad. Sci. 7: 221.1893; Gamble 1316 (920).

Erect herbs, about 50 cm tall; branchlets with white warty outgrowth. Leaves ovate-lanceolate, 3-8 × 1-2.8 cm, pubescent, base acute, apex acuminate, margin serrate. Flowers terminal, in racemes, yellowish-white. Capsules oblong, 0.6 × 0.4 cm, glandular, 3-valved; seeds 3.

Fls. & Frs. : Round the year.

Common weed of waste lands. Sangareddy tank, **BR** & **CP** 11296.

EUPHORBIA Linnaeus

1. Shrubs, > 2 m tall:
   2. Armed **E. ligularia**
   2. Unarmed **E. tirucalli**
1. Herbs, < 1 m tall:
   3. Leaves > 3 cm long **E. heterophylla**
   3. Leaves < 3 cm long :
      4. Leaves broadly oblong - elliptic :
         5. Appendage of glands apiculate **E. cristata**
         5. Appendage of glands entire:
            6. Cyathia stalked **E. indica**

6. Cyathia subsessile **E. hirta**
4. Leaves oblong - suborbicular **E. thymifolia**

Euphorbia cristata Heyne ex Roth, Nov. Pl. Spec. 226. 1821; FBI 5:247; Gamble 1274 (892). *Chamaesyce cristata* (Heyne ex Roth) Webster in J. Arnold Arbor. 48 : 424. 1967.

Ascending herbs, about 35 cm tall; branchlets striate, pubescent. Leaves broadly ovate-oblong, 0.5-1 × 0.4-0.8 cm, base obliquely cordate, apex acute, margin serrulate. Cyathia axillary and sub-terminal, 1-3 in clusters. Capsules globose, 0.4 cm across, pubescent; seeds angled, furrowed.

Fls. & Frs. : August - January.

Occasional in forests. Narsapur RF, **TP** & **CP** 11975.

Euphorbia heterophylla L., Sp. Pl. 453. 1753. *E. geniculata* Orteg., Nov. Rar. Pl. Matrit. 18. 1797; FBI 5:239; Gamble 1884 (1303).

Erect herbs, about 75 cm tall, branchlets striate, glabrous. Leaves obovate-elliptic, 3-10 × 1-6.5 cm, glabrous, base attenuate, apex acute, margin entire-serrate. Cyathia terminal, 5-8 in clusters. Capsules globose, 0.5 cm across, glabrous; seeds angled, tuberculate.

Fls. & Frs. : Round the year.

Abundant in waste lands and cultivated fields. Narsapur, **BR** & **CP** 11240; Siddipet, **CP** 11372; Choutkoor, **TP** & **CP** 11920.

Euphorbia hirta L., Sp. Pl. 454. 1753; Gamble 1275 (892). *Chamaesyce hirta* (L) Millsp. in Publ. Field Columb. Mus. Bot. Ser. 2: 303, 1909. *Euphorbia pilulifera* auct. non L.: FBI 5:250.

Ascending herbs, about 60 cm tall; branchlets hispid, Leaves oblong, 0.5-2.5 × 0.4-1.5 cm, sparsely hispid, base oblique, apex acute, margin serrate. Cyathia axillary, 10-15 in clusters. Capsules globose, 0.2 cm across, 3-lobed; seeds trigonous.

Fls. & Frs. : Round the year.

Common weed along road sides and waste places. Pillutla **RF, BR** & **CP** 11263.

Euphorbia indica Lam. Encycl. 2:423. 1786; Raju & P.N. Rao in Indian J. Bot. 2: 205. 1979. *E. hypercifolia* auct non L. 1753 : FBI 5:249; Gamble 1275 (892).

Erect herbs, about 40 cm, tall, branchlets pubescent. Leaves elliptic-broadly oblong, 1-2 × 0.5 - 1 cm, base obliquely truncate, apex acute-obtuse, margin serrulate. Cyathia axillary and terminal, 13-15 in clusters. Capsules subglobose, 1.5 × 1.3 cm, hairy; seeds 4-angled, smooth.

Fls. & Frs. : July - March.

Common in forests and in cultivated fields. Patancheru, **BR** & **CP** 10306; Ramayampet, **BR** & **CP** 11516; NSF, **BR** & **CP** 11670; Gangapur, TP & CP 12083.

Euphorbia ligularia Roxb. Fl. Ind. 2: 465. 1832. *E. nerifolia* sensu Hook. f., FBI 5:255. 1887, auct non L: Gamble 1277 (893); Khan in FFH 291.

Erect shrubs, about 5 m tall; branchlets 5-angled, with paired horns. Leaves obovate, 8-20 x 4-6.5 cm, glabrous, base narrowed, apex acute, margin entire. Cyathia in short dichotomous pedunculate cymes. Capsules 3-lobed, 1.2 x 0.8 cm, glabrous; seeds angled.

Fls. & Frs. : March - September. *Vern.* : Modu-Jamudu.

Occasional in hedges. Sivanoor, **BR** & **CP** 11549.

Euphorbia thymifolia L. Sp. Pl. 454. 1753; FBI 5:252; Gamble 1276 (893). *Chamaesyce thymifolia* (L.) Millsp. in Publ. Field Columb. Mus. Bot. Ser. 2:412. 1916.

Prostrate herbs; branchlets softly hispid. Leaves oblong - suborbicular, 0.3-0.6 × 0.2-0.4 cm, glabrous, base obliquely - rounded, apex obtuse, margin serrulate. Cyathia axillary, solitary or in clusters. Capsules globose, 0.1 cm across, hairy; seeds 4-angled, minutely tuberculate.

Fls. & Frs. : Round the year.

Common weed of all places. Narsapur, **BR** & **CP** 11231; Sangareddy tank, **BR** & **CP** 11291; Sardhana, **BR** & **CP** 11625; Poosanpalle, **TP** & **CP** 11950.

Euphorbia tirucalli L., Sp. Pl. 452. 1753; FBI 5:254; Gamble 1276 (893); Khan in FFH 290.

Erect shrubs, about 3 m tall; branchlets fleshy. Leaves linear, 1 × 0.3 cm, base cuneate, apex subacute, margin entire, deciduous. Cyathia terminal, in clusters. Capsules globose, 0.5 cm across, villous; seeds globose, smooth.

Fls. & Frs. : August - September. *Vern.* : Manchi jamudu.

Occasional in field hedges . Akkannapet, **BR** & **CP** 11556.

GIVOTIA W. Griffith

Givotia moluccana (L.) Sreem. in Taxon 24:696. 1975. *Croton moluccanum* L., Sp. Pl. 1005. 1753 excl. ref. Burman. *Givotia rottleriformis* Griff. in Calcutta Journ. Nat. Hist. 4:388. 1844; FBI 5:395; Gamble 1342 (938); Khan in FFH 300.

Trees, about 15m tall; branchlets fulvous - tomentose. Leaves orbicular, 5-16 × 5-20 cm, glabrous above, tomentose below, base cordate, apex shortly acuminate margin dentate-sinuate. Flowers axillary or terminal, in cymes, greenish-white. Drupes sub globose, 2.5 cm across, smooth.

Fls. & Frs. : April - August. *Vern.* : Tella punki.

Occasional in forests. Narsapur RF, **TP** & **CP** 14026.

JATROPHA Linnaeus

1. Leaves palmately 3-5-lobed: petals free or nearly connate **J. gossypifolia**
1. Leaves entire or slightly 3-5-lobed; petals connate to the middle or above:
   2. Leaves eglandular along margins; capsules glabrous **J. curcas**
   2. Leaves glandular-hairy along margins; capsules pubescent **J. villosa**

Jatropha curcas L., Sp. Pl. 1006. 1753; FBI 5:383; Gamble 1340(937); Khan in FFH 299.

Shrubs, about 3 m tall; branchlets glabrous. Leaves cordate, shallowly 5-lobed, 3-6 × 2-5 cm, pubescent, base cordate, apex acute, margin entire. Flowers axillary and terminal, in dichasial cymes, greenish-yellow. Capsules ellipsoid, 2.5 × 2 cm, rugose; seeds 3-4.

Fls. & Frs. : April - November. *Vern.* : Adavi-amudam.

Common along the hedges of the fields. Narsingi, **BR** & **CP** 11528; Medak, **CP** 11725.

Jatropha gossypifolia L., Sp. Pl. 1006. 1753; FBI 5:383; Gamble 1340 (937); Khan in FFH 299.

Shrubs, about 2.5 m tall; branchlets glandular-hairy. Leaves deeply 3-lobed, 3.5-12 × 2.7-9.6 cm, base cordate; lobes obovate-oblong, apex acute, margin glandular-hairy. Flowers axillary and terminal, in paniculate cymes, pink. Capsules ellipsoid, 3-lobed, 1 × 0.5 cm, glabrous; seeds 3.

Fls. & Frs. : Round the year. *Vern.* : Nalla-amudam.

Common along road sides. Ponnala, **CP** 11363; Zaheerabad, **CP** 11592.

Jatropha villosa Wight, Icon. Pl. Ind. Orient. 4(1). 1848. *J. wightiana* Muell-Arg. in DC. Prodr. 15 (2): 1080. 1866; FBI 5: 383; Gamble 1340 (937).

Undershrubs, about 2 m tall; branchlets pubescent. Leaves cordate, 6-16 × 5.5-15 cm, pubescent along nerves above, base cordate, apex acute, margin entire. Flowers terminal, in cymes. Capsules 3-lobed, 1.5 × 1 cm, thinly pubescent.

Fls. & Frs. : April - July.

Occasional in field hedges. Zaheerabad, **CP** 11594.

PHYLLANTHUS Linnaeus

1. Herbs or undershrubs:
   2. Anthers erect, dehiscence vertical **P. maderaspatensis**

2. Anthers globose, dehiscence transverse:
    3. Stipules peltate, subsagittate; filaments free **P. virgatus**
    3. Stipules not peltate, lanceolate; filaments connate:
        4. Leaves oblong **P. amarus**
        4. Leaves elliptic or obovate:
            5. Disc of male flowers with star shaped glands **P. debilis**
            5. Disc of male flowers with peltate glands **P. missionis**

1. Shrubs or trees.
    6. Trees; fruit a drupe **p. emblica**
    6. Shrubs; fruit a berry **P. reticulatus**

Phyllanthus amarus Schum. & Thonn. in Kongl. Danske Vidensk. Selsk. Skr. 4:195. 1829. *P. niruri* acut. non L.: Roxb., Fl. Ind. 3:659. 1832; FBI 5:298. *p.p.*; Gamble 1290 (903).

Erect herbs, about 25 cm tall; branchlets glabrous. leaves oblong, 0.3 - 0.7 × 0.2-0.5 cm, glabrous, base obtuse, apex acute, margin entire. Flowers axillary, solitary or in clusters, green. Capsules globose, 0.2 cm across; seeds trigonous.

Fls. & Frs. : Round the year.

Common in all places. Narsapur, **BR** & **CP** 11230.

Note: This taxon was previously reported as *P. niruri* L. Webster (1957) has shown that *P. niruri* is an American taxon, endemic to West Indies and not represented in India. *P. amarus* can easily be distinguished by having equilateral leaf bases, bisexual cymules, ribbed seeds from *P. niruri*.

Phyllanthus debilis Klein ex Willd., Sp. Pl. 4:582. 1805; FBI 5:299; Gamble 1290 (903).

Erect herbs; about 70 cm tall; branchlets glabrous. Leaves elliptic-ovate, 1-1.8 × 0.5-1 cm, glabrous, base acute-rounded, apex broadly acute, margin entire. Flowers axillary, solitary or in clusters, green. Capsules globose, 0.4 cm across, 3-valved seeds trigonous.

Fls. & Frs. : June - December.

Occasional in forests and waste lands. Narsapur RF, **TP** & **CP** 14025.

Phyllanthus emblica L., Sp. Pl. 982. 1753; FBI 5:289; Khan in FFH 295. *Emblica officinalis* Gaertn., Fruct. 2: 122. t. 108. 1790; Gamble 1295 (906).

Trees, about 10 m tall; branchlets glabrous. Leaves oblong, 1-1.8 × 0.2 - 0.4 cm, glabrous above, glaucous below, base truncate, apex apiculate, margin entire. Flowers axillary, solitary or in clusters, green. Drupes depressed-globose, 0.3 cm across, fleshy; seeds trigonous.

Fls. & Frs. : March - November. *Vern.* : Usiri.

Occasional in forests. Narsapur RF, **BR** & **CP** 14104.

Phyllanthus maderaspatensis L., Sp. Pl. 982. 1753; FBI 5:292; Gamble 1289 (902).

Erect herbs, about 40 cm tall; branchlets glabrous. Leaves obovate-spathulate, 0.5-2.5 × 0.3-0.7 cm, glabrous, base cuneate, apex obtuse-rounded, margin entire. Flowers axillary, solitary or in clusters, green. Capsules globose, 0.4 cm across, glabrous; seeds wedge shaped.

Fls. & Frs. : Round the year.

Common weed of cultivated fields, road sides and moist places. Toopran, **BR** & **CP** 11282; Pochammaralu, **BR** & **CP** 11477.

Phyllanthus missionis Hook.f., in FBI 5:297; Gamble 1290 (903).

Narsapur town, **KMS** 7964.

Note: This taxon was reported by K.M. Sebastine *et al.* (1960). However in the present study the specimen was not collected and even not traced in MH.

Phyllanthus reticulatus Poir. in Lam., Encycl. 5:298. 1804; FBI 5:288. *Kirganelia reticulata* (Poir.) Baill., Etude Euphorb. 613. 1874; Gamble 1294 (905).

Straggling shrubs, about 4 m tall; branchlets pubescent. Leaves oblong-elliptic, 1-2 × 0.5-1.2 cm, glabrous, base rounded, acute, apex acute, margin entire. Flowers axillary, in clusters, green. Berries globose, 0.7 cm across; seeds 5-10.

Fls. & Frs. : Round the year. *Vern.* : Nalla-pulsare.

Common in hedges, along streams and waste lands. Marrikunta, **CP** 11387; Medak, **BR** & **CP** 11407; Akkannapet, **BR** & **CP** 11568.

Phyllanthus virgatus Forster f., Fl. Ins. Austr. 65. 1786. *P. simplex* Retz., Obs. Bot. 5:29. 1788; FBI 5: 295; Gamble 1289 (902).

Ascending herbs, about 50 cm tall; all branches arise from the root stock. Leaves oblong-lanceolate, 1-1.5 × 0.2-0.4 cm, glabrous, base rounded, apex acute, margin entire. Flowers axillary, solitary or in clusters, green. Capsules globose, 0.3 across, 3-5-celled; seeds trigonous.

Fls. & Frs. : Round the year.

Common near water courses and in the edges of fields. Nagulabanda, **CP** 11359; Sardhana, **BR** & **CP** 11623; Pocharam RF, **TP** & **CP** 12059.

SEBASTIANIA Sprengel

Sebastiania chamaelea (L.) Muell.-Arg. in DC., Prodr. 15:1175. 1866; FBI 5:475; Gamble 1344 (940). *Tragia chamaelea* L., Sp. Pl. 981. 1753.

Erect herbs, about 60 cm tall; branchlets angled, glabrous. Leaves lanceolate, 2-3.5 × 0.2-0.4 cm, glabrous, base rounded, apex acute, margin serrate. Flowers axillary, in spikes, yellow; male

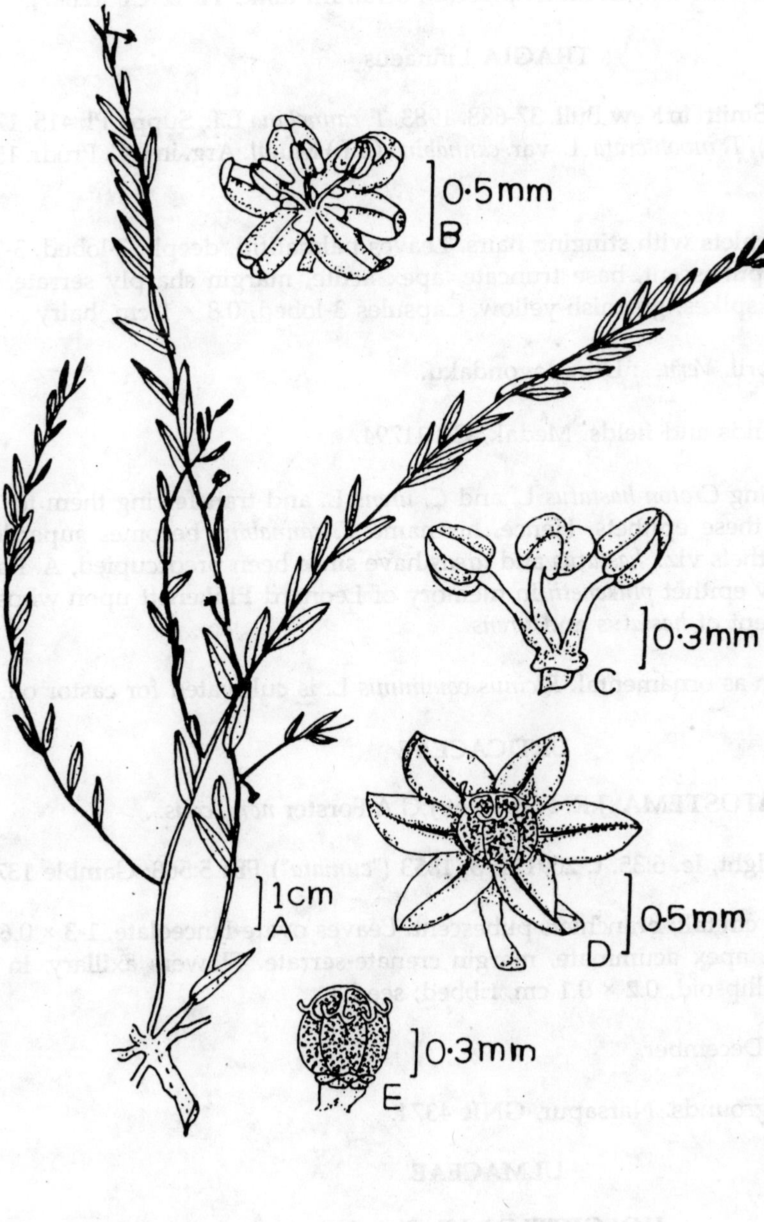

Fig. 6.26: *Phyllanthus virgatus Forst.* A. Habit, B. Male Flower, C. Stamens, D. Female flower, E. Pistil

flowers 2-3 in clusters, female flowers solitary. Capsules oblong, 0.3 × 0.2 cm, 3-lobed; each lobe with 2 rows of soft prickles: seeds carunculate.

Fls. & Frs. : Round the year.

Occasional in waste lands and in moist places. Pocharam tank, **TP & CP** 12047.

TRAGIA Linnaeus

Tragia plukenetii Radc.-Smith in Kew Bull. 37-688. 1983. *T. cannabina* L.f., Suppl. Pl. 415. 1781. *nom. illegit.*: Gamble 1333 (932). *T. involucrata* L. var. *cannabina* (L.f.) Muell. Arg. in DC. Prodr. 15 (2) 944. 1862; FBI 5: 465.

Twining herbs; branchlets with stinging hairs. Leaves palmatifid, deeply 3-lobed, 3-7.5 × 2-4 cm, obovate-lanceolate, pubescent, base truncate, apex acute, margin sharply serrate. Flowers axillary and terminal, in spikes, greenish-yellow. Capsules 3-lobed, 0.8 × 1 cm, hairy.

Fls. & Frs. : July - April. *Vern.* : Duradagondaku.

Common in waste lands and fields. Medak, **CP** 11794.

Note : Linn.f. when uniting *Croton hastatus* L. and *C. urens* L. and transferring them to *Tragia* L. failed to choose one of these epithets. Hence, his name *T. cannabina* becomes superflous and illegitimate. As these epithets viz., *hastatus* and *urens* have since been preoccupied, A. Radcliffe - Smith has selected a new epithet *plukenetii* in memory of Leonard Plukenett upon whose plates Linnaeus based his concept of *hastatus* and *urens*.

Jatropha rosea L., is grown as ornamental. *Ricinus communis* L. is cultivated for castor oil.

URTICACEAE

ELATOSTEMA J.R. Forster et J.G.A Forster *nom. cons.* .

Elatostema cuneatum Wight, Ic. 6:35. t. 2091. f. 3. 1853 ("*cuniata*") FBI 5:568; Gamble 1379 (963).

Erect herbs, about 10 cm tall; branchlets pubescent. Leaves ovate-lanceolate, 1-3 × 0.6-1.2 cm, pubescent, base cuneate, apex acuminate, margin crenate-serrate. Flowers axillary, in globose heads, yellow. Achenes ellipsoid, 0.2 × 0.1 cm, ribbed; seed 1.

Fls. & Frs. : August - December.

Occasional in rocky grounds. Narsapur, **GNR** 4374.

ULMACEAE

HOLOPTELEA J.E. Planchon

Holoptelea integrifolia (Roxb.) Planch. in Ann. Sci. Nat. Bot. (Ser.3) 10: 266. 1848; FBI 5: 481;

Gamble 1348 (943); Khan in FFH 305. *Ulmus integrifolia* Roxb. Fl. Ind. 2: 68: 1832.

Trees, about 12 m tall, branchlets pubescent. Leaves ovate - elliptic, 4-6 × 2.7-3.5 cm, glabrous, base rounded, apex acute- acuminate, margin entire. Flowers axillary, in racemes, yellow. Samaras orbicular, glabrous.

Fls. & Frs. : March - September. *Vern.* : Namli, Namlinara.

Occasional in forests and along road sides. Narsingi, **BR & CP** 11529.

MORACEAE

1. Flowers enclosed in receptacle; fruit achene **Ficus**
1. Flowers exposed; fruit berry **Streblus**

FICUS Linnaeus

1. Shrubs; leaves opposite **F.hispida**
1. Trees; leaves alternate:
   2. Receptacles sessile:
      3. Leaf apex caudate **F.religiosa**
      3. Leaf apex obtuse:
         4. Receptacles glabrous **F. benghalensis**
         4. Receptacles tomentose **F. mollis**
   2. Receptacles peduncled:
      5. Peduncles > 2 cm long **F. racemosa**
      5. Peduncles < 1 cm long **F. virens**

Ficus benghalensis L., Sp. Pl. 1059; FBI 5: 499; Gamble 1361 (952); Khan in FFH 308.

Trees, about 15 m tall; branchts with prop roots; branchlets pubescent. Leaves ovate, 10-15 × 7-10 cm, coriaceous, base rounded-subcordate, apex obtuse, margin entire. Receptacles axillary, in pairs, sessile, globose, 1.5 cm across, glabrous, red when ripe.

Fls. & Frs. : March - September. *Vern.* : Marri.

Common along road sides. Medak, **BR & CP** 11413.

Ficus hispida L.f., Suppl. Pl. 442. 1781; FBI 5:522; Gamble 1367 (956); Khan in FFH 311.

Shrubs; about 4 m tall; branchlets hispid. Leaves obovate- oblong, 10-20 × 7-10 cm, hispid-pubescent, base truncate, apex acuminate, margin minutely dentate. Receptacles on branchlets, in fasicles, depressed globose, 1.2 cm across, hispid.

Fls. & Frs. : March - November. *Vern.* : Boma-medi.

Common in forests and field hedges. Narsapur RF, **TP & CP** 12015.

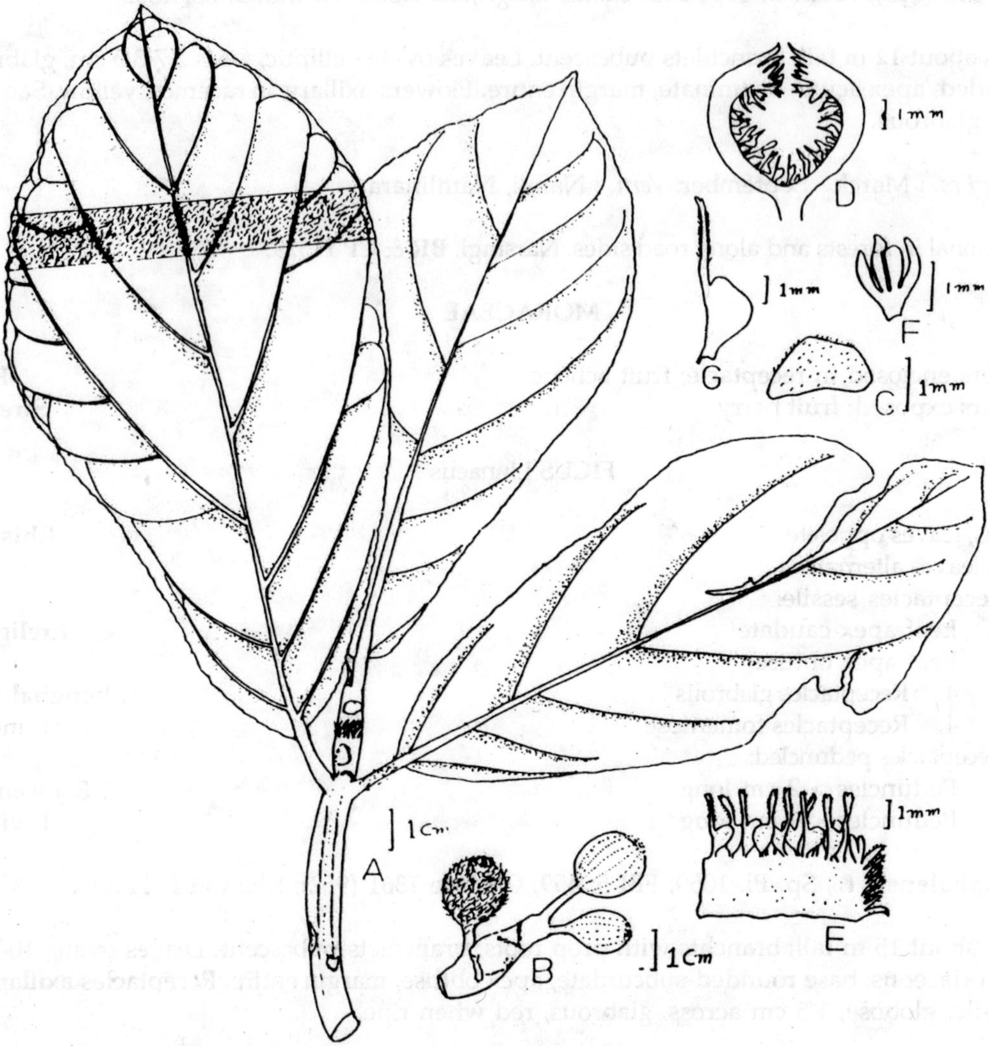

Fig. 6.27: *Ficus hispida* L.f. A. Twig, B. Fruiting twig, C. Bract, D. & E. Receptacle L.S. and part enlarged, F. Male flower, G. Female flower

Ficus mollis Vahl, symb. Bot. 1: 82. 1790. *F. tomentosa* Roxb. ex Willd., Sp. Pl. 4: 1136. 1806; FBI 5: 501; Gamble 1361 (952).

Trees, about 15 m tall; branchlets fulvous-tomentose. Leaves elliptic-obovate, 8-13 × 4.5-9 cm, coriaceous, tomentose, base cordate, apex obtuse, margin entire. Receptacles axillary, in pairs, globose. 0.4 cm across, tomentose, dull white when ripe.

Fls. & Frs. : October - March.

Occasional in forests. Gummadidala, **BR & CP** 14127.

Ficus racemosa L., Sp. Pl. 1060. 1753. *F. glomerata* Roxb., Pl. Corom. 2:13 123. 1798; FBI 5:535; Gamble 1364 (954); Khan in FFH 312.

Trees about 12 m tall; branchlets pubescent. Leaves elliptic- lanceolate, 5-20 × 2-6 cm, subcoriaceous, glabrous, base rounded, apex acute, margin entire. Receptacles on branchlets. subglobose, 3 × 2.5 cm pubescent, red when ripe.

Fls. & *Frs.* : March - September. *Vern.* : Medi.

Common along road sides near villages. Annaram, **GNR** 4145.

Ficus religiosa L., Sp. Pl. 1059. 1753; FBI 5:513; Gamble 1363 (953); Khan in FFH 309.

Trees, about 25 m tall; branchlets glabrous. Leaves broadly ovate, 5-10 × 4.5-8.5 cm, glabrous, base cordate, apex caudate, margin undulate, Receptacles axillary, in pairs, globose, 0.6 cm across, glabrous reddish-yellow when ripe.

Fls. & *Frs.* : March - August. *Vern.* : Raavi.

Occasional in villages. Medak, **TP & CP** 14081.

Ficus virens Aiton, Hort. Kew. 3: 451. 1789. *F. infectoria* auct. non Willd.: Roxb., Fl. Ind. 3:551. 1832; FBI 5:515; Gamble 1362 (953); Khan in FFH 310.

Trees, about 15 m tall; branchlets pubescent, Leaves oblong- ovate, 8-13 × 4-7 cm, glabrous, base rounded-truncate, apex acuminate, margin entire. Receptacles axillary, in pairs, globose, 1.1 cm across, glabrous, red when ripe.

Fls. & *Frs.* : March - November. *Vern.* : Banda-juvvi.

Occasional in forests. Gangapur RF, **TP & CP** 12075.

STREBLUS Loureiro

Streblus asper Lour., Fl. Cochinch. 615. 1790; FBI 5:489; Gamble 1353 (947); Khan in FFH 305.

Trees, about 6 m tall; branchlets with milky juice, glabrous. Leaves elliptic-obovate, 1.5-6 × 0.8-3 cm, glabrous above, scabrous below, base truncate, apex acute, margin denticulate, Flowers axillary, male in globose heads, female solitary, white. Berries globose, 0.5 cm across, glabrous, yellow when ripe, enclosed in persistent perianth; seed 1.

Fls. & *Frs.* : February - September. *Vern.* : Barinka.

Common along road sides near villages. Pocharam sanctuary, **BR & CP** 11613.

Artocarpus integrifolia L.f., is cultivated occasionally for its edible fruits.

CERATOPHYLLACEAE

CERATOPHYLLUM Linnaeus

Ceratophyllum demersum L. Sp. Pl. 992. 1753; FBI 5: 639; Gamble 1391 (973).

Submerged, rootless herbs; branchlets glabrescent. Leaves linear, 1-2.5 cm, dichotomously forked, apex acute, margin denticulate. Flowers axillary, solitary, unisexual. Nutlets ellipsoid, 0.4 × 0.3 cm; seed 1.

Fls. & *Frs.* : January - April.

Occasional in ponds and ditches. Medak tank, **BR & CP** 11402.

HYDROCHARITACEAE

1. Stems present; leaves cauline:
    2. Leaves whorled, < 2 cm in length — **Hydrilla**
    2. Leaves not whorled, > 2 cm in length — **Nechamandra**
1. Stems absent; leaves radical:
    3. Spathes winged; leaves orbicular-ovate — **Ottelia**
    3. Spathes not winged; leaves linear — **Vallisneria**

HYDRILLA L.C. Richard

Hydrilla verticillata (L.f.) Royle. Ill. Bot. Himal. t. 376. 1839; FBI 5: 659; Gamble 1396 (977). *Serpicuia verticillata* L.f., Suppl. Pl. 416. 1781.

Submerged herbs, about 50 cm tall; branchlets slender, glabrous. Leaves whorled, oblong-linear, 0.5-0.8 x 0.2-0.3 cm, base rounded-cuneate, apex apiculate, margin serrulate. Flowers axillary, solitary, white. Fruits subulate, 0.5 x 0.3 cm; seeds 3-5.

Fls. & *Frs.* : Round the year.

Common in ponds and tanks. Medak tank, **BR & CP** 11403; Pathur, **TP & CP** 12123.

NECHAMANDRA Planchon

Nechamandra alternifolia (Roxb.) Thw. Enum. Pl. Zeyl. 332. 1864. *Vallisneria alternifolia* Roxb. Fl. Ind. 3: 750. 1832. *Lagarosiphon roxburghii* (Planch.) Benth. in Benth. & Hook.f. Gen. Pl. 3: 451. 1880; FBI 5:659. *L. alternifolia* (Roxb.) Druce in Bot. Exch. Club Soc. Brit. Isles 1916:630. 1917; Gamble 1396 (977).

Submerged herbs, about 50 cm tall; branchlets glabrous. Leaves oblong-lanceolate or elliptic, 1-7.5 × 0.2-0.5 cm, membranous, base amplexicaul, apex acute-apiculate, margin minutely serrulate-subentire. Flowers axillary, unisexual, in spathes. Fruits ovate, 0.6 x 0.2 cm; seeds many.

Fls. & Frs. : July - February.

Common in stagnant waters. Medak town, **CP** 11798.

OTTELIA Persoon

Ottelia alismoides (L.) Pers., Syn. Pl. 1: 400. 1805; FBI 5:662: Gamble 1398 (978). *Stratiotes alismoides* L., Sp. Pl. 535. 1753.

Submerged acaulescent herbs, about 30 cm tall. Leaves radical, orbicular-ovate, 4-9.5 × 1.5-11.5 cm, base shallowly rotund, apex obtuse, margin entire. Flowers axillary, solitary, in pedunculate spathe, white. Fruits oblong-ellipsoid, 4 × 2 cm; seeds many.

Fls. & Frs. October - March.

Common in ponds, tanks and ditches. Medak tank, **BR** & CP 11406; Pathur, **TP** & **CP** 12110.

VALLISNERIA Linnaeus

Vallisneria natans (Lour.) H. Hara in Journ. Jap. Bot. 49:136. 1974. *Physkium natans* Lour., Fl. Cochinch. 663. 1790. *Vallisneria spiralis* auct. non L., : FBI 5: 660; Gamble 1369 (377).

Submerged acaulescent stoloniferous herbs. Leaves linear, 7-45 × 0.2-1 cm, flat, brown-striped, base narrow, apex obtuse, margin subentire-denticulate. Flowers axillary, in oblong spathes, white, male numerous, female solitary. Fruits linear, 9 × 0.2 cm; seeds many.

Fls. & Frs. : March - December.

Occasional in stagnant waters. Narsapur tank, **GNR** 4078.

ORCHIDACEAE

HABENARIA Willdenow

1. Leaves 3, flat on the gound; flowers white **H. roxburghii**
1. Leaves more than 3, not flat on the ground; flowers yellow **H. marginata**

Habenaria marginata Coleb. in Hook., Exot. Fl. 2:17. t. 136. 1824; FBI 6:150; Gamble 1471 (1028). *H. fusifera* Hook. f. in FBI 6: 147. 1890. *H. flavescens* Hook. f. in FBI 6: 150. 1890.

Erect herbs, about 25 cm tall. Leaves 4-5, clustered near the base, elliptic-lanceolate, 4-8 × 1-2.4 cm, fleshy, base amplexicaul, apex acute, margin entire. Flowers terminal, in spikes, yellow. Capsules elliptic; seeds many.

Fls. & Frs. : August - November.

Rare in moist places. Narsapur town, **KMS** 6809.

Habenaria roxburghii (Pers.) R. Br., Prodr. 1: 312. 1810. *Orchis roxburghii* Pers., Syn. Pl. 2:503. 1807. *Habenaria platyphylla* (Willd.) Spreng. in L., Syst. Veg. ed. 16. 690. 1826; FBI 6:140; Gamble 1470 (1027).

Erect herbs, about 40 cm tall; Leaves 3, in a rosette, orbicular-oblong, 4-6.5 × 3.5-5 cm, fleshy, base amplexicaul, apex acute, margin entire. Flowers crowded at apex, in spikes, white. Capsules elliptic; seeds many.

Fls. & Frs. : September - December.

Occasional in forests, sheltered inside thorny bushes. Muttapally, **CP** 11779.

CANNACEAE

CANNA Linnaeus

Canna indica L., Sp. Pl. 1. 1753. *C. orientalis* Roscoe, Monandr. Pl. t. 12. 1828; Gamble 1496 (1045). *C. indica* L. var. *orientalis* (Roscoe) Baker in FBI 6:260. 1892.

Erect herbs, about 1 m tall; stem glabrous. Leaves ovate- lanceolate or oblong 16-50 × 8-20 cm, glabrous, base sheathing, apex acuminate, margin entire. Flowers terminal, in pairs, bright red-yellow. Capsules oblong, warty, 4 × 2 cm; seeds 3-6.

Fls. & Frs. : Round the year.

Generally as an escepe, in villages and towns, Mombojipally, **BR & CP** 11636.

Musa paradisiaca L. (MUSACEAE) is cultivated for fruits.

AMARYLLIDACEAE

CRINUM linnaeus

1. Leaves flat; flowers 15-20 in umbels **C. asiaticum**
1. Leaves channelled; flowers 5-12 in umbels **C. defixum**

Crinum asiaticum L., Sp. Pl. 292. 1753; FBI 6:280; Gamble 1504 (1051).

Acaulescent herbs; bulbs globose, 12 cm across. Leaves oblong, 70-90 × 10-15 cm, glabrous, base narrow, apex graudually tapering, margin entire. Flowers in umbels, on scapes, white. Capsules subglobose; seeds 1-6.

Fls. & Frs. : June - September.

Occasional along river banks. Achampet, **GNR** 4766.

Crinum defixum Ker-Gawl, in Journ. Sci. Arts (London)3: 105 1817; FBI 6 : 281; Gamble 1504 (1051).

Acaulescent herbs; bulbs ovate, 6 × 4 cm. Leaves linear. 30-75 × 1.5-3 cm, glabrous, base narrow, apex obtuse, margin entire. Flowers in umbels, on scapes, white. Capsules subglobose, 4 × 2.5 cm; seeds 1-many.

Fls. & Frs. : June - September.

Occasional along the banks of rivers. Dosalangunda lake, Rusthempet, **KMS** 6802.

AGAVACEAE

AGAVE Linnaeus

Agave americana L., SP. Pl. 323. 1753; FBI 6:277; Gamble 1505 (1052); Khan in FFH 319.

Shrubs, about 1 m tall. Leaves fleshy, lanceolate, 50-150 x 10-15 cm, glaucous, apex sharphy spiny, margins dentate-spiny. Flowers terminal, in cymose panicles, light yellow. Capsules trigonous-oblong.

Fls. & *Frs.* : March - July. *Vern.* : Kalabanda.

Common in hedges. Medak, **BR & CP** 11416, Zaheerabad, **CP** 11591.

HYPOXIDACEAE

CURCULIGO Gaertner

Curculigo orchioides Gaertn., Fruct. 1: 63. t. 16.f. 11. 1788 FBI 6: 279; Gamble 1502 (1050).

Rhizomatous herbs, about 30 cm tall; rhizome elongate. Leaves lanceolate-linear, 20-46 × 0.8-3.4 cm, glabrous, base sheathing, apex acuminate, margin entire. Flowers in racemes, bright yellow. Capsules oblong, 1.5 × 1 cm, glabrescent; seeds 1-4.

Fls. & *Frs.* : July - March. *Vern.* : Nelatadi.

Occasional in moist localities and in waste lands. Gananapur, **CP** 11791.

DIOSCOREACEAE

DIOSCOREA Linnaeus

1. Leaves compund, 3-5-foliolate:
    2. Leaves 3-foliolate; leaflets 5-7-nerved **D. hispida**
    2. Leaves 3-5-foliolate; leaflets 1-3-nerved **D. pentaphylla**
1. Leaves simple, 1-foliolate:
    3. Stems twining to right; leaves opposite **D. oppositifolia**
    3. Stems twining to left; leaves alternate **D. bulbifera**

Dioscorea bulbifera L., Sp. Pl. 1033. 1753; Gamble 1511 (1055). *D. sativa* Thunb., Fl. Jap. 15. 1784, non L. 1753 : FBI 6:295.

Climbing herbs; branchlets glabrous. Leaves ovate-suborbicular, 4-14 × 3-14 cm, thin coriaceous, base cordate, apex acuminate, margin entire; bulbils in axils. Flowers axillary, in spikes, green-purple. Capsules quadrately oblong, 2.5 × 1.4 cm, glabrous; seeds winged.

Fls. & Frs. : August - March. *Vern.* : Nela dumpa.

Occasional in field hedges and in deciduous forests. Pathur, **TP & CP** 12121.

Dioscorea hispida Dennst., Schluss, Hort. Malab. 15. 1818; Gamble 1511 (1055). *D. daemona* Roxb., Fl. Ind. 3: 805. 1832; FBI 6:289.

Climbing herbs; root tuberous; branchlets prickly. Leaves 3-foliolate; leaflets obovate-oblong, 12-15 × 7-9.5 cm, hispid, base cuneate, apex acuminate, margin entire. Flowers axillary, in spikes, green. Capsules quadrately oblong. 2.5 × 1.2 cm, glabrous; seeds winged.

Fls. & Frs. : August - December. *Vern.* : Puli-dumpa.

Occasional in forests. Narsapur RF, **BR & CP** 14103.

Dioscorea oppositifolia L., Sp. Pl. 1033. 1753; FBI 6:292; Gamble 1512 (1056).

Climbing herbs; branchlets glabrous. Leaves ovate-oblong, 7-10 × 4-7.5 cm, thin coriaceous, base obtuse, apex acute-acuminate, margin entire. Flowers axillary and terminal, in dense peduncled spikes, green. Capsules quadrately obovate, 2 × 1.8 cm, glabrous; seeds winged.

Fls. & Frs. : July - November.

Occasional in forests. Pocharam RF, **TP & CP** 12063.

Dioscorea pentaphylla L., Sp. Pl. 1032 1753; FBI 6:289; Gamble 1511 (1056).

Climbing herbs; branchlets glabrescent, prickly at base. Leaves 3-5-foliolate; leaflets elliptic-ovate, 2.5-5 × 1.5-3 cm, glabrous, base subacute-cuneate, apex acute-acuminate, margin entire. Flowers axillary, in pendulous spikes, green. Capsules quadrately oblong, 2 × 1 cm, glabrous; seeds winged.

Fls. & Frs. : July - November.

Common in forests, scrub jungles and field hedges. Marrikunta, **CP** 11328; Pochammaralu, **CP** 11760.

LILIACEAE

1. Climbers :
    2. Leaves scaly; cladodes present **Asparagus**

2. Leaves well developed; cladodes absent **Gloriosa**

1. Non-climbers:
   3. Roots tuberous; flowers white **Chlorophytum**
   3. Roots not tuberous; flowers purple **Iphigenia**

ASPARAGUS Linnaeus

1. Cladodes < 1.2 cm long; bracts large **A.laevissimus**
1. Cladodes > 1.2 cm long; bracts small **A.racemosus**

Asparagus laevissimus Steud. ex Baker in Journ. Linn. Soc. 5: 14.623. 1875; FBI 6:317; Gamble 1517 (1060).

Undershrubs, branchlets terete, angular. Leaves with stout conical deflexed rigid spinous spurs. Cladodes 3-6, triquetrous, 0.6-1 × 0.1 cm, base narrow, apex acuminate, margin entire. Flowers axillary, solitary or two, in racemes, white. Berries globose, 0.5 cm across; seeds 6.

Fls. & Frs. : July - December.

Occasional in forests. Narsapur, **GNR** 4133.

Asparagus racemosus Willd., Sp. Pl. 2:152. 1799; FBI 6:316; Gamble 1517 (1060); Khan in FFH 322.

Armed climbing shrubs. Leaves scaly; cladodes 2-6, linear, 1.2-2 × 0.1 cm, filiform. Flowers axillary, in racemes, white. Berries globose, 0.6 cm across; seeds 3-6.

Fls. & Frs. : July - November. *Vern.* : Pillitheegalu.

Common in forests. Nagulabanda, **CP** 11339; Medak, **BR** & **CP** 11423.

CHLOROPHYTUM Ker-Gawler

Chlorophytum laxum R.Br., Prodr. 277. 1810; FBI 6:336; Gamble 1526 (1066).

Acaulescent herbs; Leaves linear, 6.5-32 × 0.3-0.6 cm, base sheathing, apex acuminate, margin entire. Flowers on flexuous scape, distant, white. Capsules obcordate, triquetrous, 0.5 × 0.3 cm, glabrous; seeds 1-4.

Fls. & Frs. : July - November.

Rare in moist places. Sardhana, **CP** 11778.

GLORIOSA Linnaeus

Gloriosa superba L., Sp. Pl. 305. 1753; FBI 6:358; Gamble 1519 (1061); Khan in FFH 322.

Climbing herbs; branchlets terete, glabrous. Leaves oblong- lanceolate, 4-8.5 × 1-3.5 cm, glabrous, base cordate, apex tendril like, margin entire. Flowers axillary, solitary, orange scarlet. Capsules oblong, 2.5 × 1 cm; seeds many.

Fls. & Frs. : July - November. *Vern.* : Agnisikha.

Common in forests and field hedges. Nagulabanda, **CP** 11343.

IPHIGENIA Kunth *nom. cons.*

Ighigenia indica (L) A. Gray ex Kunth, Enum. Pl. 4:213. 1843; FBI 6: 357; Gamble 1528 (1067). *Melanthium indicum* L., Mant. Pl. 226. 1771.

Erect herbs, about 20 cm tall; corms small, subglobose, stem unbranched. Leaves linear, 10-25 × 0.3-0.5 cm, flat, glabrous, base sheathing, apex acuminate, margin entire. Flowers axillary and terminal, solitary, purple.Capsules oblong, 2 × 1 cm; seeds ovate.

Fls. & Frs. : July - September.

Occasional in forests and moist places. Sardhana, **CP** 11777.

Allium cepa L. and *A. sativum* L. are cultivated for their bulbs.

PONTEDERIACEAE

1. Petioles swollen in the middle; perianth zygomorphic **Eichhornia**
1. Petioles normal; perianth actinomorphic **Monochoria**

EICHHORNIA Kunth *nom. cons.*

Eichhornia crassipes (C.Martius) Solms-Laub in A. & C. DC., Monogr. Phan. 4:527. 1883; Gamble 1530 (1069). *Pontederia crassipes* C. Martius, Nov. Gen. Sp.Pl. 9. t. 4. 1823.

Free floating stoloniferous herbs. Leaves broadly ovate-rhomboid, 2.5-10 × 2-9 cm, thin-coriaceous, glabrous, base cuneate-rotund, apex obtuse, margin entire. Flowers terminal, in spikes, blue-violet. Capsules oblong-ovate; seeds many.

Fls. & Frs. : January - April. *Vern.* : Gurrapudekka.

Occasional in stagnant waters. Domadugu, **GNR** 4613.

MONOCHORIA K. B. Presl

Monochoria vaginalis (Burm.f.) C. Presl, Rel Haenk. 1:128. 1827; FBI 6: 363; Gamble 1529 (1068). *Pontederia vaginalis* Burm. f. Fl. Ind. 80. 1768.

Erect aquatic herbs, about 80 cm tall. Leaves radical, ovate, 3-9.5 × 1.5-4.5 cm, glabrous, base cordate, apex abruptly acuminate, margin entire. Flowers solitary, in racemes, blue; racemes inserted in the hollow sheath of the petiole. Capsules oblong, 1 × 0.6 cm, glabrous; seeds many.

Fls. & Frs. : August - November.

Common in ponds, ditches and marshy localities. Patancheru, **CP** 11687; Sardhana, **CP** 11768; Gangapur, **TP** & **CP** 12082.

COMMELINACEAE

1. Inflorescence enclosed in leaf-sheaths **Tonningia**
1. Inflorescence not enclosed in leaf-sheaths:
    2. Flowers in spathaceous or foliar bracts:
        3. Bracts spathaceous; fertile stamens 3 **Commelina**
        3. Bracts foliar; fertile stamens 6 **Cyanotis**
    2. Flowers not in bracts **Murdannia**

COMMELINA Linnaeus

1. Spathes ovate-lanceolate:
    2. Seeds never appendaged **C. clavata**
    2. Seeds appendaged at one or both ends **C. attenuata**
1. Spathes funnel-shaped:
    3. Underground stems with cleistogamous flowers **C. benghalensis**
    3. Underground stems without cleistogomous flowers:
        4. Capsules subequally 3-valved; seeds free in the cells **C. paludosa**
        4. Capsules 2-valved; seeds adnate to the cells **C. erecta**

Commelina attenuata J. Koenig ex Vahl, Enum. Pl. 2:168. 1806; FBI 6:372; Gamble 1539 (1074).

Diffuse herbs, about 50 cm tall; stem glabrous. Leaves linear- lanceolate, 3-6.5 × 0.5-1 cm, sparsely pubescent, base sheathing, attenuate, apex acute, margin entire. Flowers in solitary spathes, blue. Capsules oblong, 0.4 × 0.3 cm; seeds 2-4.

Fls. & Frs. : July - December.

Occasional in moist places and near streams in forests. Patancheru, **BR** & **CP** 11303.

Commelina benghalensis L., Sp. Pl. 41. 1753; FBI 6:370; Gamble 1539 (1075).

Diffuse herbs; stem glabrous. Leaves ovate-oblong, 2-6 × 1.5-3 cm, thinly pubescent, membranous, base unequally rounded, apex acute-obtuse, margin ciliate-undulate. Flowers axillary, in turbinate spathes, blue. Capsules ellipsoid, 0.6 × 0.2 cm seeds 5.

Fls. & Frs. : August - December.

Common weed of cultivated fields and waste lands. Toopran, **BR** & **CP** 11280.

Commelina clavata C.B. Clarke, Commelin. Cyrtandr. Bengal. t. 5. 1874; FBI 6:371 ; Gamble 1539 (1074).

Diffuse herbs. Leaves elliptic-lanceolate, 2-4 × 0.3-0.7 cm, glabrescent, base tapering, apex acuminate - acute, margin entire. Flowers axillary, in spathes, pale violet. Capsules oblong, constricted in the middle, 0.2 × 0.5 cm; seeds 4.

Fls. & *Frs.* June - December.

Occasional in moist places. Toopran, **BR & CP** 11283.

Commelina erecta L., Sp. Pl. 41. 1753. *C. undulata* R. Br., Prodr. 270. 1810; FBI 6: 373. *p.p.* *C. kurzii.* Clarke in Journ. Linn. Soc. Bot. 11: 444. 1869; Gamble 1539 (1075).

Herbs, about 40 cm tall, branchlets glabrous. Leaves oblong-lanceolate, 4-11.5 × 1-3.5 cm, appressed-pubescent above, glaucous below, base oblique, apex acuminate, margin entire. Flowers in cucullate spathes, pale violet. Capsules ellipsoid, 0.5 × 0.2 cm; seeds 3.

Fls. & *Frs.* : July - October.

Occasional in hilly areas. Narsapur RF, **BR & CP** 14109.

Commelina paludosa Blume, Enum.Pl. Jav. 1: 2. 1827. *C. obliqua* Buch-Ham. ex D. Don, Prodr. Fl. Nepal. 45.1825. non Vahl, 1805: FBI 6:372; Gamble 1539 (1075).

Erect herbs, about 50 cm tall; branchlets glabrous. Leaves elliptic-lanceolate, 4-17 × 1-4 cm, membranous, sparsely scabrous, base acute-attenuate, apex acuminate, margin entire. Flowers in funnel-shaped spathes, blue. Capsules trigonous, 1.2 × 0.5 cm; seeds 3-5.

Fls. & *Frs.* : August - October.

Occasional in moist places. Narsapur, **GNR** 4514.

CYANOTIS D. Don *nom. cons.*

1. Herbs woolly or cottony **C. fasciculata**
1. Herbs glabrous or not cottony:
    2. Roots tuberous; bracts equal to the inflorescence **C. tuberosa**
    2. Roots not tuberous; bracts longer than the inflorescence **C. cristata**

Cyanotis cristata (L.) D.Don, Prodr. Fl. Nepal. 46 1825; FBI 6: 385; Gamble 1550 (1081). *Commelina cristata* L., Sp. Pl. 42. 1753.

Diffuse herbs; branchlets ascending. Leaves elliptic-lanceolate, 4-7.5 × 1.5-3 cm, sparsely pubescent, base cordate, apex obtuse, margin ciliate, Flowers in scorpioid cymes, bluish purple. Capsules triangular, 0.2 × 0.1 cm; seeds 6.

Fls. & *Frs.* : August - December.

Occasional in moist places. Nagulabanda, **CP** 11350.

Cyanotis fasciculata (Roth) Schultes & Schultes f. in L. Syst. Veg. 7: 1152. 1830; FBI 6: 387; Gamble 1550 (1082). *Tradescantia fasciculata* Heyne ex Roth., Nov. Pl. Spec. 189. 1821.

Erect herbs; branchlets cottony. Leaves linear-lanceolate, 1-5 × 0.2-0.8 cm, silk-cobwebby, base cordate, apex acute, margin entire. Flowers axillary and terminal, in cobwebby cymes, pink. Capsules oblong, 0.3 x 0.1 cm; seeds 6.

Fls. & *Frs.* : August - January.

Common in waste lands and forest areas. Sangareddy, **BR** & **CP** 11286; Sardhana **CP** 11773.

Cyanotis tuberosa (Roxb.) Schultes & Schultes f. in Roemer & Schultes in L., Syst. Veg. 7:1153. 1830; FBI 6:386; Gamble 1549 (1081). *Tradescantia tuberosa* Roxb. Pl. Corom. t. 108. 1799 & Fl. Ind. 2:119. 1832.

Erect herbs, about 40 cm tall; branchlets softly villous. Leaves radical and cauline, linear-ensiform, 3-12 × 0.6-1 cm, pubescent, base sheathing, apex subacute, margin ciliate. Flowers axillary and terminal, in corymbose cymes, purple. Capsules oblong, 0.3 × 0.1 cm; seeds 5.

Fls. & *Frs.* : June - December.

Common in shady places. Nagulabanda, **CP** 11345.

MURDANNIA Royle *nom. cons.*

1. Leaves linear, > 5 cm long:
   2. Diffuse herbs; flowers in subcorymbose cymes, pink-blue **M. nudiflora**
   2. Erect herbs; flowers in terminal panicles, rose **M. semiteres**
1. Leaves ovate- lanceolate, < 3 cm long **M spirata**

Murdannia nudiflora (L.) Brenan in Kew Bull. 7: 189. 1952. *Commelina nudiflora* L., Sp. Pl. 41. 1753. *p.p. Aneilema nudiflorum* (L.) R. Br., Prodr. 271. 1810; FBI 6:378; Gamble 1545 (1078).

Diffuse herbs, about 30 cm tall; stem glabrous. Leaves linear, 2-5.5 × 0.2-0.4 cm, glabrous, base sheathing, apex acute- acuminate, margin entire. Flowers terminal, in subcorymbose lax cymes, pink-blue. Capsules subglobose, 0.4 × 0.3 cm, glabrous; seeds 6.

Fls. & *Frs.* : Round the year.

Common in moist places. Thornal, **CP** 11704.

Murdannia semiteres (Dalz.) Sant. in Poona Agri. Coll. Mag. 41: 284. 1951. *Aneliema semiteres* Dalz.

in Hook. Journ. Bot. Kew Gard. Misc 3:138. 1851. *A. paniculatum* Wall. ex C.B. Clarke in A. & C. DC. Monogr. Phan. 3:215. 1881. non Wight 1853; FBI 6:381; Gamble 1546 (1079).

Erect herbs, about 40 cm tall; branchlets reddish. Leaves linear, 2-5.5 × 0.1 cm, filiform, glabrous. Flowers terminal, in panicles, deep violet. Capsules subglobose, 0.2 cm across; seeds many.

Fls. & *Frs.* : July - September.

Occasional near streams. Bonthapally, **GNR** 4451.

Murdannia spirata (L.) Brueckner in Engl. Nat. Pflanzenfam. (ed.2) 15 a: 173. 1930. *Commelina spirata* L., Mant. Pl. 176. 1767. *Aneilema spiratum* R. Br., Prodr. 271. 1810; FBI 6:377; Gamble 1545 (1078).

Diffuse herbs, about 50 cm tall; branchlets flexuous. Leaves ovate-lanceolate, 1-2.5 × 0.4-0.8 cm, glabrous, base amplexicaul, apex acute, margin entire. Flowers terminal, in panicles, pink - blue. Capsules oblong, 0.4 × 0.2 cm; seeds 6-10.

Fls. & *Frs.* : July - December.

Occasional in moist places. Pegarikutta East, **KMS** 6690.

TONNINGIA Necker ex. A.H.L. Jussieu

1. Leaves linear-lanceolate, succulent; capsules ellipsoid **T. axillaris**
1. Leaves oblong-lanceolate, glabrous; capsules oblong **T. cucullata**

Tonningia axillaris (L.) Kuntze, Rev. Gen. Pl. 2: 721. 1891. *Commelina axillaris* L., Sp. Pl. 42. 1753. *Amischophacelus axillaris* (L) R.S. Rao & R.V. Kammathy in Journ. Linn. Soc. Bot. London 59 : 306. 1966. *Cyanotis axillaris* (L.) D.Don, Prodr. Fl. Nepal 46. 1825; FBI 6:388; Gamble 1550 (1082).

Herbs about 60 cm tall; branchlets glabrous. Leaves linear- lanceolate, 3-8 x 0.4-0.8 cm, fleshy, base sheathing, apex acute, margin entire. Flowers axillary, in cymes, pink-purple. Capsules ellipsoid, 0.5 x 0.2 cm, glabrous; Seeds 6.

Fls. & *Frs.* : Round the year.

Common in moist places. Narsapur RF, **BR** & **CP** 14111.

Tonningia cucullata (Roth) Kuntze, Rev. Gen. Pl. 2:1722. 1891. *Tradescantia cucullata* Roth Nov. Pl. Sp. 189. 1821. *Amischophacelus cucullata* (Roth) R.S. Rao & R. V. Kammathy in Journ. Linn. Soc. Bot. London 59:306. 1966. *Cyanotis cucullata* (Roth) Kunth, Enum 4:107. 1843; FBI 6:389; Gamble 1550 (1082).

Herbs, about 30 cm tall; branchlets glabrous. Leaves oblong- lanceolate, 2.5-7.5 x 0.5-1 cm, glabrous, base obtuse, apex acute, margin entire. Flowers axillary, in fascicles, light violet. Capsules depressed in the centre with 3 projections; seeds pitted.

Fls. & *Frs.* : August - December. : *Vern.* Gorlaku.

Occasional weed of cultivated fields. Gummadidala, **BR & CP** 14115.

NOTE : Farr *et al.* (1979) treatment of *Tonningia* is followed in the present work.

ARECACEAE

1. Leaves simple, fan-like **Borassus**
1. Leaves compound, feather-like **Phoenix**

BORASSUS Linnaeus

Borassus flabellifer L., Sp. Pl. 1187. 1753; FBI 6:482; Gamble 1562 (1090); Khan in FFH 328.

Trees, about 20 m tall; stem grey-black. Leaves plicately multifid, orbicular, 1 x 1.5cm, margins split into lanceolate, acuminate segments. Flowers in spikes, pale-green. Drupes subglobose, 10 x 12 cm, yellow when ripe; pyrenes 1-3.

Fls. & *Frs.* : February - August. *Vern.* Thati, Tadi.

Rare in waste lands. Narsapur, BR & CP 14132.

PHOENIX Linnaeus

Phoenix loureirii Kunth, Enum. Pl. 3:257. 1841 *P. humilis* Royle var. *pedunculata* Becc. in Malesia 3: 379 & 387. t. 44.f. 13-15, 18-21 & 25-27, 1890; FBI 6: 427; Gamble 1560 (1088); Khan in FFH 327.

Shrubs, about 3 m tall. unbranched. Leaves unipinnate; leaflets linear, 15-30 x 0.4-1 cm, plicate, base narrow, apex acuminate, spinous, margin entire. Flowers in spikes, yellow. Drupes oblong, 1.5 x 0.8 cm,. green yellow; seed 1.

Fls. & *Frs.* : January - May. *Vern.* : Adavi etha.

Common in waste lands, scrub jungles and hedges. NSF fence, BR & CP 11649.

Cocos nucifera L. is cultivated for its fruits.

PANDANACEAE

PANDANUS Parkinson

Pandanus fascicularis Lam., Encycl. 1:372. 1785; FBI 6:780. *P. odoratissimus* L.f., Suppl. Pl. 64. 1781 *nom. superfl.* : Khan in FFH 332, *P. tectorius auct. non* Sol. ex Park, : Gamble 1570 (1095).

Shrubs, about 4 m tall; branches speading from the base, supported by stilt roots. Leaves oblong-linear or ensiform, 80- 150 x 3-9 cm, apex caudate-acuminate, margin spinulate. Flowers in cylindric spikes, white, fragrant. Drupes globose.

Fls. & Frs. : July - December. *Vern.* : Mogili.

Occasional in hedges. Pocharam RF, **BR & CP** 11601.

TYPHACEAE

TYPHA Linnaeus

Typha angustata Bory & Chaub., Exp. Sci. Morre, Bot. 1:338. 1833; FBI 6:489; Gamble 1571 (1096).

Aquatic, robust shrubs, about 2 m tall. Leaves linear, 150 x 1 cm, base sheating, apex acute, margin entire. Flowers in cylindric spikes, brown; spikes monoecious, male and female flowers separated by sterile rachis. Nutlets fusiform, brown.

Fls. & Frs. : August - December. *Vern.* : Thunga.

Common in marshy localities along water courses. Medak, **CP** 11735.

ARACEAE

PISTIA Linnaeus

Pistia stratiotes L., Sp. Pl. 963. 1753; FBI 6: 497; Gamble 1573 (1097).

Floating stoloniferous herbs. Leaves obovate, 2-8 x 1.5-8 cm, white-pubescent, base cuneate, apex rounded-truncate, margin entire. Flowers in oblong spathes, pale yellow. Berries ovate; seeds few.

Fls. & Frs. : Round the year. *Vern* : Antharathamara.

Occasional in ditches forming mats. Gummadidala, **GNR** 4498.

LEMNACEAE

1. Fronds 3-nerved; root 1 per frond **Lemna**
1. Fronds 5-more-nerved; roots several per frond **Spirodela**

LEMNA Linnaeus

Lemna perpusilla Torrey, Fl. New York 2:245. 1843; Hartog & Van der Plas in Blumea 18:363. 1970. *L. paucicostata* Hegelm, Lemnac. 139.t. 8. 1868; FBI 6:556; Gamble 1593 (1111).

Floating herbs. Fronds solitary or in groups of 2-3, ovate-oblong, 0.1-0.4 x 0.1-0.3 cm, flat, base obtuse, apex obtuse-rounded, margin entire. Flowers unisexual, male in pairs, female solitary in budding pouches. Utricles ellipsoid; seeds 1-4.

Fls. & *Frs.* : September - October.

Common in ponds. Patancheru, BR & CP 11301.

SPIRODELA Schleiden

Spirodela polyrhiza (L.) Schleiden in Linnaea 13:392. 1839; Hartog & Van der Plas in Blumea 18:360. 1970. *Lemna polyrhiza* L., Sp. Pl. 1753; FBI 6:657; Gamble 1593 (1111). *L. orbiculata* Roxb., Fl. Ind. 3:565. 1832.

Floating herbs. Fronds in groups of 3-5, suborbicular, 0.5-0.7 x 0.4-0.5 cm, flat above, convex below, base obtuse, apex rounded, margin entire. Flowers unisexual, in budding pouches. Utricles ellipsoid; seeds 1-4.

Fls. & *Frs.* : September - December.

Occasional in water ditches. Patancheru, **BR & CP** 11302.

ALISMATACEAE

SAGITTARIA Linnaeus

Sagittaria guayanensis Kunth sub sp.**lappula** (D. Don)Bogin in Mem. New York Bot. Gard. 9:192.f. 5.1955. *S. lappula* D.Don Prodr. Fl. Nepal 22.1825. *S. guanensis* sensu Hook. f. Fl.Brit. India 6:561. 1893 p.p., Fischer 1596 (1113).

An aquatic herb. Leaves broadly ovate. Flowers white, sepals appressed on the mature pistillate flowers. Seed prominently toothed and winged.

Fls. & *Frs.* : October - December.

Rare in ponds and lakes. Pathur fiedls, **TP & CP** 12127.

APONOGETONACEAE

APONOGETON Linnaeus f.

Aponogeton natans (L.) Engl. & K. Krause in Engl. Pflanzenr. 4. 13. 24:11. 1906; Gamble 1598 (1115). *Saururus natans* L., Mant. Pl. 227. 1771. *Aponogeton monostachyon* L.f., Suppl. Pl. 1781;FBI 6:564.

Aquatic or marshy herbs; root stock tuberous. Leaves oblong, 3-10.5 x 0.8-2.4 cm, glabrous, base subcordate, apex acute-obtuse, margin entire. Flowers in cylindric spikes, violet. Follicles oblong, 0.4 x 0.2 cm, glabrous, beaked; seeds 2-8.

Fls. & Frs. : August - February.

Common in ponds and tanks. Narsapur tank, **BR & CP** 11242 and **TP & CP** 12008.; Thornal, **CP** 11397.

POTAMOGETONACEAE

POTAMOGETON Linnaeus

Potamogeton nodosus Poir. in Lam., Encycl. Suppl 4:535. 1816 *P. indicus* Roxb., Fl. Ind. 1:471, 1820. non Roth ex Roemer & Schultes 1818: FBI 6:565; Gamble 1600 (1116).

Submerged rhizomatous herbs; branchlets stout. Leaves heterophyllous, 4-12 x 1-3.5 cm; upper ones floating, ovate- elliptic, base obtuse, apex subacute, margin entire; lower ones submerged, lanceolate, base cuneate. Flowers axillary, in spikes, pinkish-white. Drupes ovate, 0.3 x 0.2 cm, glabrous.

Fls. & Frs. : September - March.

Occasional in ditches and ponds, Medak tank, **BR & CP** 11404.

ERIOCAULACEAE

ERIOCAULON Linnaeus

1. Leaves narrowly linear, green, capillary **E. cinereum**
1. Leaves lanceolate, purplish-red, membranous **E.quinquangulare**

Eriocaulon cinereum R.Br., Prodr. 254. 1810. *E. sieboldianum* Sieb. & Zucc. ex Steudel, Syn. Pl. Glum. 2:272 1855; FBI 6:577; Gamble 1619 (1127).

Erect marshy herbs, about 15 cm tall; stem very short. Leaves narrowly linear, 1.5 x 0.1 cm, capillary, apex acuminate. Flowers terminal, in spherical heads, white. Capsules ellipsoid; seeds 2-3.

Fls. & Frs. : November - January.

Occasional in marshy localities. Ooracheru lake, **KMS** 6757.

Eriocaulon quinquangulare L., Sp. Pl. 87. 1753; FBI 6:582; Gamble 1620 (1128).

Erect marshy herbs; about 25 cm tall; stem very short, Leaves lanceolate, 1.5-5 x 0.2-0.3 cm, membranous, apex acuminate. Flowers terminal, in globose heads, grey-white. Capsules subglobose; seeds 2-3.

Fls. & Frs. : November - April.

Rare in marshy places. Way to Rusthempet, **KMS** 6806.

CYPERACEAE

1. Spikelets unisexual **Scleria**
1. Spikelets bisexual:
    2. Empty glumes at base of spikelets 3 or more **Rhynchospora**
    2. Empty glumes at base of spikelets 0-2:
        3. Glumes on spikelets distichous:
            4. Plants bulbous-thickened basally; spikelets 1-3 **Abildgaardia**
            4. Plants not bulbous-thickened basally; spikelets numerous:
                5. Stigmas 2:
                    6. Rhachilla articulated, deciduous **Kyllinga**
                    6. Rhachilla not articulated, persistent **Pycreus**
                5. Stigmas 3:
                    7. Rhachilla persistent **Cyperus**
                    7. Rhachilla deciduous:
                        8. Keel of nut bearing glumes winged **Indocourtoisia**
                        8. Keel of nut bearing glumes not winged **Mariscus**
        3. Glumes on spikelets spiral:
            9. Style base dilated and constricted or articulated above the nut:
                10. Hypogynous bristles present **Eleocharis**
                10. Hypogynous bristles absent:
                    11. Style persistent, if falling not leaving a tumour on the nut **Fimbristylis**
                    11. Style deciduous, leaving a tumour on the nut **Bulbostylis**
            9. Style base not dilated, continuous with the nut:
                12. Hypogynous bristles hyaline, enveloping the nut **Lipocarpha**
                12. Hypogynous bristles not hyaline not enveloping the nut:
                    13. Leaves reduced to sheaths (except in *S.maritimus*) **Schoenoplectus**
                    13. Leaves well developed:
                        14. Leaves with a collar around the stem **Fuirena**
                        14. Leaves without a collar around the stem **Rikliella**

ABILDGAARDIA Vahl

Abildgaardia ovata (Burm.f.) Kral in Sida 4:71. 1971. *Carex ovata* Burm.f. Fl. Ind. 194. 1768. *Cyperus monostachyos* L., Mant. 2:180. 1771. *Abildgaardia monostachyos* (L.) Vahl, Enum 2: 296. 1805. *Fimbristylis monostachyos* (L.) Hassk., Fl. Jav. Rar. 61. 1848 ("*monostachya*"); FBI 6:649; Gamble 1660 (1152).

Erect herbs, about 40 cm tall; stems tufted, filiform. Leaves filiform, crowded at the base of the stem, 20 x 0.1 cm, Spikelets ovate, solitary, compressed. Glumes ovate, yellow. Nuts obovate, stipitate, snow-white.

Fls. & Frs. : December - March.

Occasional in wet grass lands. Marrikunta, **CP** 11395; Pocharam tank, **TP & CP** 12046.

BULBOSTYLIS Kunth *nom.cons.*

Bulbostylis barbata (Rottb.) C.B. Clarke in FBI 6:651. 1893; Gamble 1662 (1153); Kern in Fl. Males. Ser. 1.7 539. 1974. *Scirpus barbatus* Rottb., Descr. Pl. Rar. 27. 1772 & Descr. Ic. Rar. 52 t. 17. f. 4. 1773.

Erect herbs,about 30 cm tall; stems tufted, striate. Leaves filiform, 4x0.1 cm, glabrous. Spinkelets ovate-oblong, in terminal heads. Glumes ovate, brown. Nuts trigonous, white.

Fls. & Frs. : Round the year.

Common in all places except forest. Choutkoor, **TP & CP** 11947.

CYPERUS linnaeus

1. Spikelets digitate or clustered:

  2. Glumes aristate or distinctly mucronate — **C. cuspidatus**

  2. Glumes minutely mucronate:

    3. Inflorescence a single head — **C. niveus**

    3. Inflorescence umbellate, not a single head:

      4. Glumes obovate, apex rounded — **C. difformis**

      4. Glumes oblong or ovate, apex obtuse-acute:

        5. Rhizomes present — **C. haspan**

        5. Rhizomes absent — **C. tenuispica**

1. Spikelets distinctly spicate or racemose:

  6. Rhachilla of spikelets distinctly 2-winged :

    7. Leaves short, rarely half as long as the stem:

      8. Wing of rhachilla wide, red — **C. pangorei**

      8. Wing of rhachilla narrow, pale — **C. corymbosus**

    7. Leaves long, often longer than the stem:

      9. Rhizomes stoloniferous — **C. rotundus**

      9. Rhizomes not stoloniferous — **C. exaltatus**

  6. Rhachilla of spikelets not winged:

    10. Glumes cuspidate — **C. compressus**

    10. Glumes obtuse, not cuspidate:

      11. Rhachilla of spike glabrous; keel 3-5-nerved — **C. iria**

      11. Rhachilla of spike scabrid; keel 7-nerved — **C. alulatus**

Cyperus alulatus Kern in Reinwardt. 1:463-466.f.1. 1952. *C. iria* var. *rectangularis* Kuekenth. in Engl., Pflanzenr. 101:152. 1935. *C. rectangularis* (Kuekenth.) Bennet in Indian For. 95:692. 1969.

Erect herbs, about 40 cm tall; stems tufted, triquetrous. Leaves linear, 40 x 0.2 cm, Spikelets oblong, in terminal umbellate spikes. Glumes ovate, greenish-yellow. Nuts obovate, reddish- brown.

Fls. & Frs. : August - November.

Occasional in dry paddy fields & ditches. Usirikapalle, **GNR** 4332.

Cyperus compressus L., Sp. Pl. 46. 1753; FBI 6:605; Gamble 1640 (1140); Kern in Fl. Males. Ser. 1. 7:617. 1974.

Erect herbs, about 40 cm tall; stems tufted, triquetrous. Leaves linear, 20 x 0.2 cm, scabrous. Spikelets oblong, in condensed umbellate spikes. Glumes broadly ovate, pale green. Nuts obovate, black.

Fls. & Frs. : June - January.

Occasional in moist places. Siddipet, **CP** 11376; Choutkoor, **TP & CP** 11922.

Cyperus corymbosus Rottb., Descr. Ic. Rar. 42. t. 7. f. 4. 1773; FBI 6:612; Gamble 1640 (1140).

Erect herbs, about 1 m tall; stems triquetrous. Leaves reduced to sheaths. Spikelets linear, in terminal corymbose spikes. Glumes ovate, brown. Nuts obovate, dark brown.

Fls. & Frs. : August - December.

Common in moist places. Medak tank, **BR & CP** 11455; Medak town, **BR & CP** 11461. 239.

Cyperus cuspidatus Kunth, Nov. Gen. Pl. 1:204. 1815; FBI 6:598; Kern in Fl. Males. Ser. 1.7.629. 1974. *C. uncinatus auct. non* Poir., Gamble 1639 (1139).

Erect herbs, about 10 cm tall; stems tufted, glabrous. Leaves filiform, 7 x 0.5 cm. Spikelets linear, in digitately - arranged spikes. Glumes mucronate, brown. Nuts obovate, dark brown.

Fls. & Frs. : June - January.

Rare in moist places. Narsapur town, **BR & CP** 11245.

Cyperus difformis L., Cent. Pl. 2:6. 1756; FBI 6:599; Gamble 1640 (1139); Kern in Fl. Males. Ser. 1.7:629. 1974.

Erect herbs, about 40 cm tall; stems tufted, weak. Leaves flat, 30 x 0.4 cm. Spikelets linear-lanceolate, in globose heads. Glumes obovate, brown. Nuts obovate, yellowish-brown.

Fls. & Frs. : Round the year.

Common in moist places. Medak tank, **BR & CP** 11421; Ramayampet, **BR & CP** 11505; Pocharam sanctuary, **BR & CP** 11610; NSF, **BR & CP** 11647.

Cyperus exaltatus Retz., Obs. Bot. 5:11. 1789, FBI 6:617; Gamble 1642 (1141); Kern in Fl. Males. Ser. 1.7:602. 1974.

Erect herbs, about 1 m tall; stems tufted, trigonous. Leaves flat, 75 x 0.7 cm. Spikelets oblong, in corymbose spikes. Glumes ovate, pale green. Nuts ellipsoid, brown.

Fls. & Frs. : August - December.

Occasional in wet fields. Narsapur, **GNR** 3968.

Cyperus haspan L., Sp. Pl. 45. 1753; FBI 6:600; Gamble 1640 (1139); Kern in Backer & Bakh.f., Fl. Java 3:476. 1968 ("*halpan*").

Erect herbs, about 50 cm tall; stems tufted or solitary, triquetrous. Leaves linear, 15 x 0.2 cm. Spikelets linear- oblong, in digitately arranged spikes. Glumes ovate, brown. Nuts obovate, cream-yellow.

Fls. & Frs. : November - March.

Rare in shallow water and marshy places. Thornal, **CP** 11701.

Cyperus iria L., Sp. Pl. 45. 1753; FBI 6:606; Gamble 1640 (1140); Kern in Fl. Males. Ser. 1.7:616. 1974.

Erect herbs, about 50 cm tall; stems tufted, triquetrous. Leaves flat, 30 x 0.3 cm. Spikelets oblong, in interrrupted spikes. Glumes broadly ovate. Nuts obovate, black-brown.

Fls. & Frs. : Round the year.

Common weed of paddy fields. Nagulabanda, **CP** 11352; Narsingi, **BR** & **CP** 11538; NSF, **BR** & **CP** 11663; Venkatapuram, **TP** & **CP** 14005.

Cyperus niveus Retz., Obs. Bot. 5:12. 1789; FBI 6:601; Gamble 1640 (1140).

Erect rhizomatous herbs, about 50 cm tall; stems tufted, glabrous. Leaves narrowly linear, 25 x 0.2 cm. Spikelets oblong-elliptic, in terminal solitary heads. Glumes ovate-lanceolate, pale brown. Nuts obovate, dark brown-black.

Fls. & Frs. : May - November.

Occasional in moist sandy localities. Nagulabanda, **CP** 11333.

Cyperus pangorei Rottb., Descr. Ic. Rar. 31. t. 7. f. 3. 1773; Gamble 1641 (1140). *C. tegetum* Roxb., Fl. Ind. 1:211. 1820; FBI 6:613.

Erect herbs, about 1.5 m tall; stems solitary or tufted, trigonous, sulcate. Leaves flat, 25 x 0.1 cm. Spikelets oblong, in corymbose spikes. Glumes ovate, purplish brown. Nuts ellipsoid, yellowish- brown.

Fls. & Frs. : September - December.

Occasional along the water streams and in paddy fields. Way to Rusthempet, **KMS** 6718.

Cyperus rotundus L., Sp. Pl. 45. 1753; FBI 6:614; Gamble 1641 (1140).

Erect stoloniferous herbs, about 30 cm tall; stems tufted, triquetrous. Leaves flat, 15 x 0.2 cm. Spikelets linear, in short simple spikes. Glumes ovate, reddish- brown. Nuts oblong, greyish black.

Fls. & Frs. : June - January.

Common weed of cultivated fields. Narsingi, **BR & CP** 11531; NSF, **BR & CP** 11659.

Cyperus tenuispica Steud., Syn. Pl. Glum. 2:11. 1855; Gamble 1640(1139). *C. flavidus auct. non* Retz.: FBI 6:600.

Erect herbs, about 20 cm tall; stems tufted, triquetrous, finely ribbed. Leaves linear, 15 x 0.3 cm. Spikelets linear-lanceolate, in stellately arranged spikes. Glumes oblong, pale brown. Nuts subglobose, pale yellow.

Fls. & Frs. September - February.

Occasional weed of paddy fields. Narsapur, **GNR** 4303.

ELEOCHARIS Robert Brown

1. Hypogynous bristles shorter than nut; stamens 2 **E. atropurpurea**
1. Hypogonous bristles longer than nut; stamens 3 **E. geniculata**

Eleocharis atropurpurea (Retz.) C. Presl, Rel. Haenk. 1:196. 1828; FBI 6:627; Gamble 1648 (1144). *Scirpus atropurpureus* Retz., Obs, Bot. 5:14. 1788.

Erect herbs, about 15 cm tall; stems tufted, sulcate. Leaves reduced to short sheaths. Spikelets ovate, solitary. Glumes ovate, purple. Nuts obovate, biconvex, black.

Fls. & Frs. : June - December.

Common in marshy areas. Manjira Barrage, **BR & CP** 11473.

Eleocharis geniculata (L.) Roemer & Schultes in L. Syst. Veg. 2:150. 1817. *Scirpus geniculatus* L., Sp. Pl. 48. 1753. *Eleocharis capitata* R. Br., Prodr. 225. 1810; FBI 6:627; Gamble 1648 (1145). non *Scirpus capitatus* L., 1753.

Erect herbs, about 10 cm tall; stems tufted, sulcate. Leaves reduced to membranous sheaths. Spikelets ovate, globose, solitary. Glumes broadly elliptic-ovate, grey. Nuts obovate, brown.

Fls. & Frs. : June - October.

Common in marshy areas and paddy fields. Pocharam tank., **TP & CP** 12046.

FIMBRISTYLIS Vahl *nom. cons.*

1. Nuts oblong-cylindrical; stigmas 2:
    2. Inflorescence corymbose with few to many spikelets **F. dispsacea**
    2. Inflorescence of a single terminai spikelet **F. tetragona**
1. Nuts obovate; stigmas 2 or 3:
    3. Spikelets few to many:
        4. All spikelets sessile **F. argentea**
        4. All spikelets stalked:
            5. Stigmas 2; nuts biconvex:
                6. Glumes glabrous:
                    7. Spikelets 1-1.5 mm wide **F. bis-umbellata**
                    7. Spiklets 2-4 mm wide **F. dichotoma**
                6. Glumes pubescent on upper half **F. ferruginea**
            5. Stigmas 3; nuts trigonous:
                8. Leaves shortly liguled **F. complanata**
                8. Leaves not liguled:
                    9. All or upper leaves surrounding stem base bladeless **F. miliacea**
                    9. Leaves surrounding stem base bladed **F. tenera**
    3. Spikelets 1-2 or 3:
        10. Glumes elliptic **F. polytrichoides**
        10. Glumes broadly ovate **F. schoenoides**

Fimbristylis argentea (Rottb.) Vahl, Enum. Pl. 2:294. 1806; FBI 6:640; Gamble 1659 (1151); Kern in Fl. Males. Ser. 1.7;586. 1974. *Scirpus argenteus* Rottb., Descr. Pl. Rar. 27. 1777 & Descr. Ic. Rar. 51. t. 17. f. 1773.

Spreading herbs, about 10 cm tall; stems filiform, trigonous. Leaves filiform, 5 x 0.1 cm, flexuous. Spikelets oblong, in terminal globose heads. Glumes ovate, brown. Nuts obovate, biconvex, fulvous.

Fls. & Frs. : June - January.

Occasional in moist localities. Medak tank, **BR & CP** 11401.

Fimbristylis bis-umbellata (Forssk.) Bubani, Dodecanthea 30. 1850; Gamble 1659 (1151). *Scirpus bis-umbellatus* Forssk., Fl. Aegypt.-Arab 15. 1775. *Fimbristylis dichotoma* C.B. Clarke, in FBI 6: 636. 1893.

Erect herbs, about 30 cm tall; stems tufted, trigonous. Leaves flat, 12 x 0.1 cm, pubescent. Spikelets oblong, in terminal umbels. Glumes ovate-oblong, reddish- brown. Nuts obovate, creamy yellow.

Fls. & Frs. : January - June.

Common in wet lands. Medak, **BR & CP** 11459; Ramayampet, **BR & CP** 11509.

Fimbristylis complanata (Retz.) Link, Enum. Pl. hort. Berol. 1:292. 1827; FBI 6:646; Gamble 1659 (1151). *Scirpus complanatus* Retz., Obs. Bot 5:14. 1788.

Erect herbs, about 50 cm tall; stems densely caespitose, trigonous. Leaves flat, 15 x 0.2 cm. Spikelets narrowly oblong, terminal , in compound umbels. Glumes ovate, brown. Nuts obovate, trigonous.

Fls. & Frs. : August - December.

Occasional in muddy areas. Gangapur RF, **TP & CP** 12096.

Fimbristylis dichotoma (L.) Vahl, Enum 2:287. 1806; FBI 6:635; Gamble 1658 (1151). *Scirpus dichotomus* L. Sp. Pl. ed. 1:50. 1753.

Erect herbs, about 40 cm tall; stems tufted, trigonous. Leaves linear, 30 x 0.3 cm, Spikelets ovate, in compressed umbels. Glumes broadly ovate, dark brown. Nuts obovate.

Fls. & Frs. : March-December.

Occasional along the sides of water courses. Way to Pegarikutta, KMS 6634.

Fimbristylis dipsacea (Rottb.) C.B. Clarke in FBI 6:635. 1893; Gamble 1658 (1151). *Scirpus dipsaceus* Rottb., Descr. Ic. Rar. 56. t. 1. 1773.

Erect herbs, about 15 cm tall; stems tufted, slender, trigonous. Leaves reduced to short-bladed sheaths. Spikelets globose, solitary, in simple corymbs. Glumes ovate, red-brown. Nuts oblong, biconvex, cream.

Fls. & Frs. : April - August.

Common in sandy fields. Near Ooracheru lake, **KMS** 7974.

Fimbristylis ferruginea (L.) Vahl, Enum. Pl. 2:291. 1806; FBI 6:638; Gamble 1658 (1151). *Scirpus ferrugineus* L., Sp. Pl. 50. 1753.

Erect, rhizomatous herbs, about 1 m tall; stems tufted, trigonous. Leaves flat, 6 x 0.2 cm. Spikelets ovate-oblong, in sub-compound umbels. Glumes ovate, brown. Nuts fulvous.

Fls. & Frs. : August - December.

Common in wet places and in brackish waters. Way to Rushtempet, **KMS** 7996.

Fimbristylis miliacea (L.) Vahl, Enum. Pl. 2:287. 1806; FBI 6:644; Gamble 1660 (1152). *Scirpus miliaceus* L., Syst. Nat. ed. 10 2:868. 1759.

Erect herbs, about 40 cm tall; stems tufted, slender, glabrous. Leaves linear, 30 x 0.3 cm. Spikelets globose-ovate, in lax panicles. Glumes ovate, brown. Nuts obovate, yellow.

Fls. & Frs. : July - December.

Common weed of paddy fields. Gangapur, **TP & CP** 12079.

Fimbristylis polytrichoides (Retz.) R.Br. Prodr. 226. 1810; FBI 6:632; Gamble 1659 (1150). *Scirpus polytrichoides* Retz., Obs. Bot. 4:11. 1786.

Erect herbs, about 25 cm tall; stems densely tufted, trigonous, glabrous. Leaves canaliculate, acicular, 10 x 0.2 cm. Spikelets ellipsoid, solitary, terminal. Glumes elliptic, brown. Nuts obovate, biconvex, grey-purple.

Fls. & *Frs.* : September - March.

Occasional on open wet grounds. Way to Rusthempet, **KMS** 7998.

Fimbristylis schoenoides (Retz.) Vahl Enum. Pl. 2:286. 1806; FBI 6:634; Gamble 1658 (1150). *Scirpus schoenoides* Retz., Obs. Bot. 5:14. 1788.

Erect herbs, about 40 cm tall; stems tufted, slender. Leaves filiform, 15 x 0.1 cm. Spikelets ovate-subglobose, solitary or in pairs. Glumes broadly ovate, pale brown. Nuts obovate, bioconvex, stipitate.

Fls. & *Frs.* : August - November.

Occasional in marshy places. Gangapur, **TP & CP** 12080.

Fimbristylis tenera Schultes, Mant. 2:57. 1825; FBI 6:642; Gamble 1660(1152). *F. monticola* Hochst. ex Steud. Syn. Pl. Cyp. 3:111. 1855.

Erect herbs, about 30 cm tall; stems caespitose, filiform, trigonous. Leaves filiform, 15 x 0.1 cm, Spikelets oblong, in sub compound umbels. Glumes triangularly ovate, dark brown. Nuts obovate, yellow-brown.

Fls. & *Frs.* August - November.

Occasional in moist low lands, sandy soils. Sardhana, **CP** 11744.

Fimbristylis tetragona R.Br., Prodr. 226. 1810; FBI 6:631; Gamble 1658 (1150).

Caespitose herbs, about 40 cm tall; stems slender, ribbed, 4- angular. Leaves reduced to membranous sheaths. Spikelets subglobose, solitary. Glumes oblong-ovate. Nuts oblong, terete, grey.

Fls. & *Frs.* : July - December.

Occasional in moist open places, borders of paddy fields. Usirikapally, **GNR** 4305.

FUIRENA Rottboll

1. Scales quadrate of suboribcular **F. ciliaris**
1. Scales lunate, trilobate **F. trilobites**

Fuirena ciliaris (L.) Roxb., Fl. Ind. 1:184. 1820; Kern. in Fl. Males. Ser. 1.7:519. 1974. *Scirpus ciliaris* L., Mant. Pl. 182. 1771. *Fuirena glomerata* Lam., Tabl. Encycl. 1:150. 1791; FBI 6:666; Gamble 1669 (1158).

Erect herbs, about 40 cm tall; stems noded, slender, sparsely hairy. Leaves linear-lanceolate, 9 x 0.4 cm, hairy. Spikelets oblong, in terminal umbellate clusters. Glumes obovate-oblong, ciliate. Nuts obovate, triquetrous, pale white.

Fls. & Frs. : June - January.

Common weed of fields and in moist localities. Pochammaralu, **BR & CP** 11474; Pocharam tank, **TP & CP** 12044

Fuirena trilobites C.B. Clarke in FBI 6:666. 1893 & iii. Cyper. t. 59. f. 5. 1909.

Erect herbs, about 40 cm tall; stems striate, sparsely hairy. Leaves linear-lanceolate, densely pubescent. Spikelets ovate- oblong, in axillary and terminal; subcapitate heads. Glumes obovate, cilitate, greyish- black. Nuts obovate, triquetrous.

Fls. & Frs. : November - March.

Rare in water logged grass lands. Donthi, **GNR** 4049.

INDOCOURTOISIA Bennet & Raizada

Indocourtoisia cyperoides (Roxb.) Bennet & Raizada in Indian For. 107:432. 1981. *Kyllinga cyperoides* Roxb., Fl. Ind. 1:187. 1820. *Courtoisia cyperoides* (Roxb.) Nees in Linnaea, 9:286. 1834; FBI 6: 625; Gamble 1645 (1143).

Erect herbs, about 30 cm tall; stems tufted, trigonous, glabrous. Leaves linear, 30 x 0.2 cm. Spikelets broadly elliptic, in terminal spikes. Glumes distichous, ovate. Nuts linear-oblong, golden brown.

Fls. & Frs. : November - January.

Occasional weed of paddy fields. Pocharam RF, **TP & CP** 12061.

KYLLINGA Rottboll

1. Rhizome long, stems solitary **K. brevifolia**
1. Rhizome short, stems tufted **K. bulbosa**

Kyllinga brevifolia Rottb., Descr. Ic. Rar. 13. t. 4. f. 3. 1773; FBI 6:588; Gamble 1624 (1130). *Cyperus brevifolius* (Rottb.) Hassk., Cat. Hort. Bot. Bog. 24. 1844; Kern in Fl. Males Ser. 1. 7:656. 1974.

Erect rhizomatous herbs, about 25 cm tall; stems solitary, glabrous. Leaves linear, 25 x 0.3

cm. Spikelets oblong, in terminal globose spikes Glumes ovate-elliptic. Nuts ellipsoid, biconvex, yellow.

Fls. & Frs. : july - December.

Common in all places. Ramayampet, **BR & CP** 11504; Patancheru, **CP** 11685.

Kyllinga bulbosa Beauv., Fl. d' Oware & Benin 1:11.t. 8.f. 1. 1804. *K. triceps* Rottb., Descr. Ic. Rar, 14. t. 4. f. 6. 1773, *nom. illegit*.: FBI 6:587; Gamble 1623 (1130). *Cyperus triceps* Endl., Cat. Hort. Vindob. 1:94. 1842; Kern in Fl. Males. Ser. 1.7:659. 1974.

Erect herbs, about 20 cm tall; stems tufted, glabrous. Leaves linear, 10 x 0.2 cm, Spikelets oblong, in terminal globose spikes. Glumes ovate-oblong. Nuts oblong, biconvex, brown.

Fls. & Frs. : Round the year.

Common weed of all localities. Nagulabanda, **CP** 11323; Venkatapuram, **TP & CP** 14006.

LIPOCARPHA Robert Brown *nom. cons.*

Lipocarpha sphacelata (Vahl) Kunth, Enum. Pl. 2:267. 1837; FBI 6:667. *Hypaelyptum sphacelatum* Vahl, Enum. Pl. 2:283. 1806. *Lipocarpha triceps* (Roxb.) Nees in Wight, Contrib. Bot. India 92. 1834; Gamble 1670 (1158). *Tunga triceps* Roxb. Fl. Ind. 1:183. 1820.

Erect herbs, about 25 cm tall; stems tufted, triquetrous. Leaves linear, canaliculate, 15 x 0.1 cm. Spikelets globose-ovate, in terminal heads. Glumes elliptic, purple-striped. Nuts ellipsoid, trigonous, light brown.

Fls. & Frs. : August - January.

Occasional in wet fields and margins of swamps. Pandivagu, **BR & CP** 11203.

MARISCUS Vahl *nom. cons.*

1. Base of the stem bulbous; spikelets 1-flowered **M. clarkei**
1. Base of the stem not bulbous; spiklets 2-4-flowered **M. cyperinus**

Mariscus clarkei (Cooke) T. Koyama in J.Jap. Bot. 51:313. 1976. *Cyperus clarkei* Cooke, Fl. Bombay 2: 873 (387). 1908. *Mariscus bulbosus* C. B. Clarke in FBI 6:620 1893. non Steud. 1855; Gamble 1644 (1142).

Erect stoloniferous herbs; about 15 cm tall; stems subsolitary, bulbous. Leaves linear, 12 x 0.4 cm. Spikelets terete, in terminal lax cylindrical spikes. Glumes 2, lower ovate and upper lanceolate, green. Nuts ellipsoid, trigonous, brown.

Fls. & Frs.: September - December.

Occasional in moist places. Siddipet, **CP** 11379.

Mariscus cyperinus (Retz.) Vahl, Enum. Pl. 2:377, 1806; FBI 6:621; Gamble 1644 (1143) *Kyllinga cyperina* Retz., Obs. Bot. 6:21. 1791.

Erect herbs, about 40 cm tall; stems tufted, triquetrous. Leaves narrowly linear, 30 × 0.2 cm, Spikelets linear-lanceolate, in terminal cylindrical spikes. Glumes elliptic, straw brown. Nuts oblong, brown.

Fls. & Frs. : July - October.

Occasional in marshy areas. Nagulabanda, **CP** 11331; Gnanapur, **CP** 11788.

PYCREUS Palisot de Beauvois

1. Nuts transversely wrinkled — **P. diaphanus**
1. Nuts smooth-puncticulate:
    2. Leaves distant; glumes furrowed — **P. sanguinolentus**
    2. Leaves basal; glumes not furrowed:
        3. Rachilla winged — **P. pumilus**
        3. Rachilla not winged — **P. flavidus**

Pycreus diaphanus (Schrad. ex Roemer & Schultes) Hooper & T. Koyama in J. Jap. Bot. 51: 316. 1976. *Cyperus diaphanus* Schrad. ex Roemer & Schultes Mant. 2:477. 1824. C. *latespicatus* Boeck., in Flora 42:433. 1859. *Pycreus latespicatus* (Boeck.) C.B. Clarke in FBI 6:590.

Erect herbs, about 30 cm tall; stems tufted, Striate. Leaves linear, 20 × 0.3 cm. Spikelets oblong-lanceolate, in simple umbellate spikes. Glumes broadly ovate when opened. Nuts obovate, black.

Fls. & Frs. : August - January.

Occasional in wet places along water streams. Usirikapally, **GNR** 4319.

Pycreus flavidus (Retz.) T. Koyama in J. Jap. Bot 51:313. 1976. & in Gard. Bull. Straits Settlem. 30:150. 1977. *Cycerus flavidus* Retz., Obs. Bot. 5:13. 1788. *C. capillaris* J. Koening ex Roxb. Fl. Ind. 1:198. 1820. *Pycerus globosus* (All.) Reichb. Fl. Germ. Excurs. 1:140. (10). 1830. Gamble 1627 (1132). *P. capillaris* (Koenig ex Roxb.) Nees ex C.B. Clarke in FBI 6:591 *Cyperus globosus* All., Suctuar. Fl.Pedemont, 49 1789.

Erect herbs, about 20 cm tall; stems tufted, trigonous. Leaves linear, 15 x 0.1 cm. Spikelets linear, Compressed, in stellate spikes. Glumes ovate, greenish- yellow. Nuts obovate dark brown.

Fls. & Frs. : July - November.

Occasional in wet places of forests and marshy localities. Medak town, **BR** & **CP** 11460; Rusthempet, **KMS** 7997.

Pycreus pumilus (L.) Nees in Linnaea, 9:283. 1835; Gamble 1627 (1132). *Cyperus pumilus* L., Cent. Pl. 2:6. 1756. *Pycreus nitens* Nees in Nov. Act. Acad. Caes. Leopcarol. Nat. Cur. 19 Suppl. 1:53, 1843; FBI 6:591.

Erect herbs, about 10 cm tall; stems tufted, triquetrous. Leaves linear. 10 x 0.1 cm. Spikelets oblong, in subglobose spikes. Glumes ovate, green . Nuts elliptic, brown.

Fls. & Frs. : August - November.

Occasional along the drying beds of canals. Narsapur tank, **BR & CP** 11245a.

Pycreus sanguinolentus (Vahl) Nees ex C.B. Clarke in FBI 6:590; Gamble 1627 (1132). *Cyperus sanguinolentus* Vahl, Enum. Pl. 2:351. 1806.

Erect herbs, about 40 cm tall; stems tufted, trigonous. Leaves linear, 10 x 0.2 cm. Spikelets narrowly ovate, in clustered capitate spikes. Glumes ovate, pale brown. Nuts obovate, brown.

Fls. & Frs. : July - January.

Occasional in marshy localities and paddy fields. Dundigal, **GNR** 4669.

RHYNCHOSPORA Vahl *nom. cons.*

Rhynchospora wightiana Steud., Syn. Pl. Glum. 2:148. 1855; FBI 6:669; Gamble 1672 (1160).

Erect herbs, about 35 cm tall; stems tufted, glabrous. Leaves linear, clustered at the base of stem, 8 x 0.2 cm. Spikelets lanceolate, in terminal solitary heads. Glumes elliptic-lanceolate, brown. Nuts oblong, dark brown.

Fls. & Frs. : July - December.

Occasional in moist places. Pocharam tank, **TP & CP** 12031.

RICKLIELLA J. Raynal

Rickliella squarrosa (L.) Raynal in Adansonia (Ser.2) 13:154. 1973. *Scirpus squarrosus* L., Mant. Pl. 181. 1771; FBI 6:663; Gamble 1666 (1156); Kern in Fl. Males. Ser. 1.7:516. 1974.

Erect herbs, about 20 cm tall; stems tufted, Striate, glabrous. Leaves filiform, 10 x 0.1 cm. Spikelets ovate, in pseudolateral heads. Glumes obovate, pale green. Nuts obovate trigonous. yellow.

Fls. & Frs. : August - December.

Occasional in wet sandy river beds. Narsapur RF, **TP & CP** 14055.

SCHOENOPLECTUS (Reich.) Palla

1. Hypogynous bristles present, retrorsely scabrid:
    2. Rhizomatous herbs:
        3. Spikelets sessile **S. mucronatus**
        3. Spikelets stalked **S. maritimus**
    2. Stoloniferous herbs **S. grossus**
1. Hypogynous bristles absent:
    4. Stems septate; glumes not mucronate:
        5. Nuts with transverse ridges **S. senegalensis**
        5. Nuts smooth **S. articulatus**
    4. Stems not septate; glumes mucronate:
        6. Stems terete; nuts obovate **S. lateriflorus**
        6. Stems trigonous; nuts elliptic **S. supinus**

Schoenoplectus articulatus (L.) Palla in Bot. Jahrb. Syst. 10:299. 1888. *Scirpus articulatus* L., Sp. Pl. 47. 1753; FBI 6:656; Gamble 1666 (1156); Kern in Fl. Males. Ser. 1.7: 513. 1974.

Erect herbs, about 50 cm tall; stems tufted, terete, spongy. Leaves reduced to stramineous-brown sheaths. Spikelets ovate-oblong, in stellately-arranged globose clusters. Glumes ovate, pale green. Nuts obovate, black.

Fls. & Frs.: Round the year.

Common in marshy localities. Medak tank, BR & CP 11405; Narsapur tank, TP & CP 1198.

Schoenoplectus grossus (L.f.) Palla in Allg. Bot. Z. Syst. 17 : Beibl. 3. 1911. *Scirpus grossus* L.f., Suppl. Sp. Pl. 104. 1781; FBI 6:659; Gamble 1666 (1156).

Stoloniferous herbs; about 2 m tall; stems tufted, 3-angled, spongy. Leaves flat, 100 x 2.5 cm, septate-nodulose. Spikelets ellipsoid, solitary, in terminal corymbs. Glumes elliptic, reddish- brown. Nuts obovate, brown.

Fls. & Frs. : August - December.

Rare in marshy fields. Dosalagunda lake, Rusthempet, **KMS** 6799.

Schoenoplectus lateriflorus (J.F.Gmelin) Lye in Bot. Notis. 124:290. 1971. *Scirpus lateriflorus* J.F. Gmelin in L. Syst. Veg. ed. 13.1 :127 1791. *S. supinus* var. *uninodis* C. B. Clarke in FBI 6:656. 1893; non (Del.) Asch. & Schweinf. 1887. *S. supinus* auct. non L. : Gamble 1666 (1156).

Erect herbs, about 30 cm tall; stems tufted, terete. Leaves flat, 10 x 0.2 cm, Spikelets ovate-oblong, in terminal heads. Glumes elliptic, reddish- brown. Nuts obovate, black.

Fls. & Frs. : September - January.

Common in paddy fields. Narsapur, **GNR** 4362.

Schoenoplectus maritimus (L.) Lye in Blyttia 29:145. 1971. *Scirpus maritimus* L., Sp. Pl. 75. 1753; FBI 6: 658; Gamble 1666 (1156).

Erects herbs, about 1 m tall; stems solitary. Leaves linear, 35 x 0.7 cm, keeled. Spikelets ovate-oblong, terminal in umbellate clusters. Glumes oblong, brown. Nuts obovate, pale white.

Fls. & Frs. : October - March.

Occasional in moist waste lands. Rusthempet, **KMS** 8009.

Schoenoplectus mucronatus (L.) Palla in Bot. Jahrb. Syst. 10:299. 1888. *Scripus mucronatus* L., Sp. Pl. 50. 1753; FBI 6:657; Gamble 1666 (1156), Kern in Fl. Males. Ser. 1.7 : 510. 1974.

Erect herbs, about 1 m tall; stems sparsely tufted, spongy. Leaves flat, 60 x 0.2 cm; sheaths stramineous, brown. Spikelets ovate, pseudolateral, capitate. Glumes broadly ovate, grey. Nuts obovate, plano-convex, black.

Fls. & Frs.: September - December.

Occasional in marshy places. Pathur, **TP & CP** 12133.

Schoenoplectus senegalensis (Hochst. ex Steud.) Palla in Bot. Jahrb. 10:299. 1888. *Isolepis senegalensis* Hochst. ex Steud. Syn. Pl. Glum 2:96. 1855. *Scirpus jacobil* C. Fischer in Kew Bull. 1931:103. 1931 and in Gamble, Fl. Pres. Madras. 1666 (1156). 1931.

Erect herbs, about 30 cm tall; stems caespitose, subterete, transversely septate. Leaves reduced to obliquely truncate sheaths. Spikelets ovate-subglobose, in pseudolateral heads. Glumes broadly ovate, concave. Nuts obovate, triquetrous.

Fls. & Frs. : August - December.

Occasional in marshy places. Narsapur tank, **BR & CP** 11243 and 11244.

Schoenoplectus supinus (L.) Palla in Sitzber, Zool-Bot. Ges. Wien 38:49. 1888. *Scirpus supinus* L., Sp. Pl. 49. 1753; FBI 6:655; Gamble 1666 (1156).

Erect herbs, about 30 cm tall; stems tufted, trigonous. Leaves reduced to obliquely truncate sheaths. Spikelets ovate-oblong, in pseudolateral heads. Glumes ovate, pale with sanguineous spots. Nuts elliptic, black.

Fls. & Frs. : August - December.

Common in wet grass lands and paddy fields. Pathur, **TP & CP** 12107.

SCLERIA P. Bergius

Scleria lithosperma (L.) Sw., Prod. Fl. Ind. Occid. 18. 1788. FBI 6:685; Gamble 1677 (1163). *Scirpus lithospermus* L., Sp. Pl. 51. 1753.

Erect herbs, about 80 cm tall; stems tufted, rigid. Leaves narrowly linear, clustered at the middle of the stem, 25 x 0.3 cm, pubescent. Spikelets ovate-turbinate, solitary or clustered, in axillary and terminal panicles. Glumes lanceolate, purple. Nuts obovate, trigonous, brown.

Fls. & Frs. : Round the year.

Occasional in dry localities. Pathur, TP & CP 12118.

POACEAE (*nom. alter.* Gramineae)

1. Shrubs, culms woody **Dendrocalamus**
1. Herbs, culms not woody:
  2. Spikelets with unisexual florets **Coix**
  2. Spikelets with hermaphrodite florets:
    3 Mature spikelets falling entire from their pedicels, dorsally compressed:
      4. Spikelets usually in pairs, one sessile, the other pedicelled:
        5. Spikelets in each pair usually dissimilar, the sessile bisexual, pedicelled male or neuter:
          6. Inflorescence of solitary raceme:
            7. Spikelets awned:
              8. All spikelets heterogamous:
                9. Inflorescence not supported by spathes **Sehima**
                9. Inflorescence supported by spathes **Eremopogon**
              8. Lowest pairs of spikelets homogamous **Heteropogon**
            7. Spikelets unawned **Hackelochloa**
          6. Inflorescence of racemes more than one paired:
            10. Spikelets in threes **Chrysopogon**
            10. Spikelets in pairs:
              11. Racemes supported by the spathes:
              12. Involucral spikelets present:
                13. Involucral spikelets with strong bristles, sessile, persistent. **Themeda**
                13. Involucral spikelets glabrous, pedicelled, deciduous **Iseilema**
              12. Involucral spikelets absent:
                14. Basal joints of the racemes bulbous **Apluda**
                14. Basal joints of the racemes not bulbous:
                15. Racemes axillary, decompound **Cymbopogon**
                15. Racemes terminal, umbels **Andropogon**
              11. Racemes not supported by the spathes:
                16. Inflorescence a panicle **Vetiveria**
                16. Inflorescence not a panicle, digitate racemes:
                  17. Upper lemma awned from the back, below the middle **Arthraxon**
                  17. Upper lemma awned from the apex or from the sinus:
                    18. Upper lemma well developed **Ischaemum**
                    18. Upper lemma reduced to a transulucent base of an awn **Dichanthium**
        5. Spikelets in each pair usually similar **Saccharum**
      4. Spikelets usually solitary:

19. Spikelets usually unawned (except in *Echinochioa crusgalli*):
20. Inflorescence spiciform, digitate:
21. Ligule absent **Echinochloa**
21. Ligule present:
22. Lower glume well developed:
23. Lower glume turned away from the rachis **Paspalidium**
23. Lower glume turned towards the rachis **Brachiaria**
22. Lower glume absent or reduced to a scale:
24. Spikelets plano convex, orbicular or oblong, ovate **Paspalum**
24. Spikelets elliptic or elliptic-oblong **Digitaria**
20. Inflorescence panicle:
25. Spikelets covered by involucral bristles:
26. Bristle solitary, formed by prolongation of axis **Pseudoraphis**
26. Bristles few to many, forming an involucre:
27. Spikelets ellipsoid, bristles rigid, barbed **Setaria**
27. Spikelets lanceolate, bristles plumose **Pennisetum**
25. Spikelets not covered by an involucral bristles:
28. Spikelets with an annular callus at the base **Eriochloa**
28. Spikelets without an annular callus at the base:
29. Spikelets gibbose **Sacciolepis**
29. Spikelets not gibbose **Panicum**
19. Spikelets awned:
30. Upper lemma densely ciliate along the margins; upper glume unawned **Alloteropsis**
30. Upper lemma glabrous along the margins; upper glume awned **Oplismenus**
3. Mature spikelets breaking up bearing glumes on the pedicel, laterally compressed:
31. Spikelets with 1-fertile floret:
32. Fertile lemma and palea similar in texture:
33. Fertile lemma supported by 2 empty scales **Oryza**
33. Fertile lemma without any scales **Leersia**
32. Fertile lemma and palea not similar in texture:
34. Upper glume 5-ribbed, armed with hooked spinules **Tragus**
34. Upper glume not as above:
35. Inflorescence of solitary or digitate spikes, spikelets secund:
36. Inflorescence of solitary spike **Melanocenchris**
36. Inflorescence of 2 or more umbelled or racemously arranged spikes:
37. Fertile floret 1, lemmas unawned **Cynodon**
37. Fertile florets 1-4, lemmas awned **Chloris**
35. Inflorescence paniculate, spikelets not secund:
38. Spikelets 2-3-flowered **Arundinella**
38. Spikelets 1-flowered:
39. Lemmas hyaline at maturity, unawned **Sprobolus**
39. Lemmas indurate at maturity, awned **Aristida**
31. Spikelets with 2 or more fertile florets:
40. Slender grasses, lemmas not aristulate:
41. Inflorescence of digitate spikes or racemes:
42. Axis of the inflorescence ending in a spikelet **Eleusine**

42. Axis of the spikes terminating in a sharp point **Dactyloctenium**
41. Inflorescence panicle:
43. Glumes aristate, spikelets 2-3-flowered **Dinebra**
43. Glumes acute, spikelets more than 3-flowered:
44. Racemes not secund, upper glume 3-nerved **Eragrostis**
44. Racemes conspiculously secund, upper glume 1-nerved:
45. Spikelets unawned, not in globose clusters **Eragrostiella**
45. Spikelets awned, in dense globose clusters **Elytrophorus**
40. Tall grasses, lemmas aristulate **Pharagmites**

ALLOTEROPSIS K.B. Presl.

Alloteropis cimicina (L.) Stapf in Prain. Fl. Trop. Africa 9. 487. 1919; Gamble 1766 (1223); Khan in FFH 339; Bor 276. *Milium cimicinum* L., Mant. Alt. 184. 1771. *Axonopus cimicinus* (L.) Beauv., Ess. Agrost. 12. 1812; FBI 7:64.

Annual, tufted grass, Culms about 40 cm tall, tuberculate, pilose throughout; nodes glaborous. Leafsheaths 2-4.5 cm, loose, striate, pilose-glabrous; ligule a fimbriate rim; blades 1.5 - 4.5 x 0.6-1.2 cm, ovate-lanceolate, base-amplexicaul, tuberculate cliate on margins, apex acute, glabrous. Spiciform racemes 3-9, up to 10 cm, terminal. Spikelets 4mm, dorsally compressed, 2- flowered, awned. Caryopsis elliptic-oblong.

Fls. & Frs. : July - February.

Common grass in waste lands, fields. Nagulabanda, CP 11362; Pocharam tank, **TP & MSG** 12049.

ANDROPOGON Linnaeus

Andropogon pumilus Roxb., Fl. Ind. 1:277; 1820; FBI 7:170; Gamble 1753 (1215); Bor 93.

Annual, stoloniferous grass. Culms about 30 cm tall, tufted, slender, Leafsheaths 1-3 cm, loose; ligule membranous; blades 3-7 cm, linear, base cordate, apex acuminate, glabrous. Racemes up to 2 cm, 2, subtended by a narrow spathe, digitate. Spikelets 4 mm, 2-nate, sessile, hermaphrodite, pedicelled male or neuter.

Fls. & Frs. : September - December.

Rare grass in plains. Choutkoor. **TP & MSG 11928.**

APLUDA Linnaeus

Apluda mutica L., Sp. Pl. 82. 1753; Gamble 1750 (1212); Khan in FFH 337; Bor 93. *A. varia* Hackel, ssp. *aristata* (L.) Hackel Monogr. Androp. 6.196. 1889; FBI 7:150. *A. aristata* L., Amoen. Acad. 4. 303. 1759; Gamble 1750 (1212).

Perennial, tufted grass. Culms about 3 m tall; nodes glabrous. Leafsheaths 3.5 cm, glabrous; ligule membranous; blades 6-20 x 0.2-0.5 cm, linear-lanceolate, base narrowed, apex cuspidate, glabrous. Panicles up to 30 cm, terminal. Spikelets 3-nate, one sessile, hermaphrodite, the rest pedicelled; sessile 2-flowered, awned; pedicelled muticous. Caryopsis oblong.

Fls. & Frs.: September - December. *Vern.*: Adavi - korregaddi.

Common on the hill slopes, Ramayampet, **BR & CP** 11519; Choutkoor, **TP & MSG** 11910.

ARISTIDA Linnaeus

1. Glumes aristate **A.adscensionis** var. **adscensionis**
1. Glumes not aristate:
    2. There is an articulation between the lemma and the awn. **A.funiculata**
    2. There is no articulation between the lemma and the awn:
        3. Panicle contracted **A. setacea**
        3. Panicle effuse **A. hystrix**

Aristida adscensionis L., Sp. Pl. 82. 1753; FBI 7:224; Khan in FFH 342; var, **adscensionis;** Bor 407. *A. depressa* Retz., Obs. bot. 4:22 1786; Gamble 1809 (1252).

Perennial, tufted grass. Culms about 40 cm tall, erect or geniculately ascending; nodes glabrous. Leafsheaths 2.5-4 cm; ligule a ridge of fine soft hairs; blades 2-4 x 0.1 - 0.2 cm, narrowly linear, convolute, base cordate, apex sub acute. Panicles up to 10 cm, contracted. Spikelets 1 cm, 1-flowered, linear, 3-awned. Caryopsis linear.

Fls. & Frs. : August - September. *Vern.* Pilli-bisar gaddi.

Occasional in sandy soils. Pocharam, **TP & MSG** 12023.

Aristida funiculata Trin. & Rupr. Sp. Gram, Stip. 159. 1842; var. **funiculata;** Bor 410; FBI 7:226; Gamble 1809 (1253).

Annual, tufted grass,. Culms about 50 cm tall; geniculately ascending; nodes glabrous. Leafsheaths 4-8 cm; ligule a small ciliate membrane; blades 6-16 x 0.1-0.2 cm linear, convolute, base cordate, apex acuminate. Panicles lax, narrow, up to 22 cm. Spikelets 1 cm, pedicelled, 1-flowered, awn trifid. Caryopsis linear, spiny.

Fls. & Frs. : August - September.

Common in the hill slopes and forests. Pochammaralu, **CP** 11763.

Aristida hystrix L.f., Suppl. Pl. 113. 1781; FBI 7:225: Gamble 1809 (1252); Bor 410.

Perennial, tufted grass. Culms about 60 cm tall, with creeping rootstock, ascending; nodes glabrous. Leafsheaths 2-4 cm, densely pilose at the mouth; ligule of soft hairs; blades 3-9 x 0.1-

0.2 cm, linear, base cordate, apex acuminate, sparsely pilose on the margins. Panicles lax, up to 20 cm, pyramidal. Spikelets 2.5 cm, linear, awn-trifid. Caryopsis linear, prickly.

Fl.s. & Frs. : September - December. : *Vern* : Dolluba gaddi.

Common grass in plains and lower hill slopes. Choutkoor, **TP & MSG** 11937.

Aristida setacea Retz., Obs. Bot. 4:22. 1786; FBI 7:225; Gamble 1809 (1252); Bor 412.

Perennial, tufted grass. Culms about 2 m tall, erect or geniculately ascending; nodes glabrous. Leafsheaths 2-10 cm; ligule of short hairs; blades 10-25 x 0.1-0.3 cm, linear, convolute, base cordate, apex acuminate. Panicles up to 30 cm, contracted. Spikelets 1.5 cm, linear, 1-flowered, awn trifid. Caryopsis linear, prickly.

Fls. & Frs. : September - April. *Vern.* : Gaddiparakalu, Cheepurugaddi.

Common in plains, Pochammaralu, **BR & CP** 11472; Choutkoor, **TP & MSG** 11941; Pocharam, TP & MSG 12022.

ARTHRAXON P. Beauvois

1. Lower glume of the sessile spikelet 2-keeled, linear, enchinulate on the back; anther 3 **A.lanceolatus** var. **echinatus**
1. Lower glume of the sessile spikelet not keeled, lanceolate, scabrid on the back; anthers 2 **A. lancifolius**

Arthraxon lanceolatus (Roxb.) Hochst. var. **echinatus** (Nees) Hack. Mon. Androp. 348. 1889. *Bathratherum echinatum* Nees in Edinb. New Phil. J. 18. 181. 1835. *Arthraxon echinatus* (Nees) Hochst. in Fl. 39:188. 1856; Gamble 1729 (1198); Bor 99. *A. spathaceus* Hook.f. in FBI 7:145. 1896.

Annual tufted grass. Culms about 40 cm tall, geniculately ascending; nodes glabrous. Leafsheaths 2-3cm, tuberculate ciliate; ligule fimbriate; blades 3-4.5 x 0.5-1 cm, ovate-lanceolate, base amplexicaul, apex acuminate, margin tuberculate- ciliate. Racemes up to 5 cm. Spikelets 0.6 cm, alternate, awned. Caryopsis narrowly linear.

Fls. & Frs. : September - March.

Occasional along the streams and moist places. Pupplavagu, **TP & MSG** 12088.

Arthraxon lancifolius (Trin.) Hochst. in Flora 39. 188. 1856; Gamble 1729 (1198); Bor 100. *Andropogon lancifolius* Trin. in Mem. Acad. Sci. Petersb. Ser. 6. 2 : 271. 1832. *Arthraxon microphyllus* (Trin.) Hochst. in Flora 39. 188. 1856; FBI 7:147. *Andropogon microphyllus* Trin. in Mem. Acad. Sci. Petersb. Ser. 6. 2:275. 1832.

Annual, tufted grass. Culms about 60 cm tall, decumbent at base; nodes slightly bearded. Leafsheaths 0.1-0.15 cm; ligule membranous; blades 1.5-3 x 0.4-0.8 cm, ovate-lanceolate, base amplexicaul, apex acute. Panicle densely flowered, spicate. Spikelets 3 mm, laterally compressed, awned.

Fls. & Frs. : August - December.

Common in rock crevices and in damp habitats. Ramayampet, **BR & CP** 11506.

ARUNDINELLA Raddi

Arundinella setosa Trin. Gram. Panic, 63. 1837; var. *setosa;* FBI 7:70; Bor 424.

Perennial, stoloniferous grass. Culms about 90 cm tall; nodes glabrous. Leafsheaths 4-9 cm, loose, tuberculate pilose; ligule a ciliate membrane; blades 5-12 x 0.4-0.6 cm, base narrow, apex acuminate, densely pilose or tuberculate pilose. Panicle contracted. Spikelets 3.5 cm, awned. Caryopsis oblong.

Fls. & Frs. : July - December.

Occasional in upper hills. Narsapur RF, **TP & MSG** 11970.

BRACHIARIA Grisebach

1. Panicles linear; spikelets minute **B. eruciformis**
1. Panciles oblong; spikelets oblong **B. ramosa**

Brachiaria eruciformis (J.E. Smith) Griseb., in Ledeb. Fl. Ross. 4 : 469. 1853; Bor 283; Gamble 1769 (1226). *Panicum eruciformis* J.E. Smith in Sibth. & J.E. Smith, Fl. Graeca 1 : 44. t. 59. 1806. *P. isachne* Roth ex Roemer et Schultes Syst. Veg. 2: 458. 1817; FBI 7:28.

Annual grass. Culms about 40 cm tall, geniculately ascending; nodes densely pubescent. Leafsheaths 2-4.5 cm, slightly pubescent; ligule hairy; blades 1.5-4.5 x 0.2-0.4 cm, lanceolate, base rounded, apex acuminate. Panicles up to 5.5 cm, subsecund, racemosely arranged spikes. Spikelets elipsoid, 2 mm, densely pubescent. Caryopsis ovate, dark.

Fls. & Frs. : August - December.

Common in fields and moist localities. NSF, **BR & CP** 11669; Poosanpalle, **TP & MSG** 11956.

Brachiaria ramosa (L.) Stapf in Oliver Fl. Trop. Africa 9:542. 1919; Gamble 1770 (1226); Bor 284. *Panicum ramosum.* L. Mant. Pl. 29. 1767; FBI 7: 36.

Annual, tufted grass. Culms about 50 cm tall; nodes pubescent. Leafsheaths 0.4-0.8 cm, ligule membranous with a row of hairs; blades 6-12 x 0.4-1 cm, linear-lanceolate, base rounded, apex acuminate, margin scaberulous. Panicles up to 12 cm. Spikelets 0.3 cm, broadly elliptic-oblong. Caryopsis ovate.

Fls. & Frs. : August-September. *Vern.* : Edurugaddi.

Commons in cultivated fields and waste lands. Gangapur RF, **TP & MSG** 12072.

CHLORIS Swartz

1. Spikes racemosly arranged, ligule ciliate; perennials **C.roxburghiana**
1. Spikes umbelled, ligule membranous; annuals:
    2. Empty lemma above the fertile floret rudimentary to well developed **C virgata**
    2. Empty lemma above the fertile floret well developed **C. barbata**

Chloris barbata Sw. Fl. Ind. Occ. 1:200. 1797; FBI 7:292; Gamble 1838 (1272); Khan in FFH 344; Bor 465.

Annual, tufted grass. Culms about 60 cm tall, creeping at base; nodes glabrous. Leafsheaths 4-9 cm; ligule membranous; blades 5- 7 x 0.2-0.4 cm, linear-lanceolate, base rounded, apex acuminate. Digitate racemes up to 7 cm. Spikelets 3 mm, 3-flowered, 1-fertile lemma; 2-sterile ones. Caryopsis oblong. compressed.

Fls. & Frs. July - December.

Common in plains and forests. Pillutla RF, **BR & CP** 11268; Patancheru, **BR & CP** 11318.

Chloris roxburghiana Schultes, Syst. Veg. 2 Mant. 329. 1827; Bor 468. *C. polystachya* Roxb., Fl. Ind. 1:332. 1820, non Lagasca (1816). *nom. illeget*; FBI 7:292; Gamble 1838 (1272). C. *myriostachya* Hochst. in Fl. 38. 204. 1855.

Perennial, tufted grass. Culms about 1 m tall, occasionally geniculately ascending; nodes glabrous. Leafsheaths 3-8 cm, ciliate at mouth; ligule mambranous; blades 9-22 x 0.3-0.6 cm, linear, convolute, base rounded, apex acuminate. Spikes up to 9 cm, racemosely arranged, feathery, 12-30 spikes. Spikelets 0.25 cm, 3-flowered, sterile lemmas 2. Caryopsis linear-oblong.

Fls. & Frs. : July - December.

Occasional in plains and hill slopes. Choutkoor, TP & MSG 11912.

Chioris virgata Sw. Fl. Ind. Occ. 1:203. 1797; FBI 7: 291 : Gamble 1838 (1272); Bor 468.

Annual, stoloniferous grass. Culms about 60 cm tall, geniculately ascending, rooting at base; nodes glabrous. Leafsheaths 2-10 cm, glabrous; ligule membranous; blades 4-10 x 0.1-0.4 cm, narrowly linear, base rounded, apex acuminate, Spikes 4-8, up to 6 cm, digitate. Spikelets 3 cm long, ovate-lanceolate, one perfect. Caryopsis ellipsoid, compressed.

Fls. & Frs. : June - November.

Common in plains. Narsapur tank, **GNR** 4637 and **KMS** 6764 (CAL).

CHRYSOPOGON Trinius

Chrysopogon fulvus (Spreng.) Chiov., Fl. Somala 327. 1929; Bor 116. *Pollinia fulva* Spreng., Pl.

Minus Cognit. Pug. 2:10. 1815. *Chrysopogon montanus* Trin. ex Spreng., Neue Entd. 2:93. 1821; Gamble 1739 (1205). *Andropogon monticola* Roemer & Schultes in L., Syst. Veg. 2:665, 1827; FBI 7:192.

Perennial, tufted grass. Culms about 40 cm tall; nodes glabrous. Leafsheaths 1-6 cm, ciliate at mouth; ligule membranous; blades 6-15 x 0.2-0.5 cm, linear, base truncate, apex acuminate. Panicles up to 70 cm. Spikelets 7.5 cm, in clusters, 3-flowered, 2-pedicelled, 1-sessile; sessile spikelet 6 mm, hermaphrodite, pedicelled 6 mm, male or neuter. Caryopsis linear.

Fls. & Frs. : July - November.

Common in plains and hill slopes. Choutkoor, **TP & MSG** 11934.

COIX Linnaeus

Coix lacryma-jobi L., Sp. Pl. 972. 1753; FBI 7:100; Rang. & Tad., Handb. South Indian Grass. t. 126. 1921; Gamble 1705 (1182); Bor 264.

Perennial, monoecious, grass, Culms about 1 m, tall; rooting at lower nodes; nodes thick. Leafsheaths 4-6 cm; ligule a very narrow membrane; blades 15-30 x 1-2 cm, convolute, base cordate, apex acuminate, midrib thick. Racemes up to 5 mm, lower spikelets solitary, female, through the apex of female spikelets male spikelet protrudes; female spikelets 9 mm, enclosed in a hardened, polished involucre; male spikelets 9 mm long, 2-flowered.

Fls. & Frs. : October - February.

Rare grass near the water sources. Gangapur-Pupplavagu, **TP & MSG** 12087.

CYMBOPOGON Sprengel

1. Nodes bearded; base of the blade rounded; lower glume of the sessile spikelet not channeled **C. coloratus**
1. Nodes glabrous; base of the blade cordate; lower glume of the sessile spikelet deeply channeled **C. martinii**

Cymbopogon coloratus (Hook.f.) Stapf in Kew Bull. 1906 : 321. 1906; Gamble 1756 (1217); Bor 126. *Andropogon nardus* L. var. *coloratus* Hook.f. in FBI 7:206. 1896.

Perennial, tufted grass. Culms about 1 m tall; nodes bearded. Leafsheaths 15 cm; ligule fimbriate membrane; blades 10-40 x 0.4-0.8 cm, linear-lanceolate, base rounded, apex acuminate. Panicles up to 10 cm, dense, red. Spikelets 5 mm, 2-nate; sessile spikelet hermaphrodite, pedicelled male or neuter.

Fl.s & Frs. : August - December.

Common in the forests. Narsapur tank, **TP & MSG** 12017.

Cymbopogon martinii (Roxb.) Watson in Atkinson, E.T. Gaz. N.W. Prov. India 392. 1882; Gamble 1756 (1217); Bor 129. *Andropogon martinii* Roxb., Fl. Ind. 1:280. 1820. *A. schoenanthus* L. var. *martinii* (Roxb.) Hook.f. in FBI 7:204. 1896.

Perennial, tufted grass. Culms about 2 m tall. Leaf-sheaths 7-15 cm; ligule membranous; blades 10-25 x 1-1.5 cm, linear- lanceolate, base cordate, apex acuminate. Panicles up to 25 cm. Spikelets 4 mm, 2-nate, sessile spikelet hermaphrodite, pedicelled male or neuter.

Fls. & Frs. : July - December.

Occasional in the forests and fields. Choutkoor, **TP & MSG** 11924.

CYNODON L.C. Richard

Cynodon arcuatus J.S. Presl in Presl., Reliq. Haenk. 1:290. 1830; Bor 469. *C. dactylon* (L.) Pers. var. *intermedius* (Rang. & Tad.) Fischer in Gamble, Fl. Pres. Fl. Madras 1835 (1270) 1934. *C. intermedius* Rang. & Tad. in J. Bombay Nat. Hist. Soc. 26:304. 1918 & Handb. South Indian Grass. tt. 192 & 193. 1921.

Perennial, stoloniferous grass. Culms about 40 cm tall; nodes glabrous. Leafsheaths 2.5 cm; ligule membranous; blades 6-12 x 0.3-0.5 cm, lanceolate, base cordate, apex acute. Digitate racemes 4. Spikelets 2.5 mm. Caryopsis elliptic.

Fls. & Frs. : Round the year.

Occasional in plains and moist places in the forests. Narsapur tank, **TP & MSG** 12002.

DACTYLOCTENIUM Willdenow

Dactyloctenium aegyptium (L.) Beauv., Ess. Agrost. Expl. t. 15. 1812; Gamble 1840 (1273); Bor 489. *Cynosurus aegyptius* L., Sp. Pl. 72. 1753. *Eleusine aegyptiaca* (L.) Desf., Fl. Atlant. 1.85. 1798; FBI 7:295. *Dactyloctenium aegyptiacum* Willd., Enum. Pl. Hort. Berol. 1029. 1809.

Annual grass. Culms about 10 cm tall, creeping; nodes sparsely bearded. Leafsheaths 2-5 cm; ligule a ciliate rim; blades 2-10 x 0.3-0.8 cm, linear, base rounded, apex acuminate. Spikes 2-6, digitately radiating, up to 7 cm. Spikelets 3-5-flowered, laterally compressed.

Fls. & Frs. : July - December.

Common in plains and hills. Choutkoor, **TP & MSG** 11904.

DENDROCALAMUS C.G.D. Nees

Dendrocalamus strictus (Roxb.) Nees in Linnaea 9:476. 1834; FBI 7:404; Gamble 1858 (1286); Khan in FFH 348. *Bambusa stricta* Roxb., Pl. Corom. t. 80. 1798.

Perennial, unarmed, tufted grass. Culms about 15 m tall; culms- sheaths brown, 5-10 x 6-14 cm; nodes swollen, lower ones often rooting. Leafsheaths 3-15 cm, auricle short; ligule narrow; blades 5-15 x 1.5-3 cm, linear-lanceolate. Panicles compound. Spikelets in congested globose heads, 1 cm, spiny, 3-flowered, all hermaphrodite. Caryopsis ovoid- sub globose.

Fls. & Frs. : very rare. *Vern.* : Sadanapu-veduru.

Common in the forests. NSF, **BR & CP** 11650.

DICHANTHIUM Willemet

1. Peduncle of the raceme hairy **D. aristatum**
1. Peduncle of the raceme not hairy:
    2. Lower glume of the sessile spikelets hairy below the middle **D. pertusum**
    2. Lower glume of the sessile spikelets not hairy below the middle:
        3. Culm nodes patently hairy; lower glume 7-nerved **D. annulatum**
        3. Culm nodes apressed hairy; lower glume 11-nerved **D. caricosum**

Dichanthium annulatum (Forssk.) Stapf in Prain, Fl. Trop. Africa 9:178. 1917; Gamble 1741 (1206); Bor 133; Deshpande in Fasc. Fl. India 15:5. 1984. *Andropogon annulatus* Forssk., Fl. Aegypt. - Arab. 173. 1775; FBI 7:196.

Perennial, tufted grass. Culms about 70 cm tall; nodes bearded. Leafsheaths 5-10 cm; ligule membranous; blades 6-20 x 0.3-0.5 cm, linear, base rounded, apex acuminate. Racemes up to 6 cm, subdigitate. Spikelets 4 mm, 2-nate, sessile hermaphrodite, pedicelled male or neuter. Caryopsis oblong.

Fl.s & Frs. : Round the year.

Common in plains and forests. NSF quarters, **BR & CP** 11644; Narsapur RF , **TP & MSG** 11993.

Dichanthium aristatum (Poir.) C.E. Hubbard in Kew Bull. 1939:654 1939; Bor 134. *Andropogon aristatus* Poir. in Lam. Encycl. 1:585. 1810. *A. mollicomus* Kunth Rev. Gram. 1:385 t. 96. 1836. *A. caricosus* L. var. *mollicomus* (Kunth) Hackel in DC. Monogr. Phan. 6:569 1889; FBI 7:196. *Dichanthium nodosum* Fischer in Gamble, Fl. Pres. Madras 1741 (1207) 1934.

Perennial, tufted grass, Culms about 50 cm tall; nodes bearded. Leafsheaths 3 - 9 cm; ligule membranous; blades 6-20 x 0.3-0.5 cm, base rounded, apex acuminate. Racemes up to 9 cm, subdigitate, fascicled. Spikelets 8 mm, 2-nate, sessile, hermaphrodite, pedicelled male or neuter.

Fls. & Frs. : August - November.

Common in plains and in cultivated fields. Pathacheruvu, **BR & CP** 11317. Pochammaralu, **BR & CP** 11409; Choutkoor, **TP & MSG** 11909; Gangapur RF, **TP & MSG** 12086.

Dichanthium caricosum (L.) A. Camus in Bull. Mus. Hist. Nat. (Paris) 27:549. 1921; Gamble 1741 (1206); Bor 134; Deshpande in Fasc. Fl. Ind. 15:7. 1984. *Andropogon caricosus* L., Sp. Pl. 2:1480. 1768; FBI 7:196.

Perennial, tufted grass. Culms about 1 m tall; erect or ascending; nodes glabrous. Leafsheaths 3-10 cm; blades 5-15 x 0.2-0.4 cm, linear, sparsely pubescent, base broad. Racemes up to 4 cm, 1-5. subdigitately arranged. Sessile spikelet 4 mm, lower glumes 11-nerved. Caryopsis slightly compressed.

Fls. & Frs. : July - December.

Occasional in hedges and forest. Kondapur tank, GNR 4590.

Dichanthium pertusum (L.) Clayton in Kew Bull. 32:4. 1977; Deshpande in Fasc. Fl. India 15:22 1984. *Holcus pertusus* L. Mant. Alt. 301. 1771. *Andropogon pertusus* (L.) Willd. Sp. Pl. 4:922. 1806; FBI 7:173. *Amphilophis pertusus* (L.) Nash ex Stapf in Agric. News W. Ind. 15:179. 1916; Gamble 1751 (1200). *Bothriochloa pertusa* (L.) A. Camus in Ann. Soc. Linn. Lyon 76:164. 1931; Bor 109.

Perennial, stoloniferous grass, Culms about 50 cm tall; nodes sparsely hairy. Leafsheaths 2-6 cm; ligule membranous; blades 3-10 x 0.2-0.4 cm, linear, base truncate, apex acuminate. Racemes 2-6, up to 5 cm, digitate. Spikelets 6 mm, 2-nate, pitted, sessile hermaphrodite, pedicelled male or neuter.

Fls. & Frs. : July - January.

Common in plains and hilly areas. Choutkoor, **TP & MSG** 11918.

DIECTOMIS Kunth *nom. cons.*

Diectomis fastigiata (Sw.) Kunth in HBK Nov. Gen. Sp. 1:193. t. 64. 1816; Bor 135. *Andropogon fastigiatus* Sw. Prodr. Veg. Ind. Occ. 26. 1788; FBI 7:167.

Perennial tufted grass, culms up to 50 cm tall. Leaves 7-30 cm long, narrowly linear, apex acuminate. Spikes solitary.

Fls. & Frs. : August - December.

Rare in tank beds. Pocharam tank, **TP & MSG** 15221.

DIGITARIA Heister ex Fabricius

1. Spikelets heteromorphous; caryopsis ellipsoid **D. bicornis**
1. Spikelets homomorphous; caryopsis oblong **D. ciliaris**

Digitaria bicornis (Lam.) Roemer & Schultes in L. Syst. Veg. 2: 470. 1817; Bor 299; Veldkamp in Blumea 21 : 30. 1973. *Paspalum bicornis* Lam. Tabl. Encycl. 1 : 176. 1791. *P. heteranthum* Hook. f. in FBI 7 : 16 1896. *Digitaria biformis* Willd., Enum. Hort. Berol. 92. 1809; Bor 299.

Annual, tufted grass, Culms about 50 cm tall; nodes sparsely bearded. Leafsheaths 2-8 cm, sparsely tuberculately hairy; ligule membranous; blades 2-6 × 0.4-0.6 cm, oblong, base cordate, apex acuminate. Digitate racemes 2-9, up to 10 cm. Spikelets 3 mm. Caryopsis ellipsoid.

Fls. & Frs. : July - November.

Common in plains. Nagulabanda, **CP** 11361; Sivanoor, **BR & CP** 11550; NSF, **BR & CP** 11674.

Digitaria ciliaris (Retz.) Koel., Descr. Gram. 27. 1802; Veldkamp in Blumea 21 : 32. 1973. *Panicum ciliare* Retz., Obs. Bot. 4 : 16. 1786. *Digitaria marginata* Link var. *fimbriata* (Link) Stapf in Prain Fl. Trop. Africa 9 : 440. 1919; Gamble 1764 (1222). *D. adscendens* (Kunth) Henrard in Blumea 1:92. 1944; Bor 298. *Panicum adscendens* Kunth, Nov. Gen. Pl. 1 : 97. 1815.

Annual, tufted grass. Culms about 70 cm tall, geniculately ascending at base; nodes glabrous. Leafsheaths 3-10 cm, glabrous; ligule membranous, truncate; blades 4-10 × 0.3-0.5 cm, linear-lanceolate, base cordate, apex acute. Racemes 4-9, up to 10 cm, digitate; rachis serrate, trigonous. Spikelets 3 mm, elliptic-lanceolate, 2-nate, appresed to the rachis, glabrous-sparsely ciliate. Caryopsis oblong.

Fls. & Frs. : July - October.

Rare in moist plains and in hill slopes. Way to Rusthempet, **KMS** 6807.

DINEBRA N.J. Jacquin

Dinebra retroflexa (Vahl) Panz. in Denkschr. Acad. Wiss. Muechen. 270. t. 12. 1814; Gamble 1841 (1274); Bor 491. *Cynosurus retroflexus* Vahl, Symb. Bot. 2.20. 1791. *Dinebra arabica* Jacq., Fragm. Bot. 77. t. 121 f. 1. 1807; FBI 7 : 297.

Annual, tufted grass. Culms about 70 cm tall; nodes glabrous. Leafsheaths 5-9 cm; ligule narrow, membranous; blades 5-10 × 0.2-0.4 cm, linear, base cordate, apex acuminate. Spikes racemosely arranged, up to 20 cm. Spikelets 5 mm, 2-flowered. Caryopsis ellipsoid.

Fls. & Frs. : July - September

Common weed of plains and hill slopes. NSF, **BR & CP** 11673; Poosanpalle, **TP & MSG** 11954.

ECHINOCHLOA P. Beauvois

1. Upper glume and lower lemma acute or cuspidate, unawned **E. colona**
1. Upper glume and lower lemma acuminate-cuspidate, awned **E. crus-galli**

Echinochloa colona (L.) Link Hort. Berol. 2 : 209. 1833; Gamble 1776 (1230); Khan in FFH 340; Bor 308. *Panicum colonum* L., Syst. Nat. (ed. 10) 870. 1759; FBI 7 : 32.

Annual, tufted grass. Culms about 70 cm tall; nodes glabrous. Leafsheaths 5-15 cm; ligule obsolete; blades 5-25 × 0.3-0.6 cm, linear-lanceolate, base cordate, apex acuminate. Panicles up to 15 cm, racemes appressed to the rachis. Spikelets 2.5 mm, broadly ovate. Caryopsis broadly ellipsoid.

Fls. & Frs. : Round the year.

Frequent weed of rice fields. Ramayampet, **BR** & **CP** 11513; Gangapur, **CP** 11793.

Echinochloa crus-galli (L.) Beauv. Ess. Agrost. 53 : 161. 1812; Gamble 1777 (1231); Bor 310. *Panicum crus-galli* L., Sp. Pl. 1753; FBI 7 : 30.

Annual tufted grass. Culms about 70 cm tall; nodes glabrous. Leafsheaths 5-25 cm; ligule obsolete; blades 5-20 × 0.4-0.8 cm, linear-lanceolate, base cordate, apex acuminate. Panicles up to 10 cm, racemes spreading. Spikelets 5 mm, awned, broadly ovate. Caryopsis broadly ellipsoid.

Fls. & Frs. : Round the year.

Common in rice fields. Narsingi, **BR** & **CP** 11530

ELEUSINE Gaertner

Eleusine indica (L.) Gaertn., Fruct. 1 : 8. 1788; FBI 7 : 293; Gamble 1839 (1273); Bor 493. *Cynosurus indicus* L., Sp. Pl. 72. 1753.

Annual, tufted grass. Culms about 60 cm tall; nodes glabrous. Leafsheaths 4-10 cm; ligule hairy; blades 4-25 × 0.4-0.6 cm, linear, base truncate, apex acuminate. Spikes 2-7 or more, up to 10 cm, digitate. Spikelets 3 mm, 3-6-flowered. Caryopsis oblong.

Fls. & Frs. : June - September.

Common grass in plains and hill slopes Ponnala, **CP** 11367.

ELYTROPHORUS P. Beauvois

Elytrophorus spicatus (Willd.) A. Camus in Lecomte, Fl. Gen. Indo-Chine 7 : 547. 1923; Gamble 1843 (1275); Bor 493. *Dactylis spicata* Willd. in Ges. Naturf. Freunde Berlin Neue Schriften 3:416. 1801. *Elytrophorus articulatus* Beauv., Ess. Agrost. 67. t. 14.f. 2. 1812; FBI 7:306.

Annual, tufted grass. Culms about 30 cm tall; nodes glabrous. Leafsheaths short, overlapping; ligule fringed, hairy; blades 5-15 × 0.2-0.7 cm, narrow. Catkin like spikes with globose cluster arranged on the rachis of the inflorescence. Spikelets 5 mm, 4-6 flowered, laterally compressed. Caryopsis narrowly oblong.

Fls. & Frs. : August-January

Occasional in ditches and wet places. Pocharam RF, **TP** & **MSG** 12055.

ERAGROSTIELLA Bor

Eragrostiella bifaria (Vahl) Bor in Indian Forester 66 : 270. 1940 & Grass India 494. 1960. *Poa bifaria* Vahl,

Symb. Bot. 2:19. 1791. *Eragostis bifaria* (Vahl) Wight ex Steud., Nom. Bot. ed. 2. 1:562. 1840. FBI 7:325; Gamble 1828 (1265).

Perennial, tufted grass. Culms about 50 cm tall; nodes glabrous. Leafsheaths 4-8 cm; ligule pubescent; blades 5-20 × 1-1.5 cm, linear, filiform, base narrow, apex acute. Spikes up to 25 cm. Spikelets 5 mm, alternately arranged on the rachis, 6-40-flowered. Caryopsis globose.

Fls. & Frs. : July - January. *Vern.* : Nooligaddi

Common grass in plains and hilly areas. Nagulabanda, **CP** 11360.

ERAGROSTIS Wolf

1. Spikelets breaking from base to apex :
   2. Palea deciduous :
      3. Spikelets broadly ovate-oblong — **E. unioloides**
      3. Spikelets narrowly oblong-linear :
         4. Annuals; spikelets scattered, linear — **E. pilosa**
         4. Perennials; spikelets fascicled, oblong :
            5. Apex of lemma shortly beaked; caryopsis ellipsoid — **E. atrovirens**
            5. Apex of lemma acute; caryopsis oblong — **E. gangetica**
   2. Palea persistent :
      6. Pedicels of spikelets with a glandular band :
         7. Panicle-branches slender; leaf blades without glands — **E. tenuifolia**
         7. Panicle-branches stiff; leafblades with warty glands — **E. minor**
      6. Pedicles of spikelets without glandular band :
         8. Spikelets scattered, versatile; long pedicellate — **E. tremula**
         8. Spikelets more or less fascicled; short pedicellate — **E. nutans**
1. Spikelets breaking from apex to base :
   9. Lemmas ciliate on the margins — **E. ciliata**
   9. Lemmas not ciliate on the margins :
      10. Palea ciliolate on the keels — **E. viscosa**
      10. Palea glabrous or scabrid on the keels :
          11. Branches of the panicle more or less whorled; spikelets loosely scattered, few-flowered — **E. japonica**
          11. Branches of the panicle mostly solitary; spikelets densely scattered, 6-14-flowered — **E. diarrhena**

Eragrostis atrovirens (Desf.) Trin. ex Steud., Nom. Bot. ed. 2. 1:562. 1840; Bor 503. *Poa atrovirens* Desf., Fl. Atlant. 1:73 t. 14. 1798.

Perennial, tufted grass. Culms about 70 cm tall; nodes glabrous. Leafsheaths 4-12 cm; ligule cillate; blades 10-30 × 0.3-0.4 cm, linear, apex acuminate. Panicles up to 30 cm, effuse. Spikelets 4 mm, oblong, 4-20-flowered. Caryopsis ellipsoid.

Fls. & Frs. : August - May

Common in plains and waste lands. Pochammaralu, **BR** and **CP** 11497; Narsapur RF, **TP & MSG** 12011; Pocharam tank, **TP & MSG** 12039.

Eragrostis ciliata (Roxb.) Nees, Agrost. Bras. 512. 1829; FBI 7; 313; Gamble 1825 (1263); Bor 506. *Poa ciliata* Roxb., Fl. Ind. 1: 336. 1820.

Annual, tufted grass. Culms about 75 cm tall. Leafsheaths bearded at mouth; ligule hairy; blades 5-16 × 0.2-0.4 cm, linear-lanceolate, flat, base rounded, apex acuminate. Panicles up to 6 cm, cylindric, compact. Spikelets 4 mm, compressed, 6-12-flowered. Caryopsis ellipsoid.

Fls. & Frs. : April - September.

Occasional in waste places. Usirakapalle, **GNR** 4327.

Eragrostis diarrhena (Schultes) Steud., Syn. Pl. Glum. 1:266. 1854; Gamble 1826 (1264); Bor 507. *Poa diarrhena* Schultes, Mant. Syst. Veg. 2:616. 1827. *Eragrostis interrupta* Beauv. var. *diarrhena* Stapf in Hook.f., in FBI 7:316. 1896.

Perennial, tufted grass. Culms about 90 cm tall; nodes glabrous. Leafsheaths 5-10 cm; ligule fimbriate membrane; blades 5-20 × 0.1-0.3 cm, linear, base rounded, apex acuminate. Panicles up to 20 cm, contracted, interrupted, dense. Spikelets 4 mm, ovate linear. Caryopsis obovoid.

Fls. & Frs. : September - March.

Common along the moist places and in the cultivated fields. Choutkoor, **TP & MSG** 11905; Pathur, **TP & MSG** 12177.

Eragrostis gangetica (Roxb.) Steud., Syn. Pl. Glum. 1:266. 1854; Gamble 1826 (1264); Bor 508. *Poa gangetica* Roxb., Fl. Ind. 1:341. 1820. *Eragrostis stenophylla* Hochst. ex Miq. in Verh. Konink. Nederl. Inst. 3, 4 : 39, 1851; FBI 7: 318.*p.p*.

Perennial, tufted grass. Culms about 50 cm tall; nodes glabrous. Leafsheaths 2-10 cm; ligule a ciliate rim; blades 5-15 x 0.2-0.4 cm, very narrow, base narrow, apex acuminate. Panicles open, up to 15 cm. Spikelets 6 mm, 8-20-flowered. Caryopsis oblong.

Fls. & Frs. : September - May.

Common in moist places and in fields. Choutkoor, **TP & MSG** 11948; Pocharam tank, **TP & MSG** 12042.

Eragrostis japonica (Thunb.) Trin. in Mem. Acad. Sci. Petersb. 6. Sci. Math. 1:405. 1831; Gamble 1826 (1264); Bor 509. *Poa japonica* Thunb. Fl. Jap. 51. 1784. *Eragrostis interrupta* Beauv. var. *tenuissima* Stapf in Hook. f., FBI 7:316. 1896.

Perennial, tufted grass. Culms about 1 m tall; nodes glabrous. Leafsheaths 7-20 cm; ligule a membranous ring; blades 10-20 x 0.2-0.4 cm, linear, base broad, apex acuminate. Panicles up to 50 cm, contracted, dense. Spikelets 2 mm, ovate-oblong, 2-9- flowered. Caryopsis obovoid.

Fls. & Frs. : September - April.

Occasional in cultivated fields. Pocharam RF, **TP & MSG** 12056 & 12065.

Eragrostis minor Host, Ic. Descr. Austr. 4:15. 1809; FBI 7:321. *E. poaeoides* Beauv., Ess. Agrost. 162. 1812; Gamble 1827 (1264) ; Bor 512.

Annual, tufted grass. Culms about 45 cm tall; nodes glabrous. Leafsheaths 2-4; ligule a pubescent ring; blades 7-15 × 0.1-0.2 cm, linear, base narrow, apex acuminate. Panicles up to 10 cm, lax, open. Spikelets 3 mm, ovate-oblong, 6-12-flowered. Caryopsis broadly oblong.

Fls. & Frs. : July - December. *Vern.* : Good-bai gaddi.

Rare in fields and forests. Pocharam RF, **TP & MSG** 12062.

Eragrostis nutans (Retz.) Nees ex Steud. Nom. Bot. ed. 2. 1:563. 1840; Gamble 1826 (1264); Bor 511. *Poa nutans* Retz., Obs. Bot. 4: 19.1786. *Eragrostis stenophylla auct. non* Hochst. 1851 : FBI 7: 318.

Perennial, tufted grass. Culms about 70 cm tall, geniculately ascending; nodes glabrous. Leafsheaths 4-8 cm; ligule a frimbriate membrane; blades 6-15 × 0.4-0.6 cm, plicate, apex setaceous. Panicles up to 20 cm, contracted, effuse. Spikelets 6 mm, oblong, 8-14-flowered. Caryopsis subglobose.

Fls. & Frs. : June - March.

Occasional near marshy localities. Pegarikutta, **KMS** 6655.

Eragrostis pilosa (L.) Beauv. Ess. Agrost. 71. 162 & 175. 1812; FBI 7: 323; Gamble 1827 (1265); Bor 512. *Poa pilosa* L., Sp. Pl. 68. 1753.

Annual, tufted grass. Culms about 70 tall; nodes glabrous. Leafsheaths 7-20 cm; ligule a soft hairy ridge; blades 10-20 × 0.3-0.5 cm, linear. Panicles up to 15 cm, open. Spikelets 5 mm, linear-oblong, 5-12-flowered. Carposis ellipsoid.

Fls. & Frs. : August - November.

Common in cultivated fields. Choutkoor, **TP & MSG** 11949; Narsapur forest nursery, **TP & MSG** 12018.

Eragrostis tenuifolia Hochst. ex Steud., Syn. Pl. Glum. 1:268. 1854; FBI 7:322. Gamble 1827 (1265); Bor 514.

Perennial, tufted grass. Culms about 60 cm tall, slender; nodes glabrous. Leafsheaths 3-7 cm, glandular ciliate near mouth; ligule ciliate; blades 3-6 × 0.1-0.3 cm, linear, base rounded, apex acuminate. Panicles up to 20 cm, lax open; branches filiform, axils pilose. Spikelets 4 mm, linear-oblong. Caryopsis oblong.

Fls. & Frs. : July - October.

Rare in moist localities. Narsapur, **GNR** 4165.

Eragrsotis tremula Hochst. ex Steud. Syn. Pl. Glum. 1:269. 1854; Stapf in Hook.f., FBI 7:320; Gamble 1827 (1264); Bor 514.

Annual, tufted grass. Culms about 50 cm tall; nodes glabrous. Leafsheaths 3-10 cm; ligule sparsely ciliate; blades 5-15 x 0.2- 0.5 cm, linear, base cordate, apex acuminate. Panicles up to 20 cm, effuse, Spikelets 7 mm, linear-oblong. Caryopsis globose.

Fls. & *Frs.* : August - November.

Frequent in dry sandy places. Ramayampet, **BR & CP** 11520; Pocharam, **TP & MSG** 12034 & 12043.

Eragrostis unioloides (Retz.) Nees ex Steud., Syn. Pl. Glum. 1:264. 1854; Gamble 1820 (1264); Khan in FFH 342; Bor 515. *Poa unioloides* Retz., Obs. Bot. 5:19. 1789. *Eragrostis amabillis* auct. non L. : FBI 7:317. 1896.

Annual, tufted grass. Culms about 50 cm tall; nodes glabrous. Leafsheaths 5-10 cm, ligule obscure; blades 5-10 × 0.2-0.3 cm, linear-oblong, base cordate, apex acuminate. Panicles up to 10 cm. Spikelets 1 cm, ovoid - elliptic, 20-50-flowered. Caryopsis obovoid - ellipsoid.

Fls. & *Frs.* : July - March.

Common in plains and forests. Narsapur tank, **TP & MSG** 12001.

Eragrostis viscosa (Retz.) Trin. in Mem. Acad. Sci. Petersb. 6. Sci. Math. 1:397. 1831; Gamble 1826 (1264); Bor 515. *Poa viscosa* Retz., Obs. Bot. 4:20 1786/87. *Eragrostis tenella* var. *viscosa* (Retz.) Stapf in Hook. f., FBI 7; 315. 1896.

Perennial, tufted, viscid grass. Culms about 60 cm tall, glandular patches scattered throughout; nodes glabrous. Leafsheaths 2-5 cm; ligule a fimbriate membrane; blades 3-10 × 0.2 -0.4 cm, linear, base narrow, apex acuminate. Panicles up to 20 cm, narrowly cylindric, dense. Spikelets 4 mm, oblong, 5-20- flowered. Caryopsis ovoid.

Fls. & *Frs.* : July - March. *Vern.* : Bankagaddi.

Abundant in plains and cultivated fields. Choutkoor, **TP & MSG** 11906 & 11915.

EREMOPOGON Stapf

Eremopogon foveolatus (Del.) Stapf in Prain, Fl. Trop. Africa 9: 183. 1917; Gamble 1751 (1218); Bor 148. *Andropogon foveolatus* Del., Descr. Egypt. Hist. Nat. 16. t. 8. f. 2. 1812; FBI 7:168.

Perennial, tufted grass. Culms about 80 cm tall; nodes bearded. Leafsheaths 4-6 cm; ligule ciliate; blades 5-15 × 0.1-0.3 cm, base rounded, apex acuminate. Racemes solitary, up to 5 cm, spatheolate. Spikelets 6 mm, 2-nate, sessile hermophrodite, pedicelled male or neuter. Caryopsis ellipsoid.

Fls. & Frs. : September - January.

Common in waste lands. Choutkoor, **TP & MSG** 11930.

ERIOCHLOA Humboidt, Bonpaland & Kunth

Eriochloa procera (Retz.) C.E. Hubbard in Kew Bull. 1930; Gamble 1767 (1224); Bor 312. *Agrostis procera* Retz., Obs. Bot. 4:19. 1786. *Eriochloa polystachya* Hook.f., in FBI 7:20. 1896.

Perennial, tufted grass. Culms about 60 cm tall; nodes glabrous. Leafsheaths 4-15 cm; ligule fimbriate; blades 5-20 x 0.4-0.7 cm, linear, base rounded, apex acuminate. Panicles up to 10 cm, lax. Spikelets 3 mm, ellipsoid, lower glume absent. Caryopsis oblong-ellipsoid.

Fls. & Frs. : August - October.

Common grass near streams and marshy places. Narsingi, **BR & CP** 11542; Akkannapet, **BR & CP** 11569; Narsapur tank, **TP & MSG** 12020; Pupplavagu, **TP & MSG** 12089.

HACKELOCHLOA O. Kuntze

Hackelochloa granularis (L.) Kuntze, Rev. Gen. Pl. 2:776. 1891; Gamble 1758 (1218); Bor 159. *Cenchrus granularis* L., Mant. Pl. App. 575. 1771. *Manisuris granularis* (L.) L.f., Nov. Gram. Gen. 40. 1779; FBI 7:159.

Annual, tufted grass. Culms about 50 cm tall; nodes bearded. Leafsheaths 2-5 cm, hispid; ligule membranous, fimbriate; blades 4-10 × 0.5-1.2 cm, lanceolate, hispid, base cordate, margin ciliate. Racemes spiciform, up to 2.5 cm. Spikelets 2-nate, one sessile, other with pedicel fused to the joint; sessile hermaphrodite, pedicelled male or neuter.

Fls. & Frs.: July - September. *Vern.* : Nalakacherlagaddi.

Occasional weed in cultivated fields. Madikonakutta, **KMS** 6736.

HETEROPOGON Persoon

Heteropogon contortus (L.) Beauv. ex Roemer & Schultes in L., Syst. Veg. 2:836. 1817; Gamble 1743 (1208); Khan in FFH 336; Bor 163. *Andropogon contortus* L., Sp. Pl. 1045. 1753; FBI 7:199.

Annual, tufted grass. Culms about 60 cm tall; nodes glabrous. Leafsheaths 5-10 cm; ligule membranous; blades 4-10 × 0.2-0.4 cm, lanceolate, base rounded, apex acute. Racemes solitary up to 6 cm, awned. Spikelets: lower homogamous, unawned, either male or neuter; upper heterogamous, awned, 2-nate, sessile hermaphrodite, pedicelled male or neuter. Caryopsis linear.

Fls. & Frs. : July - November. *Vern.* : Kesari gaddi.

Common in plains and waste lands. Ramayampet, **BR & CP** 11541.

ISCHAEMUM Linnaeus

1. Base of the lower leaf blades rounded or sub-cordate, without petiole **I. rugosum**
1. Base of the lower leaf blades sagittate or deeply cordate, with distinct petiole **I. semisagittatum**

Ischaemum rugosum Salisb., Ic. Stirp. Rar. 1. t. 1. 1791; FBI 7:127; Gamble 1722 (1194); Khan in FFH 334; Bor 184.

Annual, tufted grass. Culms about 75 cm tall; nodes bearded. Leafsheaths 2-8 cm, compressed; ligule membranous; blades 8-20 × 0.2-1.2 cm, flat, lanceolate, base narrowed, apex acuminate. Racemes digitate, spiciform, up to 8 cm. Spikelets 6 mm, 2-nate, sessile and pedicelled closely pressed together. Caryopsis oblong.

Fls. & Frs. : July - December.

Occasional weed in paddy fields. Gangapur RF, **TP & MSG 12071.**

Ischaemum semisagittatum Roxb., Fl. Ind. 1: 322. 1820; FBI 7: 130; Bor 185.

Annual, tufted grass. Culms about 60 cm tall, decumbent at base; nodes glabrous. Leafsheaths loose, blades 5-8 × 1-12 cm, oblong - lanceolate, auricled base cordate - sagittate, apex acuminate, lower with filiform petioles, upper sessile. Racemes 2, up to 7 cm, villous with white or grey hairs. fragile. Spikelets 6 mm, 2- nate.

Fls. & Frs. : August - December.

Rare in moist localities. Rusthempet, **KMS 6797** (CAL).

ISEILEMA Andersson

1. Spathes broad; peduncles glabrous **I. laxum**
1. Spathes narrow; peduncles hispidulous **I. prostatum**

Iseilema laxum Hackel in A. & C. DC. Manogr. Phan. 6:682 1889; FBI 7:218; Gamble 1748 (1211); Bor 188.

Perennial, ascending grass. Culms about 80 cm tall; nodes glabrous. Leafsheaths 2-4 cm, lax; ligule a fimbriate rim; blades 4-12 x 0.2-0.5 cm, lanceolate, scaberulous at apex. Panicles up to 12 cm, interrupted. Spikelets 3-nate; involucral spikelets 4.5 mm, oblong, male or neuter; pedicelled similar to involucral spikelets; sessile hermaphrodite. Caryopsis ellipsoid.

Fls. & Frs. : July - December.

Occasional in marshy areas and in forests. Pegarikutta, **KMS 6702.**

Iseilema prostratum (L.) Anders, in Nov. Act. Soc. Sci. Upsal. ser. 3. 2:251. 1856; Gamble 1747

(1211); Bor 188. *Andropogon prostratum* L., Mant. Alt. 304. 1771. *Iseilema wightii* Anders. *loc. cit.* 1856; FBI 7:218.

Perennial, tufted grass. Culms about 70 cm tall; nodes more or less bearded. Leafsheaths 3-5 cm; ligule membranous; blades 4-12 x 0.2-0.4 cm, linear, base rounded, apex acute. Panicles up to 30 cm, lax; spathes linear-lanceolate. Spikelets 3-nate; involucral spikelets 4 mm, male or neuter; pedicelled spikelets similar to involucral spikelets; sessile spikelets hermaphrodite. Caryopsis oblong.

Fls. & Frs. : November - June.

Common in plains and occasional in forests. Posanpalle, **TP** & **MSG** 11962.

LEERSIA Solander ex Swartz. *nom. cons.*

Leersia hexandra Sw. Prod. 21. 1788; FBI 7:94; Gamble 1845 (1277); Bor 599.

Annual, aquatic, floating grass. Culms about 40 cm tall; nodes bearded. Leafsheaths 4-10 cm; ligule membranous; blades 5-15 × 0.3-0.6 cm, lanceolate, apex acute, scabrous on both surfaces. Panicles up to 70 cm. Spikelets 4 mm. solitary, oblong, laterally compressed. Caryopsis oblong.

Fls. & Frs. : September - February.

Rare in ponds and marshy areas. Dundigal, **GNR** 4668.

MELANOCENCHRIS C.G.D. Nees

Melanocenchris jacquemontii Jaub. & Spach, Illus. Pl. Or. 4:36. 325. 1851; Bor 473. *M. royleana* Nees ex Steud., Syn. Pl. Glum. 1:218. 1854; Gamble 1821 (1268). *Gracilea royleana* Hook. f., in FBI 7:284. 1896.

Annual, tufted grass. Culms about 20 cm tall; nodes glabrous. Leafsheaths 1-2 cm; ligule hairy; blades 2-4 × 0.1-0.5 cm, base rounded, apex acuminate. Spikes up to 6 cm, filiform. Spikelets 8 mm. clustered on a simple axis, 2-flowered.

Fls. & Frs. : July - September.

Common in dry, sandy soils and waste lands. Pandivaagu, **BR** & **CP** 11206; Choutkoor, **TP** & **MSG** 11931.

OPLISMENUS *P. Beauvois*

1. Annual, tufted, blades ovate-lanceolate, raceme branches capillary **O. burmannii**
1. Perennials, stoloniferous, blades elliptic - lanceolate, raceme branches stiff **O. compositus**

Oplismenus burmannii (Retz.) Beauv. Ess. Agrost. 54. 168 & 169. 1812; FBI 7:68; Gamble 1778 (1232); Bor 317. *Panicum burmannii* Retz. Obs. Bot. 3:16. 1783.

Annual grass. Culms about 50 cm tall, creeping; nodes glabrous- sparsely hairy. Leafsheaths 1.5 cm; ligule a ridge of hairs; blades 2-5 × 0.5-1 cm, ovate-lanceolate, base amplexicaul, apex acuminate. Speciform racemes, secund, up to 6 cm; branches distant. Spikelets elliptic ovate, 3 mm, solitary. Caryposis linear.

Fls. & Frs. : August - December. *Vern.* : Kodijuttu gaddi.

Common grass in cultivated fields and shades of tall trees. Narsapur forest nursery, **BR & CP** 11226; Narsapur RF, **TP & MSG** 11995; Gangapur RF, **TP & MSG** 12069.

Oplismenus compositus (L.) Beauv. Ess. Agrost. 54. 168. 169. 1812; FBI 7:66; Gamble 1778 (1231); Bor 317.

Perennial, stoloniferous grass. Culms about 60 cm tall, creeping; nodes glabrous. Leafsheaths 2-9 cm; ligule a fimbriate membrane; blades 3-10 × 1-2 cm, elliptic-lanceolate, base amplexicaul, apex acuminate. Racemes spiciform, up to 20 cm; branches distant. Spikelets 4 mm, elliptic, 2-flowered, awned. Caryopsis linear.

Fls. & Frs. : November - January. *Vern.* : Kodijuttugaddi.

Common in damp and shady places of forests. Gangapur, **TP & MSG** 12094.

ORYZA Linnaeus

Oryza rufipogon Girff., Notul. 3:5. 1851; Bor 605.

Perennial, stoloniferous, aquatic grass. Culms about 1 m tall, rooting at nodes; nodes glabrous. Leafsheathes overlapping; ligule membranous; blades 20-30 × 0.2-0.6 cm, linear, base rounded, apex acuminate, Panicles up to 15 cm, drooping. Spikelets linear-oblong, 7 mm, awned, stamens 6. Caryopsis narrowly oblong.

Fls. & Frs. : August - December.

Occasional aquatic grass found near the pools and ditches. Narayanapur RF, **BR & CP** 11273; Sardhana, **CP** 11769; Pocharam tank, **TP & MSG** 12050.

NOTE : This taxon is closely related to cultivated paddy, *Oryza sativa* L. var. *sativa*. It can be esily distinguished from the cultivated paddy by caducous spikelets.

PANICUM Linnaeus

1. Lower glume more than half or as long as the spikelet **P. trypheron**
1. Lower glume less than half the length of the spikelet:
    2. Leafsheaths margin densely ciliate; lower glume suborbicular, apex truncate **P. repens**
    2. Leafsheaths margin glabrous; lower glume ovate, apex acute or acuminate **P. psilopodium**

Panicum psilopodium Trin. Gram. Panic. 217. 1826; FBI 7:46, Gamble 1782 (1234); Bor 329.

Perennial, tufted grass, culms about 50 cm tall; nodes glabrous. Leafsheaths 4-8 cm; ligule a fimbriate membrane; blades 4-10 × 0.3-0.9 cm, elliptic-lanceolate, base cordate, apex acuminate. Panicles up to 12 cm, ovate-oblong, effuse. Caryopsis ellipsoid.

Fls. & Frs. : July - September.

Common in shady and wet places. Ramayampet, **BR & CP** 11521; Gnanapur, **CP** 11785; Poosanpalle, **TP & MSG** 11959.

Panicum repens L., Sp. Pl. ed. 2. 87. 1762; FBI 7:49; Gamble 1783 (1235); Bor 330.

Perennial, stoloniferous grass. Culms about 45 cm tall, creeping; nodes glabrous. Leafsheaths 2-6 cm, densely ciliate on the margin; ligule membranous; blades 5-15 × 0.4-0.8 cm, linear-lanceolate, base rounded, apex acute. Panicles up to 10 cm. Spikelets 3 mm. Caryopsis ellipsoid.

Fls. & Frs. : March - December. *Vern.* : Arikelu.

Common in the fields and waste lands. Choutkoor, **TP & MSG** 11907.

Panicum trypheron Schultes, Mant. Syst. Veg. 2: 244. 1824; FBI 7: 47 Gamble 1783 (1235); Bor 331.

Annual, tufted grass. Culms about 60 cm tall; nodes glabrous. Leafsheaths 2-8 cm; ligule membranous; blades 2-12 × 0.2-0.6 cm, linear-narrowly lanceolate, base truncate, apex acuminate. Panicles up to 15 cm, open. Spikelets 3 mm, ellipsoid. Caryopsis ovoid.

Fls. & Frs. : October - March.

Common grass in moist places. Pocharam tank, **TP & MSG** 12035, 12036 & 12041.

PASPALIDIUM Stapf

1. Annuals; ligule absent; spikelets gibbose **P. flavidum**
1. Perennials; ligule present; spikelets not gibbose **P. geminatum**

Paspalidium flavidum (Retz.) A. Camus in Lecomte, Fl. Gen. Indo-chine 7: 419. 1922; Gamble 1744 (1229); Khan in FFH 340; Bor 333. *Panicum flavidum* Retz., Obs. Bot 4:15. 1786; FBI 7:28.

Annual, tufted grass. Culms about 20 cm tall; nodes glabrous. Leafsheaths 6-15 cm; ligule fringed membranous; blades 5-20 × 0.5-0.8 cm, linear-lanceolate, base rounded, apex acuminate. Panicle racemes up to 20 cm, secund, in 2 rows, 2-floret. Caryopsis ovate, oblong.

Fls. & Frs. : July - October.

Common in cultivated fields. Marrikunta, **CP** 11391.

Paspalidium geminatum (Forssk.) Stapf in Prain, Fl. Trop. Africa 9:583, 1920; Gamble 1774 (1229);

Bor 333. *Panicum geminatum* Forssk., Fl. Aegypt - Arab. 18. 1775. *P. paspaloides* Pers., Syn. Pl. 81. 1801; FBI 7:30.

Perennial, marshy grass. Culms about 50 cm tall; creeping below; nodes glabrous. Leafsheaths 4-10 cm; ligule a ciliate rim; blades 5-15 × 0.3-0.6 cm, linear, base rounded, apex acuminate. Panicles up to 15 cm, racemes 2 cm, ovoid. Spikelets 3 mm, 2-flowered. Caryopsis ellipsoid.

Fls. & Frs. : December - July.

Occasional grass near water sources. Narsapur tank, **TP** & **MSG** 12006.

PASPALUM Linnaeus

Paspalum scrobiculatum L., Mant. Pl. 29. 1767; FBI 7: 10. Gamble 1772 (1227); Bor 340.

Annual, tufted grass. Culms about 70 cm tall; nodes glabrous. Leafsheaths 5-15 cm; ligule a ciliate rim; blades 10-40 × 0.3-0.7 cm, linear, base rounded, apex acute. Racemes 2-5, up to 10 cm, axis flattened. Spikelets 2 mm, glabrous, in 2-rows, lower glumes obscure. Caryopsis plano convex.

Fls. & *Frs.* : Round the year. *Vern*: Arikelu.

Occasionally cultivated in black cotton soils and rarely as an escape. Pocharam, tank, **TP** & **MSG** 12037.

PENNISETUM L. Richard

1. Lemmas similar; rachis not winged on the ribs, **P. hohenackeri**
1. Lemmas dissimilar; rachis winged on the ribs **P. pedicellatum**

Pennisetum hohenackeri Hochst. ex Steud., Syn. Pl. Glum. 1:103 1854; Gamble 1792 (1241); Bor 344. *P. alopecuros* Steud., Syn. Pl. Glum. 1:102 1854; FBI 7:84.

Perennial, tufted grass. Culms about 70 cm tall; nodes glabrous. Leafsheaths overlapping; ligule a small fairy ring; blades 15-40 × 0.2-0.5 cm, linear, apex acuminate. Spikelet racemes up to 12 cm, solitary; involucral bristles 1.5 cm, spikelets elliptic - lanceolate, 8 mm. Caryopsis narrowly oblong.

Fls. & Frs. : August - October.

Occasional in sandy soils near streams and bunds of rice field. Medak, **CP** 11734; Rayampalli canal. **TP & MSG** 12111.

Pennisetum pedicellatum Trin. in Mem., Acad. Imp. Sci. Petersp. Ser. 6. Sci. Math. Nat. 3:184. 1834; FBI 7:86; Gamble 1792 (1241); Bor. 346.

Annual, tufted grass. Culms about 1 m tall; nodes glabrous. Leafsheaths 3-15 cm; ligule a fimbriate membrane; blades 5-30 × 0.3-1.5 cm, flat, linear-lanceolate, apex acuminate, margin scabrid. Panicles spiciform, up to 10 cm, dense. Spikelets 6 mm, solitary or 2-3 together. Caryopsis oblong.

Fls. & Frs. : September - December.

Occasional in open grass lands. Narsapur, **GNR** 3961.

PHRAGMITES Adanson

Phragmites karka (Retz.) Trin. ex Steud., Nom, Bot. ed. 2.2:2 324. 1841; FBI 7: 304; Gamble 1807 (1251); Khan in FFH 341; Bor 416. *Arundo karka* Retz., Obs. Bot. 4:21. 1786.

Perennial, stoloniferous shrub. Culms about 5 m tall, stout, woody below, nodes glabrous. Leafsheaths 1.5-20 cm; ligule a ciliate rim; blades 15-60 × 0.7 - 4 cm, linear-lanceolate, stiff, base rounded, apex acuminate. Panicles up to 60 cm, lax, open. Spikelets 1.5 cm, lanceolate, 4-flowered. Caryopsis oblong.

Fls. & Frs. : November - March. *Vern.* : Kulpa-gaddi.

Occasional robust grass near water sources. Pochammaralu, **BR & CP** 11470.

PSEUDORAPHIS W. Griffith

Pseudoraphis spinescens (R.Br.) Vickery in Proc. Roy. Soc. Queensland 62. n. 7:69. 1952; Bor 353. *Panicum spinescens* R.Br., Prodr. 193. 1810. *Chamaeraphis spinescens* (R. Br.) Poir. in Lam., Encycl. Suppl. 2:189. 1811; FBI 7:62. *Pseudoraphis aspera* (Koen.) Pilger, Notizbl. Bot. Gart. Berlin 10. 210 : 1928; Gamble 1790 (1239).

Perennial, marshy, floating grass. Culms about 60 cm tall; nodes woolly. Leafsheaths 4-8 cm; ligule membranous; blades 5-10 x 0.2-0.5 cm, linear, base trunate, apex acuminate. Panicles up to 14 cm, effuse or contracted. Spikelets narrowly linear, 6 mm. Caryopsis linear-oblong.

Fls. & Frs. October - March.

Occasional grass in ponds and marshy areas. Narsapur tank, **TP & MSG** 12012.

SACCHARUM Linnaeus

Saccharum spontaneum L., Mant. Pl. 183. 1771; FBI 7:118; Gamble 1709 (1185); Bor 214.

Perennial, tufted grass. Culms about 6 m tall; nodes waxy. Leafsheaths 5-20 cm; ligule membranous with fringe of hairs; blades 15-90 x 0.3-0.8 cm, apex acuminate, midrib prominent. Panicles up to 60 cm, contracted. Spikelets linear, 5 mm, 2-nate, sessile spikelets 4 mm, hermaphrodite, pedicelled 4 mm, male or neuter.

Fls. & Frs. : October - February. *Vern.* : Rellugaddi.

Very common along stream and banks of river. Gangapur, **TP & MSG** 12085.

SACCIOLEPIS Nash

1. Culms spongy, robust; spikelets in interrupted panicles, glabrous **S. interrupta**
1. Culms slender; spikelets in continuous crowded panicles; tuberculate pilose:
   2. Spikelets cylindric-elliptic, acute **S. indica**
   2. Spikelets subglobose, blunt **S. myosuroides**

Sacciolepis indica (L.) Chase in Proc. Biol. Soc. Wash. 21:8. 1908; Gamble 1787 (1238); Bor 357. *Aira indica* L., Sp. Pl. 63. 1753 (in "Errata") *Panicum indicum* L., Mant. Pl. 184. 1771 non Miller 1768; FBI 7: 41.

Annual, marshy, tufted grass. Culms about 70 cm tall; nodes glabrous. Leafsheaths 2-5 cm; ligule absent; blades 5-15 x 0.1 - 0.4 cm, linear, filiform, base truncate, apex acute. Spike like panicles up to 7 cm, contracted, cylindric. Spikelets 3.5 mm, linear. Caryopsis linear.

Fls. & Frs. : August - October.

Occasional in marshy areas. Pocharam tank, **TP & MSG** 12038 & 12040.

Sacciolepis interrupta (Willd.) Stapf in Prain, Fl. Trop. Africa 9:757. 1920 Gamble; 1787 (1238); Bor 358. *Panicum interruptum* Willd., Sp. Pl. 1:341. 1798; FBI 7:40.

Perennial, aquatic tufted grass. Culms about 1 m tall; nodes glabrous. Leafsheaths 5-15 cm, linear-lanceolate, base truncate, apex acuminate, midrib prominent. Spike like panicles up to 20 cm, cylindric, contracted, interrupted. Spikelets 6 mm, elliptic- lanceolate. Caryopsis ovoid.

Fls. & Frs. : January - December.

Occasional grass in marshy areas. Narayanapur RF, **BR & CP** 11272.

Sacciolepis myosuriodes (R.Br.) A. Camus in Lecomte, Fl. Indo- Chine 7:460. 1922; Gamble 1787 (1238); Bor 358. *Panicum myosuroides* R.Br., Prodr. 189:1810; FBI 7:42.

Annual, tufted grass. Culms about 60 cm tall, terete, rooting at lower nodes; nodes glabrous. Leafsheaths 4-8 cm, compressed; ligule membranous; blades 10-30 × 0.2-0.6 cm, flat, linear, apex acuminate, margin scaberulus. Panicles spiciform, up to 20 cm. Spikelets 2 mm, ovate-oblong. Caryopsis ellipsoid.

Fls. & Frs. : August - December.

Occasional in wet paddy fields. Usirakapalle, **GNR** 4309.

SEHIMA Forsskal

Sehima sulcatum (Hack.) A. Camus in Bull. Mus. Hist. Nat. Paris 27:373. 1921; Gamble 1724 (1195); Bor 219. *Ischaemum sulcatum* Hack., in DC. Monogr. Phan. 6:248. 1889; FBI 7: 137.

Perennial, tufted grass. Culms about 50 cm tall, geniculately ascending; nodes glabrous. Leafsheaths lax, glabrous, constricted at mouth; ligule a ciliate membrane; blades 10-18 x 0.2-0.3 cm, linear-laneeolate, apex acuminate. Racemes up to 8 cm, solitary. Spikelets 2-nate, 1 cm long.

Fls. & Frs. : August - December.

Rare in open forests and plains. Narsapur, **GNR** 4026.

SETARIA Beauvois

1. Panicle not continuous, interrupted; involucral bristles 3-4; spikelets up to 2 mm long **S. intermedia**
1. Panicle continuous; involucral bristles 6-11; spikelets more than 2 mm long **S. pumila**

Setaria intermedia Roemer & Schultes in L., Syst. Veg. 2:489. 1817; FBI 7:79. Gamble 1789 (1239). *Panicum tomentosum* Roxb., Fl. Ind. 1:303. 1820. *Setaria tomentosa* (Roxb.) Kunth, Rev. Gram. 147. 1829; Bor 365.

Annual, tufted grass. Culms about 80 cm tall; nodes glabrous. Leafsheaths 4-10 cm; ligule ciliate, blades 7-15 x 0.4-1 cm, linear - lanceolate, base rounded, apex acuminate. Panicles up to 10 cm, lax, interrupted; involucral bristles antrorsely barbed. Spikelets broadly ovate, 2 mm. Caryopsis ellipsoid.

Fls. & Frs. : July - November.

Common grass in moist localities, Ramayampet, **BR** & **CP** 11515.

Setaria pumila (Poir.) Roemer & Schultes in L., Syst. Veg. 2:891. 1817. *Panicum pumilum* Poir. Encycl. Suppl. 4: 273. 1816. *P. pallidefuscum* Schum., Beskr. Guin. Pl. 58. 1827. *Setaria pallidefusca* (Schum.) Stapf & C.E. Hubbard in Kew Bull. 1930. 259. 1930; Gamble 1789 (1239); Bor 363. *Setaria glauca* auct. non (L.) : Beauv., Ess. Agrost. 51 169 & 178. 1812; FBI 7: 78.

Annual, tufted grass. Culms about 60 cm tall; nodes glabrous. Leafsheaths 2-5 cm; ligule a ridge of hairs; blades 5-15 x 0.2 - 0.6 cm, linear-lanceolate, base cordate, apex acuminate. Panicles spiciform, up to 8 cm, cylindric, densely flowered; involucral bristles antrorsely barbed. Spikelets 2 mm, ovoid. Caryopsis ovoid - ellipsoid.

Fls. & Frs. : July - November. *Vern.* Nakkathoka-gaddi.

Common in plains and in the forests. Pillutla RF, **BR** & **CP** 11269; Nagulabanda, **CP** 11346; Ramayampet, **BR** & **CP** 11502; 8 NS**BR** & **CP** 11658; Choutkoor, TP & MSG 11929.

SPOROBOLUS Robert Brown

1. Spikelets crowded; stamens 2; branches of panicle not capillary **S. indicus var. diander**
1. Spikelets scattered; stamens 3; branches of panicle capillary **S. tenuissimus**

Sporobolus indicus (L.) R.Br., var. **diander (Retz.)** Jovet & Guedes in Taxon 22:163 1973. *Agrostis diandra* Retz., Obs. Bot. 5:19. 1789. *Sporobolus diander* (Retz.) Beauv., Ess. Agrost. 26 : 147 & 178. 1812; FBI 7:247; Gamble 1817 (1258); Bor 629.

Perennial, tufted grass. Culms about 60 cm tall; nodes glabrous. Leafsheaths 3-10 cm; ligule a ridge of minute hairs; blades 5-12 × 0.1-0.2 cm, linear, base truncate, apex acuminate. Panicles up to 25 cm, lax. Spikelets 1.5 mm, linear. Caryopsis obovoid.

Fls. & Frs. : July - November.

Common in moist plains and upper ghats. Pocharam sanctuary, **CP** 11757; Choutkoor, **TP** & **MSG** 11919.

Sporobolus tenuissimus (Schrank) Kuntze, Rev. Gen. Pl. 3 : 369. 1893; Bor 633. *Panicum tenuissimum* Schrank in Denkschr. Bot. Ges. Regensb. 2:26 1822 *Sporobolus minutiflorus* (Trin.) Link, Enum. Hort. Berol. 1:88. 1827; FBI 7:248. Gamble 1817 (1258).

Annual, tufted grass. Culms about 60 cm tall; nodes glabrous. Leafsheaths 3-8 cm, ligule a membranous ring; blades 3-6 x 0.05- 0.15 cm, linear, base rounded, apex acuminate, Panicles up to 25 cm, effuse. Spikelets very minute. 1 mm, linear. Caryopsis lanceolate.

Fls. & Frs. : July - January.

Occasional grass of moist localities and under shades of trees. Narsapur forest nursery, **TP** & **MSG** 12079.

THEMEDA Forsskal

1. Perennials; culms slender; lower glume with stiff tubercle based bristles **T. quadrivalvis**
1. Annuals; culms robust; lower glumes glabrous **T. triandra**

Themeda quadrivalvis (L.) O. Kuntze, Rev. Gen. Pl. 2:794. 1891, var. **quadrivalvis;** Gamble 1746 (1210); Bor 252 *Andropogon quadrivalvis* L. in Murr., Syst., Veg. ed. 13. 758. 1774. *Anthistiria ciliata* L.f. Suppl. Sp. 113. 1781; FBI 7:213.

Annual, slender grass. Culms about 1 m tall, rooting from the lower nodes; nodes glabrous. Leafsheaths 3-7 cm; ligule a narrow ciliate membrane; blades 7-15 × 0.3-0.5 cm, lanceolate, base cordate, apex acuminate. Spatheolate panicles up to 20 cm, dense; racemes 5 cm. Spikelets 7 mm, 3-nate, involural spikelets 4 mm, male; Pedicelled spikelets narrow; sessile spikelets 7 mm, hermaphrodite. Caryopsis linear.

Fls. & Frs. : August - December.

Common on slopes of hills. Narsapur, **TP** & **MSG** 11996.

Themeda triandra Forssk. Fl. Aegypt. - Arab. 123 & 178. 1775; Gamble 1746 (1209); Bor 254.

Anthistiria imberbis Retz., Obs. Bot. 3:11. 1783; FBI 7:211. *Themeda imberbis* (Retz.) T. Cook, Fl. Pres. Bombay 2:993. 1908.

Perennial, tufted grass. Culms about 2 m tall, robust; nodes glabrous. Leafsheaths 4-12 cm; ligule a ciliolate membrane; blades 8-25 × 0.3-0.5 cm, linear, base rounded, apex acuminate, scabraous. Spatheolate panicles up to 30 cm, dense. Spikelets 3- nate; involucral spikeletes 4 mm, male; pedicelled spikelets similar to involucral spikelets; sessile spikelets 6.5 mm, hermaphrodite. Caryopsis linear.

Fls. & Frs. : August - December.

Occasional grass in hilly areas and plains near streams. Patancheru, **BR & CP** 11315.

TRAGUS A. Haller

Tragus roxburghii Panigrahi in Kew Bull. 29:495. 1974. *T. racemosus* Hook.f. in FBI 7:97. 1896 non Scop. 1777. *T. biflorus* auct. non Schultes in Mant. Syst. Veg. 2:205. 1824; Gamble 1813 (1255); Bor 682.

Perennial, tufted grass. Culms about 25 cm tall; nodes glabrous. Leafsheaths 2-4 cm, glabrous; ligule a membranous ciliate rim; blades 1-3 x 0.2-0.4 cm, lanceolate, base rounded, apex acute, margin pectinately ciliate. Spiciform racemes up to 4 cm, solitary. Spikelets 3 mm, lanceolate, 2, 1-flowered; lower glume scarious. Caryopsis ellipsoid.

Fls. & Frs. : July - September.

Common in waste plains and lower hill slopes. Choutkoor, **TP & MSG** 11945.

VETIVERIA Bory de St. Vincent.

Vetiveria lawsonii (Hook.f.) Blatter & McCann. in Journ. Bombay Nat. Hist. Soc. 32:409. 1828; Gamble 1733 (1201); Bor 258. *Andropogon lawsonii* Hook.f., in FBI 7:187. 1896.

Perennial, tufted grass. Culms about 2 m tall; nodes glabrous. Leafsheaths 5-20 cm; ligule a small fimbriate membrane with long hairs; blades 10-20 x 0.4-0.6 cm, linear, base truncate, apex acuminate. Panicles up to 20 cm. Racemes up to 6 cm. Spikelets lanceolate, 2-nate; sessile spikelet 5 mm, hermaphrodite; pedicelled spikelet 6 mm, male. Caryopsis linear-oblong.

Fls. & Frs. : August - January.

Occasional grass in marshy areas. Gangapur, **TP & MSG** 12093.

Oryza sativa L. var. *sativa, Eleusine coracana* (L.) Gaertn., *Pennisetum americanum* (L.) K. Schum., *Setaria italica* Beavu., *Sorghum bicolor* (L.) Moench, *Zea mays* L., are cultivated for grains. *Saccharum officinarum* L. is cultivated for the cane.

PTERIDOPHYTES

SELAGINELLACEAE

SELAGINELLA P. Beauvois *nom. cons.*

Saleginella bryopteris (L.) Baker in Journ. Bot. Brit. & For. 22:376. 1884. *Lycopodium bryopteris* L., Sp. Pl. 1103. 1753.

Erect herbs, about 10 cm tall, Rhizome long with tufts of roots. Stems dichotomously branched, heterphyllous. Leaves in four rows, central two rows smaller than the lateral two rows; blades ovate-elliptic, green above, yellowish- with white tomentose below, apex sharply mucronate, margin serrate strobili up to 0.4 cm.

Sporangia : August - October.

Occasional in moist and shady localities. Pocharam tank, **TP & CP** 12029.

MARSILEACEAE

Marsilea Linnaeus

Marsilea quadrifolia L., Sp. Pl. 1099. 1753; Gupta Marsilea Monogr. 22 1962

Aquatic creeping herbs; rhizome dichotomously branched, rooting at nodes downwards. Petioles up to 15 cm, lamina divided into 4 leaflets; leaflets obovate, 1.5-2 x 0.8-2 cm, base cuneate, apex entire. Sporocarps stalked, been-shaped, arising near the petiole, brown.

Sporocarps : July - December.

Occasional in ponds and ditches. Patancheru, **BR & CP** 11314. Gangapur RF, **TP & CP** 12066.

CHEILANTHACEAE

Cheilanthes Swartz *nom. cons.*

Cheilanthes tenuifolia (Burm.f.) Swartz, Syn. Fil. 129. 332. 1806; Bedd., Hand. Ferns. Brit. Ind. Suppl. 92 1892 *Trichamanes tenuifolia* Burm, f, Fl. Ind. 237. 1768.

Rhizomatous herbs, about 20 cm tall. Leaves ovate-lanceolate; pinules linear-acuminate. Sori marginal, protected by reflexed margin.

Sori : July - November.

Occasional near moist localities. Narsapur RF, **TP & CP** 14052

ACTINOPTERIDACEAE

Actinopteris Link

Actinopteris radiata (Swartz.) Link in Fil. Spec. 80. 1841; Beddome, Ferns of South India t. 124. 1864. *Asplenium radiatum* Swartz in Journ. Bot. Schard. 1800. 2 : 50. 1801.

Erect marshy fern, about 10 cm tall. Rhizome short, covered with fine hair-like roots. Lamina fan-shaped, up to 3 cm, arise from the rhizome, bipinnate; pinnae 3-5, radiating, deeply dissected, lobes linear, glabrous. Sori arranged in two rows on the lower side of the pinnae lobes, globose, brown.

Sori : June - September.

Common in rock crevices on hills. Medak, **CP** 11724.

BIBLIOGRAPHY

Aiton, W. 1789. *Hortus Kewensis* (ed.1) London.

Aiton, W.T. 1810-1813. *Hortus Kewensis*. (ed.2) London.

Ali, S.I. 1968. Revision of the genus *Indigofera* L. from west Pakistan and north-west Himalayas. *Bot. Notiser* 3: 543-577.

Ali, S.I. 1967. The genus *Goniogyna* DC. *Taxon* 16 : 463-464.

Anderson, T. 1867. An enumeration to Indian species of Acanthaceae. *Journ. Linn. Soc. Bot*. 9: 425-526.

Austin, D.F. 1977. *Ipomoea carnea* Jacq. Vs. *Ipomoea fistulosa* Mart. ex Choisy. *Taxon* 26: 235-238.

Backeb, C. V. 1960. Die Cactaceae. Band IV. Jena.

Backer, C.A. and R.C. Bakhuizen van den Brink. 1963-68. *Flora of Java*. 3 Vols Netherlands.

Balakrishnan, N.P. 1964. New plant records from south India. *Bull. Bot. Surv. India* 6: 81-92.

Banerjee, S.P. 1962. Identity of *Trichodesma amplexicaule* DC. (non Roth) (Boraginaceae). *Bull. Bot. Surv. India* 16: 10.

Beauvois, A.M.F. J.P. de. 1812. *Essai d'une nouvelle Agrostographie*. Pairs.

Beddome, R.H. 1868-74. *Icones Plantarum Indiae Orientalis*. Madras.

Beddome, R.H. 1869-74. *The Flora Sylvatica for Southern India*. 2 Vols. Madras.

Beddome, R.H. 1883. *Handbook of the Ferns of British India, Ceylon and the Malay Peninsula*. Calcutta.

Bennet, S.S.R. 1986. *Name changes in flowering plants of India and adjoining regions*. Dehra Dun.

Bentham, G. 1835. *Scrophularineae Indicae : A synopsis of the Eastern Indian Scrophulariaceae*. London.

Bentham, G. and J.D. Hooker. 1862-1883. *Genera plantarum*. 3 Vols. London.

Biswas, K. and C. Calder. 1937. *Handbook of common water and marshy plants of India and Burma*. Delhi.(ed. 2. 1955. Calcutta).

Blatter, E. and C.McCann. 1935. *The Bombay grasses*. Delhi..

Blume, C.L. Von. 1836-1849. *Rumphia, sive Commentationes botanicae imprimis de plantis Indiae Orientalis*. Leiden.

Bogle, A.L. 1970. The genera of Molluginaceae and Aizoaceae in the South Eastern United States. *Journ. Arnold Arbor*. 51: 431-462.

Bor, N.L. 1960. *The grasses of Burma Ceylon, India and Pakistan* (excluding Bambuseae). London (ed. 2. 1973).

Bradley, H. 1849. Statistics of the circar of Dowlutabad, Hyderabad, Deccan. *Madras Journ. Lit and Sci* 15: 481-55.

Compbell, A.C.1898. *Glimpses of the Nizam's Dominions*. London.

Chandrabose, M. and N.C. Nair. 1981. The genus *Polygala* L. (Polygalaceae) in Andhra Pradesh, Kerala and Tamil Nadu (South India). *Proc. Indian Acad. Sci*. (Plant. Sci.) 90: 107-127.

Clarke, C.B. 1876. *Compositae Indica descriptae et Secus genera Benthamii ordinatae*. Calcutta.

Cooke, T. 1901-1908. *The Flora of the Presidency of Bombay*. London. (repr. ed. 1958, Calcutta).

Cronquist, A.J.1981. *An integrated system of classification of flowering plants*. London.

Dahlgren, R.M.T. 1980. A revised system of classification of the Angiosperms. *Bot. Journ. Linn. Soc*. 80: 91-124.

Davis. P.H. and V.H. Heywood. 1963. *Principles of Angiosperm Taxonomy*. London.
Dellie. A.R. 1813. *Description de I Egypte Historoin naturelle*. Vol.2. Paris.
Desfontaines, R.L. 1798-1799. *Flora Atlantica* etc. 2 Vols and Atlas. Paris.
Ding Hou, L. 1962. Celastraceae-I; in *Fl. Males*. Ser. 1.6: 227-291.
Don, D. 1825. *Prodromous Florae Nepalensis*. London.
Don, G. 1831-1838. *A General History of Dichlamydeous plants*. London.
Elliot, W. 1859. *Flora Andhrica*. Madras.
Ellis, J.L. 1987-1990. *Flora of Nallamalais*, Calcutta.
Engler, H.G.A. 1900-1937. *Das Pflanzenreich Regni vegetabilis Conspectus*. Leipzig.
Engler, H.G.A. and K.A.E. Prantl. 1887-1915. *Die Naturalichen Pflanzenfamilien etc*. 23 Vols. ed 1. Leipzig.
Forster, J.R. and J.G.A. Forster. 1775. *Characteres Generum Plantarum* etc. London.
Forsskal, P. 1775. Flora *Aegyptiaco-Arabica*. Copenhagen.
Gaertner, J. 1788-1807. *De fructibus et Siminibus Plantarum*. 3.Vols. Stuttgart.
Gamble J.S. and C.E.C. Fischer. 1915-1935. *Flora of the Presidency of Madras*, London. (repr. ed. 1957, Calcutta).
Gmelin, J.F. 1791-1792. *Linnaeus Systema Naturae* ed. 13. Leipzig.
Gray, A. 1848. *A Manual of Botany of the Northern United States*. Boston and Cambridge (ed.2 New York).
Greuter, W. (ed.) 1988. *International Code of Botanical Nomenclature*. Konigstein (Germany).
Hackel, E. 1889. *Monographia Andropogonearum*. Paris.
Haines, H.M. 1921-25. *The Botany of Bihar and Orissa* 3 Vols. Calcutta. (repr. ed. 1961).
Harley, R.M. 1983. *Becium* Lindl., a new genus of Labiatae new to India and Sri Lanka. *Kew Bull*. 38: 56.
Haworth, A.H. 1812. *Synopsis Plantarum Succulentarum*. London.
Heine, H. 1962-63. *Tropical African plants*, XXVI. Some west African Acanthaceae. *Kew Bull*. 16: 173-78.
Henry, A.N. K. Vivekananthan and N.C. Nair 1978. Rare and threatened flowering plants of South India. *Journ. Bombay Nat. Hist. Soc*. 75: 684-697.
Hooker, J.D. *et. al*. 1872-1897. *Flora Indica*. London.
Hooker, J.D. and T. Thomson. 1855. *Flora Indica*. London.
Hooker, W.J. and G.A.W. Arnott. 1830-1841. *The Botany of Captain Beechey's voyage*. London.
Huber, H. 1973. In Abeywickrama, B.A. (ed.) *A revised handbook of Flora of Ceylon*. Vol.1.Part 1. University of Ceylon. Peradenia.
Humboldt, F.H.A., A.J.A. Bonpland and C.B. Kunth. 1816-1825. *Nova Genera et Species Plantarum* etc. Paris.
Hutchinson, J. 1959. *The families of flowering plants*. 2 Vols. London.
Iltis, H.H. 1960. Studies in the Capparidaceae VII; Old world Cleomes adventive in the new world. *Brittonia*. 12: 279-294.
Jacobs, M. 1960. Capparidaceae; in *Fl. Males*. Ser. 1.6: 61-226.
Jacquin, N.J. 1787-1797. *Collectanea and Botanicam. Chemicam et Historium naturalem Spectantia*. 5 Vols. Wien.
Jacquin, 1781-1795. *Icones Plantarum rariorum*. Wein.
Jain, S.K. 1982. *Flora of India—Tasks before 2000 A.D*. Panchanam Maheshwari Memorial Medal lecture, Indian Botanical Society.
Jain, S.K. and U.R.Deshpande. 1978. Transfer of some Indian species of *Bothriochloa* and *Capillipedium* to *Dichanthium*. *Bull. Bot. Surv. India* 20: 133-135.

Jain, S.K. and R.R. Rao 1977. *Field and Herbarium methods*. New Delhi.

Jeffrey, C. 1980. Further notes on Cucurbitaceae of the Indian Subcontinent. *Kew Bull*. 34: 789-810.

Jussieu, A.L. de. 1789. *Genera Plantarum Secundum Ordines Naturales Disposita*. (ed.2.1791). Paris.

Kapoor, S.L. and L.D. Kapoor 1973. Further contribution to the flora of the Karimnagar district of Andhra Pradesh. *Bull.Bot. Surv. India* 15: 76-84.

Karthikeyan, S.M.P. Nayar and R.S. Raghavan. 1981. *An annotated Bibliography of Taxonomic Botany of Peninsular India 1959-1978*. Botanical Survey of India. Howrah.

Kern, J.H. 1974. Cyperaceae; in Fl. Males. ser 1.7: 510-742.

Khan, M.S. 1953. *Forest Flora of Hyderbad state*. Hyderabad.

Krishnamurthy, A.V.R.G. 1960. *Forest types of Andhra Pradesh*. Mss.

Kunth, C.S. 1833-1850. *Enumeratio Plantarum omnium hucusque cognitarum*. Stuttgart.

Kuntze, O. 1891-1898. *Revisio Generum Plantarum Vascularium Omnium*. Leipzig.

Lamarck, J.B.A. P.M. de. 1783-1817. *Encyclopedie methodique: Botanique*. 8 Vols. and 5 suppl., Paris. (Vol.5-13 by J.L.M. Poiret).

Lamarck, J.B.A. P.M. de. 1791-1823. *Tableau Encyclopedique et methodique: Botanique*. 3 Vols. and Suppl. Paris (Vol.2, part 5, 2 and Vol. 3 by J.L.M. Poiret).

Lawrence, G.H.M. 1951. *Taxonomy of vascular plants*. New york.

Lecomte, P.H. 1907-1950. *Flora Indo-Chine*. 7 Vols. Paris.

Lewis, W.H. 1961. Merger of North American *Houstonia* and *Oldenlandia* under *Hedyotis*. *Rhodora* 63: 216-223.

Linnaeus, C. 1753. *Species Plantarum*. 2 Vols. Stockholm.

Linnaeus, C. 1755-56. *Centuria Plantarum*. 2 Vols. Stockholm.

Linnaeus, C. 1758-59. *Systema Naturae*. ed. 10. Vol.2. Botany. Stockholm.

Linnaeus, C. 1767. *Mantissa Plantarum* Pages 1-143, Stockholm.

Linnaeus, C. 1771. *Mantissa Plantarum Altera*. Pages 143-587. Stockholm.

Linnaeus, f., 1781. *Supplementum Plantarum Systematis Vegetabillum* etc. Braunschweig.

Maheshwari. J.K.1963. *The flora of Delhi*. New Delhi.

Matthew, K.M. 1981-1983. *The flora of Tamilnadu Carnatic*. 3 Vols. Tiruchirapalli.

McCann, C. 1942. Observations on Indian Duckweeds, Lemnaceae *Journ. Bombay Nat. Hist. Soc.* 43: 148-162.

Mears, J.A. 1980. The Linnaean species of *Gomphrena* L. (Amaranthaceae) *Taxon* 29: 85-95.

Merrill, E.D. 1917. *An Interpretation of Rumphius Herbarium Amboinense*. Manila.

Michaux, A. 1803. *Flora boreali-americana*, 2 Vols. Paris.

Miller, P. 1768. *The Gardeners Dictonary*. (ed.8). London.

Mitra, R.L. and S.K. Jain. 1985. Concept of *Phyllanthus niruri* (Euphorbiaceae) in Indian floras. *Bull. Bot. Surv. India* 27: 161-176.

Moore, A.H. 1907. Revision of the genus *Spilanthus*. Proc. Amer Acd. Arts. Sci. 42-521-569.

Moulali, A.A. 1990. *Acanthaceae Verbenaceae and Lamiaceae in Andhra Pradesh*. Ph.D. thesis. S.K. University, Anantapur.

Mukerjee, S.K. 1945. A Revision of the Indo-Burmese species of *Lidernia* All. *Jour. Ind. Bot. Soc.* 24: 127-134.

Nair, N.C. and A.N. Henry 1983. *Flora of Tamil Nadu*, India. Vol. 1. Coimbatore.

Narasimha Rao, G.1986. *Studies on the floristics of Angiosperms and the dermotypes of Acanthaceae of Narsapur taluk (Medak district, Andhra Pradesh)*. Ph.D. thesis, Osmania University. Hyderabad.

Nees, C.G.D. 1841. *Flora Africae Australioris Illustrationes monographiae*. Glogan.

Nees, 1847. Acanthaceae in DC., Prodr.Vol. 11: 46-519.

Ohwi, J. 1947. New or Noteworthy Grasses from Asia. *Bull. Tokyo Sci. Mus.* n. 18.

Oza, G.M. 1966. Correct name for *Ventilago calyculata* Tulasne. *Journ. Bombay Nat. Hist. Soc.* 63: 455.

Panigrahi, G. 1977. Proposal to reject the name *Limonia acidissima* L.*nom. ambio. rejic.* in favour of *L. elephantum* (Corr.) *comb. nov. Taxon* 26: 576-577.

Panigrahi, G.1987. *Echites antidysenterica* (L.) Roxb. ex Fleming and *Holarhena antidysenterica* (L.) Wall., validly published synonyms of *Wallida antidysenterica. Taxon* 36: 464-467.

Patridge, E.A. 1911. *Forest Flora of H.H. the Nizam's Dominions. Hyderabad. Deccan.* Hyderabad.

Pennel, F.W. 1935. Scrophulariaceae of Eastern temperate North America. *Acad. Nat. Sci. Philad. Monogr.* 1: 137-159.

Pennel, F.W. 1943. A second summary of the Scrophulariaceae of New Guinea. *Journ. Arnold Arbor.* 24: 243-274.

Prain, D. 1963. *Bengal plants* (repr. ed.) 2 Vols. Calcutta.

Prain, L.C. 1909. Botany in *"Imperial Gazetterer of India."* Provincial series, Hyderabad state. Calcutta.

Prasanna, P.V. 1987. *Poaceae in Rayalaseema. Andhra Pradesh* Ph.D. thesis. S.K. University, Anantapur.

Pullaiah, T. 1987. Materials for the Flora of Andhra Pradesh, Ranunculaceae-Dipterocarpaceae. *Journ. Swamy Bot. Club* 4: 145-155.

Pullaiah, T. 1989. Orobanchaceae, Lentiblariaceae, Gesneriaceae, Bignoniaceae and Pedaliaceae in Andhra Pradesh. *Journ Econ. Tax. Bot.* 13: 353-356.

Pullaiah, T., P.V. Prasanna and G. Obulesu 1992. *Flora of Adilabad District*, C.B.S. Publishers, New Delhi.

Pullaiah, T., and B. Ravi Prasad Rao. 1995. *Flora of Nizamabad district*. Dehra Dun.

Pullaiah, T., and N. Yesoda, 1989. *Flora of Anantapur district*. Dehra Dun.

Radford, A.E. *et al.* 1974. *Vascular plant Systematics.* New York.

Raghava Rao, S.R. 1989. *Flora of Mahaboobnagar district, Andhra Pradesh.* Ph.D. thesis. Osmania University, Hyderabad.

Raghavaswamy, B.V. and P.N. Rao. 1987. Further additions to the Flora of Andhra Pradesh. *Journ. Econ. Tax. Bot.* 9: 489-492.

Raizada, M.B. 1966. Nomenclatural changes in Indian Plants. *Indian Forester* 92: 299-339.

Rajagopal, T. 1973. *Flora of Hydereabad including study of the foliar epidermal charecters of species as an aid to taxonomy.* Ph.D. thesis. Osmania University, Hyderabad.

Raju , D.C.S. 1966. Plants from Papi hill in Andhra Pradesh. *Indian Forester* 92: 482-492.

Raju, R.R.V., E. Chennaiah and T. Pullaiah. 1988. Capparaceae in Andhra Pradesh. *Journ Swamy Bot. Club* 5: 137-141.

Raju, R.R.V. and T. Pullaiah 1995. *Flora of Kurnool district.* Dehra Dun.

Raju, V.S. 1984. New angiospermous taxa described from the state of Andhra Pradesh after its formation. *Journ Econ. Tax. Bot.* 5: 469-472.

Raju, V.S. and A. Ragan. 1988. Additions of the Cyperaceae of Andhra Pradesh. *Jorun. Swamy Bot. Club* 5: 115.

Raju, V.S. and A. Ragan 1989. The genus *Eleocharis* in Andhra Pradesh, India. In *Plant Science Research in India* (ed. M.L. Trivedi *et al.,*). New Delhi.

Raju, V.S., and A. Ragan 1990. Additions to the genus *Cyperus* L. in Andhra Pradesh. *Journ Econ. Tax. Bot.* 14: 467-468.

Ramachandrachary. S.T. 1980. *Floristic study of Achampet taluk* (Mahboobnagar Dist., Andhra Pradesh Ph.D. thesis. Osmania Univeristy, Hyderabad.

Ramachandrachary, S.T. and N. Ramayya. 1986. Additions to the forest flora of Telangana region in Andhra Pradesh. *Indian Journ. Bot.* 9: 156-158.

Ramayya, N. and T. Rajagopal, 1973. Taxonomy, forest ecology and anatomy of the flowering plants of the erstwhile state of Hyderabad, a review. *Ind. Nat. Sci. Acad.* 45: 302-313.

Randeria, A.F. 1960. The Compositae genus *Blumea,* a taxonomic revision. *Blumea* 10: 173-317.

Rao, P.N., V.S. Raju and R.S. Rao. 1983. Floristic Research work in Andhra Pradesh. *Bull. Bot. Surv. India* 23: 90-95.

Rao, R.S. 1956. *Parthenium hystercophorus* L., A new record for India. *Journ. Bombay Nat. Hist. Soc.* 54: 218-220.

Rao, R.S. and R.V. Kammathy. 1966. Notes on Indian Commelinaceae V. *Journ. Linn Soc. (London) Bot.* 59: 305-308.

Raven, P.H. 1964. The old world species of *Ludwigia* including *Jussiaea* with a synopsis of the genus (Onagraceae). *Reinwardtia* 6: 327-427.

Retzius. A.J. 1779-1791. *Observationes botanicae.* Leipzig.

Ridsdale, C.E. 1978. A revision of the tribe Naucleae s.s. (Rubiaceae). *Blumea* 24: 307-366.

Roemer, J.J. and J.A. Schultes. 1817-1830. *Linnaeus Sytema Vegetabilium* ed. 15. 7 Vols. Stuttgart.

Roth, A.W. 1827. *Enumeratio Plantarum Phanerogamarum* etc. Leipzig.

Roxburgh, W. 1795-1819. *Plants of the coast of Coromandel.* 3. Vols London.

Roxburgh, W. 1832. *Flora Indica.* 3 Vols. Calcutta.

Saldanha, C.J. 1963. The genus *Striga* Lour. in Western India. *Bull. Bot. Surv. India* 5: 67-70.

Saldanha, C.J. 1986. *Flora of Karnataka.*Vol.1. New Delhi.

Saldanha, C.J. and D.H. Nicolson. 1976. *Flora of Hassan district.* New Delhi.

Santapur, H. 1954. Annual excursion of the Indian Botanical Society. Hyderabad, Deccan. *Journ. Ind. Bot. Soc.* 33: 78-92.

Santapau, H. 1955. *Botanicl collector's manual.* Calcutta.

Santapau, H. 1958. *Flora of Purandhar.* New Delhi.

Santapau, H. 1967. The Flora of Khandala on the Western ghats of India (ed. 3) *Rec. Bot. Surv. India* 16:, 1-372.

Santapau, H. and S.K. Wagh. 1964. Nomenclatural changes in Indian Plants. *Bull. Bot. Surv. India* 5: 105-109.

Sayeeduddin, M. 1935. Some of the common flowering plants of Hyderabad state, their distribution, economic and medicinal importance. *Journ. Asiat. Soc. Bengal Sci.* 1: 9-22.

Sayeeduddin. M. 1938. A further contribution to some of the common flowering plants of Hyderabad state, their distribution and economic importance. *Journ. Bombay Nat. Hist. Soc.* 40: 191-212.

Sayeeduddin, M. 1941 a. Preliminary notes of a recent tour to Amrabad forest reserve. *Journ. Bombay Nat. Hist. Soc.* 41: 907-910.

Sayeeduddin, M. 1941 b. Additions to our knowledge of the flowering plants of H.E.H. the Nizam dominations, Hyderabad, Deccan. *Journ. Bombay Nat. Hist. Soc.* 42: 903-924.

Sayeeduddin, M. 1954. A brief sketch of the flora of Hyderabad, Deccan. Souvenir. 41st session. Indian Sceience Congress. Hyderabad.

Sebastine, K.M. and A. N. Henry. 1966. A Contribution to the flora of Pakhal and sorrounding areas of Narasampet taluk, Warangal district, Andhra Pradesh. *Bull. Bot. Surv. India.* 8: 304-311.

Sevastine, K.M., K. Thotathri and N.P. Balakrishnan. 1960. Observation on the flora of Narsapur, Medak district, Andhra Pradesh, *Bull. Bot. Surv. India.* 2: 275-285.

Sedgwick, L.J. 1919. On the *Alyscripus rugosus* D.C. and its allied species. *Journ. Ind. Bot. Soc.* 1: 14-18.

Silar Mohammed, M. 1991. *Flora of Mohammadabad reserve Forest, Ranga reddy district.* M. Phill. dissertation. S.K. University, Anantapur.

Sivaraajan, V.V. and P. Mathew. 1983. The genus *Lindernia* All. (*Scrophulariaceae*) in India. *Journ. Bombay Nat. Hist. Soc.* 80-131-140.

Sprengel, C.P.J. 1824-1828. *Linnaeus Systema Vegetabilium* ed.16. Gottingen.

Sreemadhavan, C.P. 1968. *Indoneesiella—a* substitute name in Acanthaceae. *Phytologia* 16: 466.

Stapf, O. 1917-1930. Gramineae; in Prain, *Fl. Trop. Africa* 9: 1-768.

Subba Rao, G.V. and G.R. Kumari. 1967. Contriburtion to the flora of Karimnagar district, A.P. *Bull. Bot. India* 9: 95-113.

Subramanyam, K. 1962. A study of the grasses of the H.E.H. the Nizam's dominations, Hyderabad, Deccan. Osmania Univ. Journ. Res. 12: 28-65.

Tadulingam, M.C. and G. Venkata Narayana. 1932. A *Handbook of some south Indian weeds.* Madras.

Takhtajan, A.L. 1980. Outline of the classification of flowering plants (Magnollophyta). *Bot. Rev.* 46: 225-259.

Thothathri, K. 1961. Studies in Leguminosae: 2. The Genus *Pongamia* Vent. in India. *Bull. Bot. Surv. India.* 3: 418.

Thothathri, K. 1964. The flora of Nagarjunakonda Valley and the surrounding hills. Indian For 90: 539-546.

Thunberg, C.P. 1794-1800. *Prodroums Plantarum Capensium.* Upsala.

Thwaites, G.H.W. 1858-64. *Enumeratio Plantarum Zeylaniae.* London.

Tirvengadum D.D. 1978 a. Reestablishment of genus *Catunaregam* Wolf. (Rubiaceae). A Contribution towards the plants described in Rhee de *Hortus Malabraicus. Taxon* 27: 512-17.

Tirvengadum, D.D. 1978 b. A synopsis of the Rubiaceae Gradenieae of Ceylon (Sri Lanka). *Bull Mus. Hist. Nat (Paris)* Ser. 3.35: 33.

Tirvengadum, D.D. 1983. New taxa and name changes in Tropical Asiatic Rubiaceae. *Nord Journ. Bot.* 3: 45-496.

Trimen, M. 1894. *A handbook to the flora of Ceylon.* London.

Trimen, M. and J.D. Hooker. 1983-1900. *A handbook to the flora of Ceylon.* Vol.6. (Suppl.) London. (1931 ed. by Alston).

Torrey, J. 1943. *A Flora of the state of New York* 2 Vols. Albany.

Vahl, M. 1790-1794. *Symbolae botanicae* 9 Vols. Copenhagen.

Van Steenis, C.B.B.J. (ed.) 1950-1984. *Flora Malesiana* 10 Vols. Netherlands.

Verdcourt, B. 1976. Rubiaceae: in *Flora of Tropical East Africa : Rubiaceae* I. 339-379. (ed. R.M. Polhill). London.

Wagh, S.K. 1960. *Studies on the flora of Andhra State.* Ph.D. thesis. Bombay University.

Walker, A.L. 1849 Statistical report on the circar of Warangal (Hyderabad, Deccan). Madras Journ. Lit. and Sci. 15: 219-301.

Wallich, N. 1829-32. *Plantae Asiaticae. Rariores.* 3 Vols. London.

Wight, R. 1834. *Contribution to the Botany of India.* London.

Wight, R. 1838-1853. Icones Plantarum Orientalis 6 Vols. Madars.

Wight, R. 1840-50. *Illustrations to Indian Botany,* London.

Wight, R. and G.A. Walker-Arnott. 1834. Prodromus Florae Peninsulae Indiae Orientalis. London.

Willdenow, C.L. 1797-1830. *Species Plantarum* (ed. 4 of L., *Sp.Pl.)* 4 Vols. Berlin.

Willis, J.C. 1973. A dictionary of flowering plants and ferns (rev. ed. 8th by H.K. Airy Shaw). Cambridge.

Yesoda, N., P.V. Prasanna, R.R.V. Raju and T. Pullaiah. 1987. Two new records of grasses from Andhra Pradesh. *Journ. Bombay Nat. Hist. Soc.* 84: 265-267.

INDEX OF FAMILIES

INDEX OF GENERA AND SPECIES

Note : Index is provided to the taxa dealt is Systematic Enumeration only. Basionyms & synonyms are in *Italics*.